Network Cabling for Contractors

Network Cabling for Contractors

Daniel E. Capano
Diversified Technical Services, Inc.

McGraw-Hill
New York San Francisco Washington, D.C. Auckland Bogotá Caracas Lisbon London Madrid Mexico City Milan Montreal New Delhi San Juan Singapore Sydney Tokyo Toronto

Library of Congress Cataloging-in-Publication Data
Capano, Daniel E.
Network cabling for contractors / Daniel E. Capano.
p. cm.
ISBN 0-07-012011-0
1. Telecommunication wiring. 2. Telecommunication systems. I. Title.
TK5103.12.C37 2000
621.382′3—dc21 00-056876

McGraw-Hill
A Division of The ***McGraw-Hill*** *Companies*

1 2 3 4 5 6 7 8 9 0 DOC/DOC 0 6 5 4 3 2 1 0

ISBN 0-07-012011-0

The sponsoring editor for this book was Zoe Foundotos and the production supervisor was Sherri Souffrance.

It was set in 10/12 Century Schoolbook by The PRD Group and was printed and bound by R. R. Donnelley & Sons, Inc.

This book is printed on recycled, acid-free paper containing a minimum of 50% recycled, de-inked fiber.

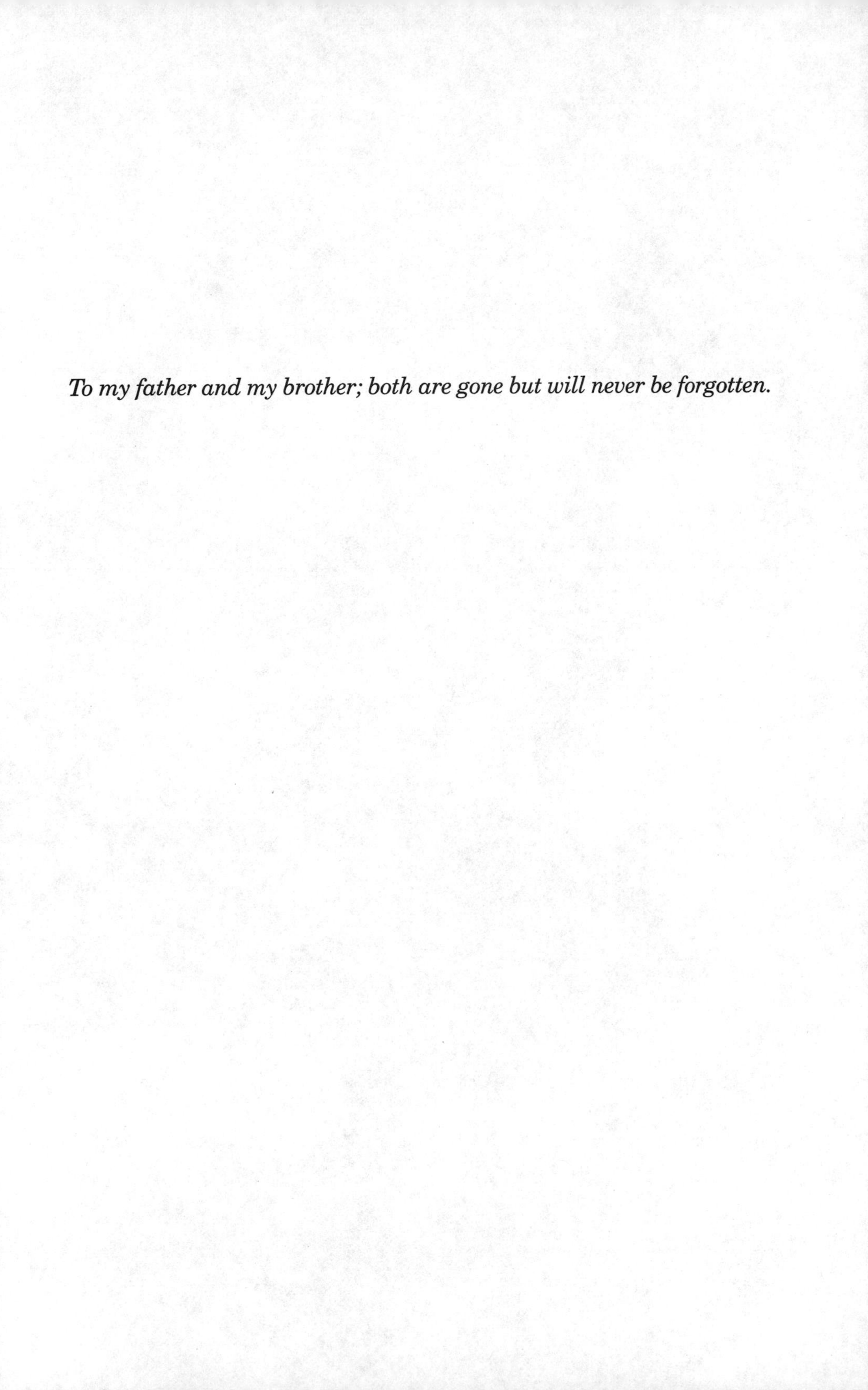

To my father and my brother; both are gone but will never be forgotten.

Contents

Introduction

When this book was first proposed, it was envisioned as a vehicle for those persons unfamiliar with the finer details of data communication technology. The explicit purpose of this book is to guide the reader through the development of data communication from the earliest system of transferring information (data) from one location to another. Data is just a buzzword for any information deemed useful. The terms information and data will be used interchangeably throughout the following discussions. This book will attempt to present, in everyday language, the background, development, and practical use of the technology. The following pages were written with the nontechnician in mind. The technically oriented and adept need not put this book aside, however, as I have endeavored to apply useful reference information liberally throughout which should prove useful. In any event, I intend that this book be used as a handy reference as well as a basic tutorial.

The book has been broken down into four broad chapters. Chapter 1 introduces the reader to the basic concepts of Data Communications. Certain basic concepts will be defined, leading to a discussion of basic computer network topics. Emphasis will be placed upon short-range communication, such as serial and parallel data communication, and telephony. Telephony in itself is a fascinating and complex subject. A thorough discussion, however, would comprise a book twice the size of this one. I wish to give the reader a thorough enough understanding of the telephone to make informed decisions and to apply the technology to the best advantage. The ISO Model, the basis for most Data Communication schemes and the framework by which most systems are applied, will be dealt with in detail in this chapter, as will Local Area Network architectures.

Chapter 2 will attempt to demystify the world of communication protocols. There is a jumble of words and letters making up this, the actual intelligence, if you will, of transferring the information from one point to another. Sophisticated sets of rules and regulations exist and have been applied for the orderly transfer of error-free data between users. Protocols are agreed upon sets of rules. Rules of conduct, so to speak, for senders and receivers of information. The latest protocols will be discussed, with frequent reference to developers and applications. It would be ultimately rewarding to the reader to pay particular attention to this chapter, which is easily the most technical. This chapter and the concepts contained in it will have a major effect on how the reader chooses and uses the technology. It will also form a basis for the chapters that follow.

The last two chapters will deal with the practical side of the application of the technology. Chapter 3, The Medium, delves into the nuts and bolts of the subject. This chapter speaks of wires and fibers and obscure, but soon to be familiar, concepts such as Bandwidth and Throughput. This chapter will further build upon the information presented in the previous two. It contains a wealth of useful data to illustrate and aid the reader in whatever part of the data communication enterprise they may be involved with, whether it be signing the purchase order or supervising or performing the installation or use. This section contains useful sidebars presenting relevant NEC articles. It is intended that this chapter become a useful reference for the installer as well as the user.

The last chapter will put all of the previous four into useful perspective by applying what has been learned to a fictitious office and factory complex data communication network. The reader will be guided through the design and installation of an office and manufacturing network linked to home and branch offices.

I have included a comprehensive Glossary, which should be used frequently. New terms will be explained when initially introduced. It is recommended that the reader refer back to the Glossary rather than attempt to find the term in the original text. A Bibliography of selected text follows.

The technology that is explained in this book is quite complex. The men and women who conceived the concepts and developed the means by which it became real are giants in the field, attaining near legend status. All were freethinking, independent individuals who realized the potential for global community and enhanced personal freedom, which would result from the developments of universal communication. The Internet has been growing steadily for many years. The many foolish attempts to restrict and censor the new medium have failed ignominiously. The development of this technology has quietly revolutionized the way we live and interact with each other; it has created enormous amounts of wealth across the entire social strata; it will continue to dramatically change every facet of our personal and professional lives.

I thank my wife, Loreen, for her patience and support. I also thank my family and my many friends for their interest in this project. In addition, thanks to my good friend Bill O'Hara for giving me the opportunity and encouragement. Finally, many thanks to Zoe Foundotous of McGraw-Hill for her incredible patience.

—Daniel E. Capano
Stamford, 2000

Chapter

1

Data Communications

Concepts

Data communications defined

Data, as defined by Webster, is any information that can be known and/or communicated. Information can be practically anything. Words, sounds, and pictures are information. Unfortunately, most information is inaccessible to the great majority of us. How many of us have traveled to Europe to view the works of the great artists? Great works of art can be digitized and transmitted to anywhere on the face of the earth. The information can be processed at the other end and then reproduced or stored for future enjoyment. Entire libraries have been digitized and placed into storage. In short, practically anything that can be known can be converted into electronic format and communicated to another location. Communication is the movement of information from one person or place to another person or place. Communication takes many forms. Communication is the primary means by which we express our wants, needs, and desires to the world. We learn from communication. Communication brings the world into our living rooms and cars. This last mode of communication is what we are interested in. It is this revolution in information technology and communication that has many people scratching their heads in amazement and wondering aloud.

Data communication is the technology by which data is transferred or conveyed from one location to another. When you operate an ATM machine, you are communicating with a computer somewhere that is accepting the data you are sending and then sending data back. Much of this is automated, of course, but this common form of data communication is taken for granted. Another very common form of data communication is the supermarket checkout. The clerk swipes the package, the laser reads the bar code, and a query goes out to the mainframe for the correct price, which is then sent back. The information is sent to another computer, which tracks the inventory and issues a buy order for that item when the minimum quantity is reached. In all probability, this order is sent to another computer at the vendor's location.

These are minor examples of the way data communication affects your life. If you use a computer on a network, you have participated in this marvelous technology. If you have gone on-line and surfed the Web, you have also had the pleasure of tapping into the vast resources available via the advanced high-speed data communication networks in existence worldwide, interconnected and untamed.

Types of data

There are two types of data that can be communicated over an electronic media. The first is known as analog, the second as digital. Each has been used extensively over the years, requiring completely different methods of communication. Each type of data uses different equipment and differs greatly in the speed at which data can be transferred. Each type has advantages and disadvantages.

Analog data. Analog data is so named because it is analogous to the data it represents. That is, it is an electronic translation of the data from which it was translated. An analog signal can vary over time. The signal can reach a minimum and maximum or any value in between depending upon the magnitude of the input signal. Sound is an analog signal, its pitch and volume being components. Light is also analog in that it is continuously changeable. Any sensation felt can be considered an analog signal. Each one of these examples can, with the correct equipment, be translated into an electrical signal. The electrical signals will vary with the intensity of the sensation. The sensation of heat applied to the proper device will output a voltage, which will increase in proportion to the intensity of the heat. A microphone is a very common analog input device, directly translating sound waves into electrical signals. These signals are proportional to the air pressure variations caused by the variations in the sounds received.

This information is useful if an industrial process is being monitored or controlled. In the case of communication between computers, the situation is somewhat different. Information starts out in digital form and may end up as analog. Computers always communicate digitally. The medium over which they communicate may not be capable of transmitting digital data, however. This situation requires a device to translate the digital information into analog and then back into digital form on the other end. This device is known as a *Modem* and will be discussed in a later section.

Digital data. Digital data exists in one of two forms. In the realm of data communication, the two options are 0 and 1. The data can be considered on or off. Each piece of information in a digital system is called a *Bit*. A bit can be either a one or a zero. These states can be thought of as on or off, or as high or low. A bit cannot exist in more than one state. Groups of bits are combined to form an entity known as a *Byte* (Figure 1). A byte is eight bits wide and is the smallest intelligible digital element used in data communication. Bytes are further combined into sixteen and thirty-two bit wide *Words* (Figure 2). The significance of the width of the entity is that it correlates with the speed in which the data is transferred. The wider the entity, the more information can be transferred at one time.

Byte = 8 bits

Figure 1

Word = 2 or more Bytes

Figure 2

Bits are arranged in specific order to represent the alphabet and characters used to control the operation of the computer and communication equipment. Each arrangement conforms to a standardized coding scheme, which has been agreed upon by the communicating parties. The most commonly used coding scheme is known as ASCII (pronounced ASK-EE) which stands for American Standard Code for Information Interchange. ASCII is a 7-bit code in that it uses 7 bits to represent each character. As stated previously, a byte is 8-bits wide. The eighth bit is used for error checking and will be explained later. For an ASCII code table, see Table 1.

Asynchronous data communication

The first type of communication we will discuss will be that known as *Asynchronous*. Asynchronous means that data is sent irregularly or without *Synchronizing* the data. Each character is sent following a signal from the sending equipment, which also signals the end of the

TABLE 1 ASCII Table

Char	Decimal	Hex	Char	Decimal	Hex
NUL	0	0	SCH	1	1
STX	2	2	ETX	3	3
EOT	4	4	ENQ	5	5
ACK	6	6	BEL	7	7
BS	8	8	HT	9	9
NL	10	a	VT	11	b
NP	12	c	CR	13	d
SO	14	e	SI	15	f
DLE	16	10	DC1	17	11
DC2	18	12	DC3	19	13
DC4	20	14	NAK	21	15
SYN	22	16	ET	23	17
CAN	24	18	EM	25	19
SUB	26	1a	ESC	27	1b
FS	28	1c	GS	29	1d
RS	30	1e	US	31	1f
SP	32	20	!	33	21
“	34	22	#	35	23
$	36	24	%	37	25
&	38	26	‘	39	27
(	40	28	)	41	29
*	42	2a	+	43	2b
,	44	2c	–	45	2d
.	46	2e	/	47	2f
0	48	30	1	49	31
2	50	32	3	51	33
4	52	34	5	53	35
6	54	36	7	55	37
8	56	38	9	57	39
:	58	3a	;	59	3b
<	60	3c		61	3d
>	62	3e	?	63	3f
@	64	40	A	65	41
B	66	42	C	67	43
D	68	44	E	69	45
F	70	46	G	71	47
H	72	48	I	73	49
J	74	4a	K	75	4b
L	76	4c	M	77	4d
N	78	4e	O	79	4f
P	80	50	Q	81	51
R	82	52	S	83	53
T	84	54	U	85	55
V	86	56	W	87	57
X	88	58	Y	89	59
Z	90	5a	[	91	5b
	92	5c	]	93	5d
^	94	5e	_	95	5f
`	96	60	a	97	61
b	98	62	c	99	63
d	100	64	@	101	65
f	102	66	g	103	67
h	104	68	i	105	69
j	106	6a	k	107	6b
l	108	6c	m	109	6d
n	110	6e	o	111	6f
p	112	70	g	113	71
r	114	72	s	115	73
t	116	74	u	117	75
v	118	76	w	119	77
x	120	78	y	121	79
z	122	7a	{	123	7b

|	124	7c	}	125	7d
~	126	7e	DEL	127	7f
--	128	80	--	129	81
--	130	82	--	131	83
--	132	84	--	133	85
--	134	86	--	135	87
--	136	88	--	137	89
--	138	8a	--	139	8b
--	140	8c	--	141	8d
--	142	8e	--	143	8f
--	144	90	--	145	91
--	146	92	--	147	93
--	148	94	--	149	95
--	150	96	--	151	97
--	152	98	--	153	99
--	154	9a	--	155	9b
--	156	9c	--	159	9f
	160	a0	i	161	a1
¢	162	a2	ϵ	163	a3
	164	a4	Y	165	a5
|	166	a6	§	167	a7
"	168	a8	©	169	a9
▲	170	aa	«	171	ab
¬	172	ac	-	173	ad
®	174	ae	¯	175	af
°	176	b0	±	177	b1
²	178	b2	³	179	b3
´	180	b4	μ	181	b5
¶	182	b6	'	183	b7
¹	184	b8		185	b9
°	186	ba	»	187	bb
$^1/_4$	188	bc	$^1/_2$	189	bd
$^3/_4$	190	be	¿	191	bf
À	192	c0	Á	193	c1
Å	194	c2	Ã	195	c3
A	196	c4	A	197	c5
Æ	198	c6	Ç	199	c7
È	200	c8	É	201	c9
Ê	202	ca	Ë	203	cb
Ì	204	cc	Í	205	cd
Î	206	ce	Ï	207	cf
D	208	d0	Ñ	209	d1
Ò	210	d2	Ó	211	d3
Ô	212	d1	Õ	213	d5
Ö	214	d6	X	215	d7
Ø	216	d8	Ù	217	d9
Ú	218	da	Û	219	db
Ü	220	dc	Ý	221	dd
b	222	de	β	223	df
à	224	e0	á	225	e1
â	226	e2	ã	227	e3
ä	228	e4	â	229	e5
æ	230	e6	ç	231	e7
è	232	e8	é	233	e9
ê	234	ea	ĕ	235	eb
ì	236	cc	í	237	ed
î	238	ee	ï	239	ef
ŏ	240	f0	ñ	241	f1
ò	242	f2	ó	243	f3
ô	244	f4	õ	245	f5
ö	246	f6	÷	247	f7
ø	248	f8	ù	249	f9
ú	250	fa	û	251	fb
ü	252	fc	ý	253	fd
p	254	fe	ÿ	255	ff

character transmission. Asynchronous transmission differs from synchronous transmission in that there is no regular timing signal associated with the transfer of data, so data may be transferred at any time.

Imagine you are sitting at home watching the television. You see something on the tube that you think might interest a friend. You call the friend to tell him. This call was not a scheduled communication and happened without any prior agreement between the two of you. You have communicated asynchronously. Computers communicate similarly. Computer A is sitting quietly when computer B sends a request for data. Computer A responds by sending the requested data. The computers have communicated asynchronously. But the similarity between the computers and your telephone call ends there. The computers in the above example have no innate language skills and must be set up to communicate with each other using a common set of rules. All communicating parties must agree upon this set of rules so that any one machine can communicate with any other machine. This agreed upon set of rules is known as a *Protocol*. Protocols are what make data communication possible and reliable. Protocols, however, are used almost exclusively on digital communication equipment.

Telecommunications basics

Before the use of computers became commonplace, the most technologically evolved equipment used (besides color television) was the telephone system. Through the telephone system, we could reach all points on the globe. All sorts of data could be transmitted using this system, albeit slowly. Facsimile was the most common form of widely used data transfer available if one desired to communicate with anything other than voice. For large organizations, the Teletype was a standard fixture, allowing rapid communication to all points of the world.

The telephone system in use today evolved from a giant monopoly. The Bell System was closely held, equipment on the system conforming to a single set of specifications. Through court challenges and the evolution of technology, this slowly changed and now the market is full of products that are designed to communicate with the system. The telephone system has been called the largest single time-shared network in the world, the Internet aiding and abetting this distinction.

The local loop. A user point of interface with the telephone system is the telephone set. From this device, the user accesses the network. What happens to the user's traffic after the receiver is picked up and the number dialed is usually not of any concern to the user, so long as the call goes through. The simplest telephone system is two telephone sets connected together with an integral power source. This arrangement works well over short distances, and is used extensively in small installations, particularly home intercom systems. To communicate with someone in another town or city requires a somewhat more sophisticated apparatus.

At the user, the interface consists of a telephone set or computer. The telephone set is a device that converts acoustic energy into electrical energy for transmission over the telephone network. Telephone sets are available with many different features and enhancements. Basically, all one needs to access the system is a simple handset with dialing capability. The telephone set consists of several parts: the microphone, the speaker, the dialing apparatus, the ringer, and the switchhook (Figure 3). These elements have been integral to telephone instruments almost from the inception of the technology. All telephone instruments must interface with the two-wire systems that serve all but a minute portion of the network. The microphone and speaker (transmitter and receiver) are the primary interface with the network. Until recently, microphones in a telephone instrument conformed to a century-old design that relied upon carbon granules to perform the conversion from acoustic to electrical energy. Today, it is rare to find such a set in regular use. Inexpensive telephone sets using electronic microphones have largely supplanted the older technology. Speaker technology has not changed significantly. The speakers used in modern telephone sets closely resemble those in use for close to a century, with the exception of size.

The microphone and speaker are usually in the handset. In order to enhance sound quality, handsets are designed to provide Sidetone, also referred to as local echo or feedback. This feedback results from a lack

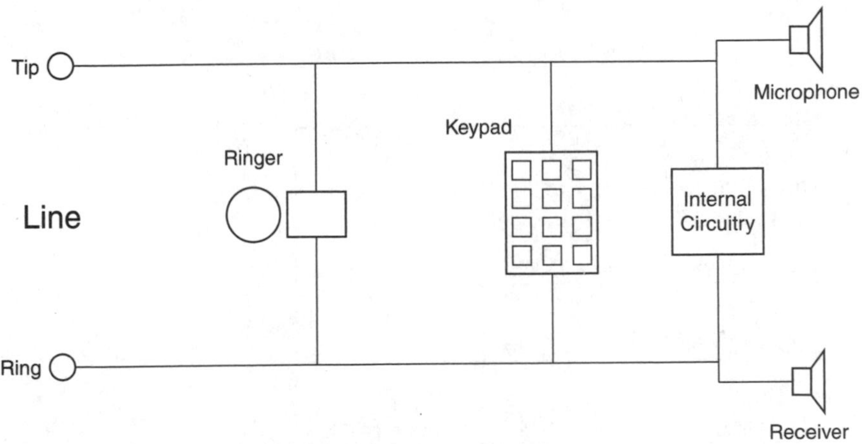

Figure 3 A Telephone Set Schematic

of perfect isolation between the transmitter and receiver. Sidetone permits a user to regulate voice level. If no feedback is provided, users tend to shout into the telephone set, causing distortion. Given too much feedback, users do not speak loudly enough.

The dialing apparatus still exists in two forms. The first is the rotary dial, which utilizes pulses to send the desired connection information to the telephone network's central office for proper routing. The rotary dial operates by alternately opening and closing a series of electrical contacts as the dial spins. This system produces regular interruptions to the system current and produces a series of electrical pulses that sequence switching equipment. The other type of dialing apparatus is the tone keypad. This system uses a standard set of tones that are assigned to each key. Tone dialing is called DTMF, or Dual Tone Multi-Frequency (Figure 4). Each key is assigned a pair of frequencies which makes it unique to the other keys. These tones are transmitted to a tone receiver in the central office for switching and routing. The advantages of tone dialing over pulse dialing are taken for granted. To dial a seven-digit number on a rotary dial required that each number be "dialed," which took up to three seconds for a "0." In tone dialing, a tone duration of less than one quarter of a second is adequate for the receiver to adequately process the tone signal. This means that a seven-digit number could be punched in a fraction of the time a rotary would require, allowing many more calls per unit time.

The switchhook is an important part of the telephone set. This device provides isolation from the network when the set is not in use and signals the system that a call is to be made when the handset is lifted. When the handset is lifted, the telephone set goes "off-hook." This term

1 697hz + 1209hz	ABC 2 697hz + 1336hz	DEF 3 697hz + 1477hz
GHI 4 770hz + 1209hz	JKL 5 770hz + 1336hz	MNO 6 770hz + 1477hz
PRS 7 852hz + 1209hz	TUV 8 852hz + 1336hz	WXY 9 852hz + 1477hz
* 941hz + 1209hz	OPER 0 941hz + 1336hz	# 941hz + 1477hz

Figure 4 DTMF Keypad & Frequencies

is a holdover from the early days of the telephone when the telephone transmitter, or microphone, was hung on a hook which was attached to a switch inside of the telephone set. When the set is off-hook, current flows through the device and to the central office. The central office then provides the user with a line, which is signaled by the presence of a dialtone. Even if the set is "on-hook" or "hung-up," the ringer is still connected to the system. An important distinction is that the voice, or communication, circuit is based upon Direct Current (DC), while the ringer uses Alternating Current (AC).

Ringers are the devices which signal an incoming telephone call. While there may be some older-type ringers still around, the predominant type of ringer is electronic, providing all sorts of unique tones. Older-style ringers used electromagnetic "clapper" type devices with gongs to signal a call.

The telephone circuit consists of two wires. On these wires, called Tip and Ring, after the nomenclature assigned to the corresponding parts of the cord plugs used by switchboard operators, a single Voice Grade channel is carried. The wire that connects the user, or subscriber, to the local office is called the Local Loop. The local loop consists of the wire, poles, terminals, conduit, and ancillary equipment used to connect the subscriber to the local office. Local loops terminate at the local office. The local office is also called the End office or Local switching office.

A local loop may be of two types (Figure 5). The first type is most common and is called a dial-up line or connection. The subscriber, who dials a number and connects through the local switching office to another subscriber, initiates this type of connection at will. When the call is completed, the connection is terminated. The second type of con-

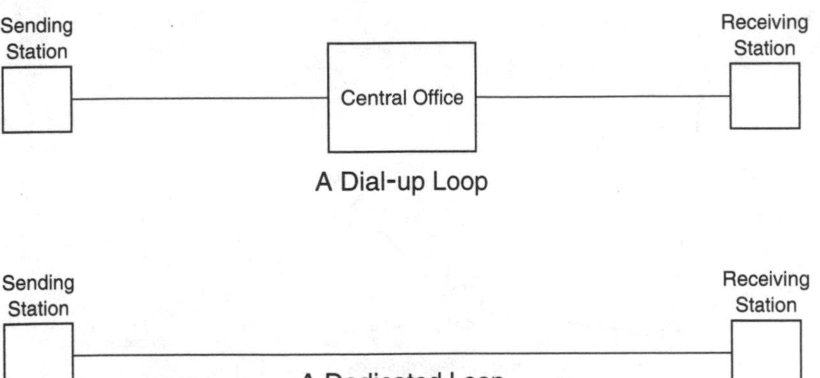

Figure 5 Local Loops: Dedicated and Dial-up

nection is called a dedicated line. A dedicated line is wired to remain connected to another subscriber or location until physically disconnected. These dedicated lines are used by organizations that need to maintain open and uninterrupted communication between branches or locations. This type of connection is also called a leased or private line.

Where several thousand local subscribers are serviced, it becomes impractical to use one central office to service all of them. Smaller central offices are used to concentrate the local loops and are then connected using a cable called a trunk. Trunk cables connect central offices together in order to more effectively serve subscribers. Each central office can set up connections between subscribers who are serviced by different central offices.

Trunks are classified by their function in the telephone network switching hierarchy (Figure 6). In large population centers, a large number of central offices are needed, requiring a large number of interconnecting trunks. This situation is also impractical, and would be impossible to reliably manage. Tandem Trunks connect central offices to Tandem Switches, which serve to switch traffic between central offices. This allows a minimum of cabling, some connections being made through interoffice trunks and others going to central offices outside of the local calling area.

Local trunks are those which connect local central offices through a local Tandem Switch. Traffic may be switched through several local offices to establish a reliable end-to-end connection. This eliminates the need for trunks from every central office, particularly those with small traffic volumes.

For calls outside of the local calling area, Toll, or Interexchange Tandem Switches, connect the central offices to Interexchange (IEC)

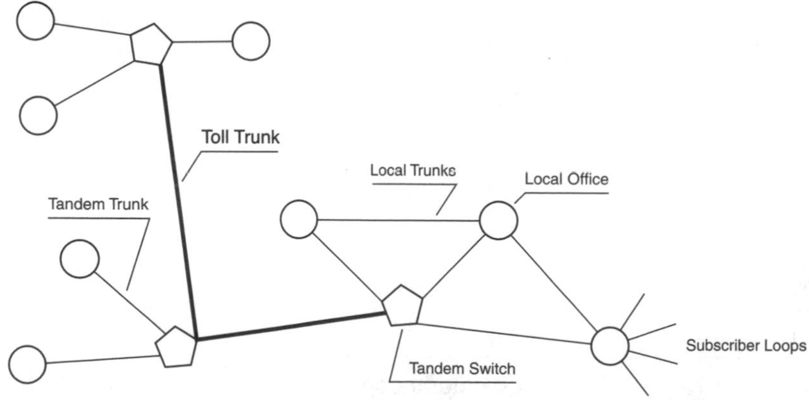

Figure 6 Trunks

Trunks for establishment of the connection. The telephone network is divided into service areas called Exchanges. Exchanges serve specific geographical locales, and are usually defined by the bounds of a population center. IEC Trunks are also called Intertoll trunks.

While there are several good reasons to establish a dedicated connection between locations, these connections can become quite expensive. Unless there is justification on the basis of protecting assets or safety, a dedicated line should not be considered. The majority of connections are made over a dial-up connection. This type of connection is more powerful, and takes advantage of the wide reach of the network rather than being limited to two points. In order to access the entire network, and access subscribers wherever they are located, the network must be capable of switching.

Call switching. Early telephone systems were manually switched by operators. Local loops and trunks terminated at jacks, which were located in banks on Switchboards. An operator connected a call using cords with plug ends. When a telephone set went off-hook, the operator was signaled. The operator would plug a headset and one end of a cord set into the signaling line jacks; after getting the desired number from the caller, the operator signaled the receiving line. After connecting with the called party, the operator would plug the other end of the cord set into that subscribers jack, releasing the call and the headset. The cycle would then start over.

This system worked well in the early days of telephony, when the volume of traffic was low and most calls were locally placed. A cadre of female operators (chosen for the pleasing "melodious" quality of the female voice) quickly developed. Unfortunately, human interaction with and manipulation of the call introduced significant delays and, worse, interference for personal reasons. The development of automatic call switching may not have occurred as it did had just such an incident not occurred.

Almon B. Strowger, an undertaker in Kansas City, noticed that his business began to drop off unexpectedly. Investigating, he found that the wife of his competitor was an operator at the local exchange. Whenever a caller asked to be connected to Strowger, this operator switched the call over to his competitor. He resolved to remove the human part of the switching process. With the help of his nephew, Walter Strowger, he devised an electromechanical switch that depended upon the use of electromagnets and movable switch contacts. He was issued a patent in 1891 for his device. He did not invent the idea of automatic switching, however, as this concept was devised by Messrs. Connolly and McTighe. Nonetheless, the development of the "Strowger Switch" revolutionized the industry.

Automatic switching allowed the elimination of human operators at a greatly improved connection speed. A greater number of subscribers would be possible, leading to larger revenue. Unfortunately, the public did not readily take to the new system. The telephone was "high-tech" in the late nineteenth century, and human operators were a ready source of technical support. Also, the new telephones, which used push buttons to place a call, were expensive and not widely installed. Before consolidation in the telephone industry, there would have been several competing telephone companies serving the same area, requiring possibly more than one telephone at the subscriber's residence or place of business.

The first large-scale implementation of the Strowger switch occurred with its installation in La Porte, Indiana, in 1892. The new system served about ninety subscribers. As will be described below, the switch was relatively unsophisticated. Strowger had formed a company to manufacture his switches and subsequently sold his interests to his backers for $1800. Less than thirty years later, his patents were sold to the Bell System for 2.5 million dollars. He died in 1902.

The Strowger switch is an electromechanical device, utilizing electromagnets, wiper arms, ratchets, and rotating shafts to connect one subscriber to another. Switch contacts are arranged in an arc, in rows stacked one atop the other and allowing space between rows for the movement of the wiper arm. The original switch is described as having ten rows of ten contacts each. The wiper arm was a metal brush or strip that made contact with one of the switch contacts. Strowger's first model was made from a circular "collar" box through which pins were inserted around the periphery. A pencil was used to rotate around and through the pins.

The calling subscriber's line is connected to the wiper arm and the other subscriber's are connected to the switch contacts. For a system such as the small one in La Porte, Indiana, this would have required ninety of these switches, one for each subcriber. The switch worked by stepping the wiper arm through various positions until it reached the called subscriber's number. For instance, if the called subscriber's number was 5-2-3, the wiper arm would move up five levels, then across two switch contacts, and then again across three more contacts. Each subcriber, including the calling subscriber, was assigned to a switch contact. The wiper arm mechanism was controlled by a series of pulses sent from the caller. Early telephones on this system used four push buttons. One button initiated the call, moving the first available wiper arm to its neutral or "home" position. This position rereferenced the wiper arm at the beginning of each call. To enter a number, the caller pressed the first, the "hundreds," button the number of times indicated by the called party's number. In the previous example, the button would be pushed five times. The next, the "tens," button was pushed twice, and the "ones" button three times. The effect of the buttons was first to "seize" the telephone line and the first available wiper arm. Then each button would make and break the circuit, generating pulses that activated an electromagnet. This electromagnet operated a ratchet attached to the wiper-arm shaft. This method of switching was called step-by-step because of the way in which the call was connected. A Strowger switch is shown in Figure 7.

The switch in the illustration is the product of the Automatic Electric Company (AE) circa 1902. These are front and side views of the switch, of which there was one for each subscriber line. The upper half of the illustration shows the electromagnets and driven mechanical linkages. The lower half shows the banks of switch contacts. Several of these switches were connected together to provide a connection. Improved switches and switching techniques allowed a reduction in the size of the switch from the one shown. Also, improved telephone numbering systems were developed producing more effective routing of calls between trunks.

A drawback of the push-button system was the requirement of having to use five wires to place a call. This problem was alleviated through the development of the rotary dial system. The rotary dial

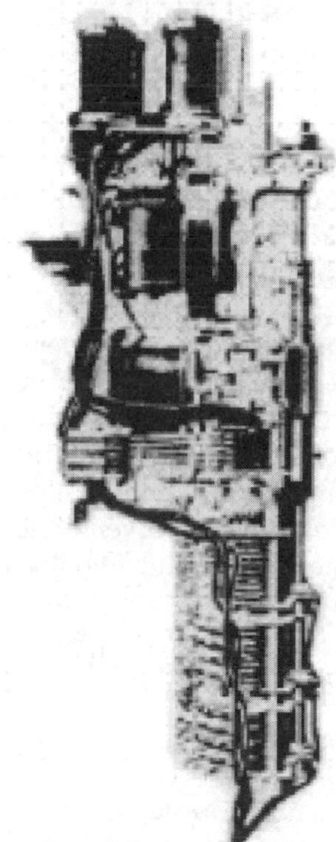

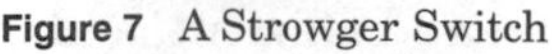

Figure 7 A Strowger Switch

used a system of Timed Pulse (TP) dialing. This system used only two wires for transmission, those being for voice transmission. The dial produced pulses that also produced interruptions in the transmitting line, setting the wiper arm to the proper digits. By placing the finger stop at the extreme travel of the dial, a certain minimum amount of time was required to dial the next number, producing a recognizable break between successive digits.

Without a human operator connecting a call, it is necessary to produce several progress indicators, or Tones by which to signal the subscriber to the condition of the connection. This convention continues today, though several changes have been made. In modern systems, a voice message typically informs a caller of a problem. The most familiar tone is the Dial Tone. The dial tone indicates that the line is in good condition and that there is an available line. This tone also signals the calling subscriber to enter a number. The next signal provided is the Busy Signal or Tone. This tone signals the caller that the called party is connected to another line. A "fast" busy indicates that there are no available lines. These tones are generated by tone generators located at the switching or central office. When a subscriber manages to successfully connect to another subscriber, a Ring tone is sent to the called subscriber's bell. The Ring Generator develops this tone.

It is interesting to note several of the ironies occurring during the development and use of this system. The original push-button telephone set is the precursor of the modern touch-tone telephone set. The function was quite different from the system in use today. Making a call using the button or dial apparatus was initially rejected by the public because it was customary for a human operator to handle the entire transaction. Many people felt that the mechanics of placing a call was "the telephone company's job." Originally, the Bell System rejected the use of dial phones, preferring human operators. When threatened by the possibility of a strike by operators, Bell moved to upgrade the system; dial telephones were introduced in Norfolk, Virginia, in 1919.

With the development and implementation of Digital Telephony, use of the Strowger Switch has all but disappeared. Several museums have dedicated themselves to the recovery and restoration of old switches. Until the development of digital telephony, the development of automatic switching was the most revolutionary advance in telephony to that point. Savings in time and labor were enormous, and a greater revenue was realized by any operator who installed this system.

Digital telephony

The Transistor was introduced to the world in 1948. In less than twenty years, this device has largely supplanted most older vacuum-tube based equipment. The transistor was rugged and efficient, and it generated very little heat, using very little power. It also opened the door to the digital age. As previously discussed, digital data differs from analog data in that digital data represents only two electrical states, whereas analog data represents a continuous, though varying, voltage level for a given input. The conversion to all-digital communication is not complete, however. In the intervening fifty-odd years from the introduction of the transistor, steady gains have been made.

The first area to realize the benefits of the transistor was the computer industry. These devices lent themselves eminently to the application of transistorized circuitry. Recall that the early computers were based either on electromechanical devices or vacuum tubes. These devices were prone to break down and the earliest large-scale computer, ENIAC, employed a staff of twelve people just to change vacuum tubes. Transistors solved many of the problems associated with these devices. The first fully transistorized computer was the CDC 1604, manufactured by Control Data Corporation, and was introduced in 1958.

Analog to digital conversion. The first and most basic problem in digital communication, particularly telephony, is the translation of analog data into digital data. We live an analog world. The sights and sounds surrounding us are of an analog nature. If we were to instantaneously translate the data we see or hear directly into digital pulses, we would approximate the visual by flashing the scene at regular intervals. You could possibly recognize something in one of the flashes. The same is true of sound. One could vary the volume of the sound in a staccato fashion. Again, one could possibly recognize parts of the sound or even a melody, but this would, at best, only present an approximate idea of the sight or sound. Yet, this is the underlying principle behind a technique called Analog to Digital Conversion, or A/D.

A/D is necessary to allow analog data to be transmitted over a digital system. At the receiving end, the data must be reconverted to analog. This step is called Digital to Analog Conversion, or D/A. Analog data consists of a constantly changing voltage level. Digital consists of pulses. The most common method of conversion between the two types of data is called Pulse Code Modulation (PCM). PCM "samples" the analog data and produces an equivalent binary value. Each sample represents the level of the analog signal at the instant of sampling. Each sample is represented by an eight-bit word. Henry Nyquist was an electrical engineer who worked for AT&T in the early part of this century. He became interested in the speed of telephone and telegraph transmissions and the distortion that resulted from attenuation and delay of the transmitted signal. He was also interested in expanding the available bandwidth of the transmission media and reducing transmission costs overall. He was primarily concerned with what he called "Line Speed," which he defined as the speed of transmission of a linear circuit. In 1928, he offered the concept of "sampling" as a way of increasing the bandwidth and transmission speed of a transmission media. Sampling is the technique by which an analog signal is broken into smaller pieces of a fixed value. Nyquist theorized that an analog signal could be perfectly reproduced in this manner by sampling at twice the frequency of the highest frequency of the signal. Another way to say this is that the highest frequency that can be represented in this manner is one-half the frequency of the sampling frequency. The sampling frequency is also called the "Nyquist Frequency." Figure 8 shows a representation of the conversion.

The range of the human voice, for instance, is 0 to 4000 Hz. According to Nyquist, it would take 8000 samples, 2 × 4000, to perfectly capture the entire range of the human voice. In practice, the signal is "quantized into 256 samples." This is not an arbitrary value, but 256 is the number of values that can be represented by one 8-bit word. If we take 8000 eight-bit samples per second, the required bandwidth for a voice-grade signal is 64000 bits per second, or 64 Kbps. Again, these numbers are not arbitrary and fit neatly into the hierarchy of telecommunication circuitry. The basis for high-speed communication channels worldwide is the single 64-Kbps channel. The basic 64-Kbps channel is called a DS-0 channel. To make up a T1 circuit, 24 DS-0 channels are multiplexed into 192 bit frames. To each frame is added a "framing" bit, making each frame 193 bits. Multiplying this figure by 8000 samples gives us 1,544,000 bits or 1.54 Mbps, the speed of a T1 carrier. In Europe and Mexico, the carriers are given the designation E1 and consist of 30 DS-0 channels or 2.048 Mbps. These carriers are explained in more detail as follows.

Fractional T1 is built-up of increments of DS-0 channels to a maximum of 768 Kbps. These circuits are typically used in dedicated circuit situations. DS stands for Digital Service. It is the telecommunication service used in most of North America and defines a four-level hierarchy of service, increasing in bandwidth. All DS makes use of the sampling technique described above. Digital service is built-up from the single DS channel into larger capacity channels. The largest bandwidth available at this writing is the DS-4 channel (T4), which has a bandwidth of 274.176 Mbps. Bandwidth is a term used to describe the

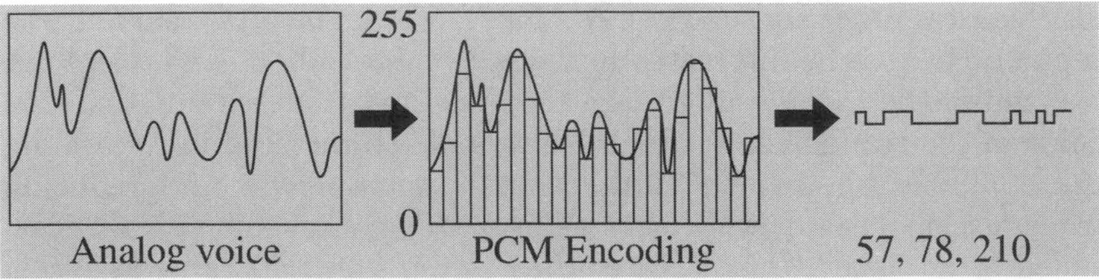

Figure 8 Analog to Digital Conversion

throughput capability of the circuit. To illustrate, picture a DS-0 channel as a path through the woods that only allows one person to walk in each direction. A DS-4 or T4 circuit would be comparable to 20 lanes of superhighway.

Channels are not simply added together to make up larger capacity circuits. A technique known as multiplexing is used to combine channels and data into larger circuits. This assures the subscriber timely delivery of transmitted data. Multiplexing further samples and parses a series of data frames into smaller units and then interleaves these units together. At the sampling rate used, the multiplexing procedure is not noticeable and data is received apparently in one piece. Another example of multiplexing is to picture a railroad company that must ship goods from four different customers with the stipulation that the goods must reach their destination at the same time. The customers are paying for this service, so the railroad must figure out a way to do this in the most cost-effective manner.

One solution is to send four trains and boxcars on four separate tracks. This would theoretically get the goods to the destination in the prescribed time frame, but would be unnecessarily expensive. Another solution would be to send the same four trains along the same track. This would be less expensive, and would introduce a minor delay in transmission, though not necessarily in delivery. The third and most practical solution is to make up one train consisting of the four boxcars and one engine. All of the boxcars would arrive at the same time using the same track. At the receiving end, data posted on each boxcar would indicate further routing to the correct customer. In effect, the goods in the boxcars are interleaved with other goods and boxcars. This is essentially what happens in multiplexing.

Multiplexing is used extensively to extend the capacity of a transmitting channel. As described in the example, multiple messages are impressed upon the transmitting medium to allow them to be sent at the same time. Multiplexers are devices located on the transmitting end of the circuit and are referred to as a "MUX." On the receiving end, a demultiplexer, or "DEMUX," is used to reverse the process. There are two major methods of multiplexing in common use. The first is called Time Division Multiplexing, or TDM. The other is called Frequency Division Multiplexing, or FDM.

Time Division Multiplexing is basically the breaking up of a message into much smaller parts for transmission with other similarly broken up signals. The multiplexer performs the actual "breaking up" of the transmitted signal. Each piece of the signal is assigned a reference number and impressed upon the medium along with other signal fragments. The signals form a composite signal and contain fragments which are assigned in a repeating, rotating fashion. Figure 9 illustrates the strategy used for parsing and transmitting signal fragments.

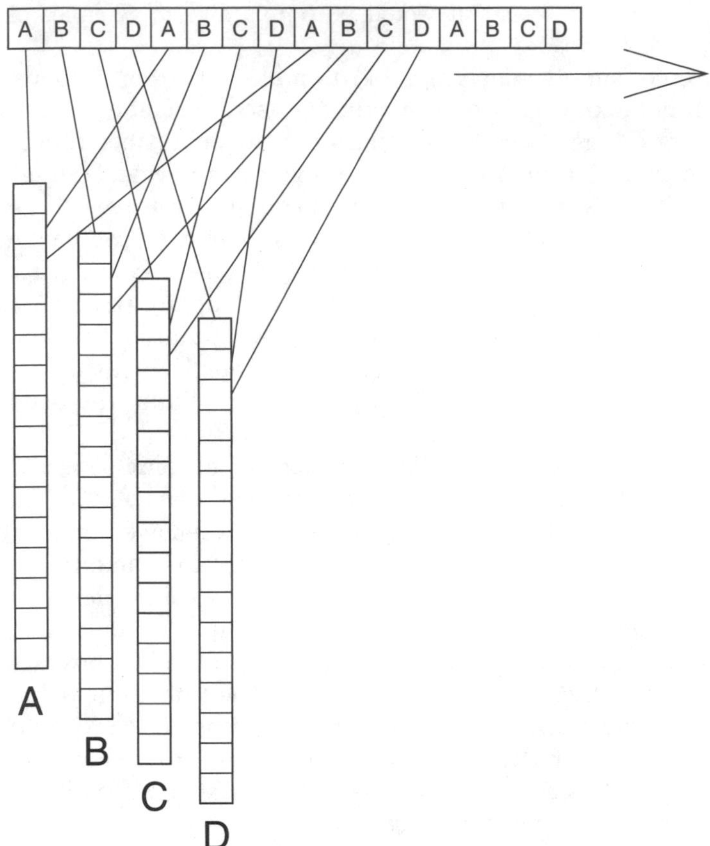

Figure 9 Time Division Multiplexing

Three variants on TDM are defined:

- Asynchronous Time Division Multiplexing (ATDM). Multiplexing in which data is transmitted asynchronously. A discussion of asynchronous and synchronous transmission follows this section. Briefly, asynchronous transmission indicates that data is transmitted in an irregular fashion. There is no regular schedule for the data to be sent and transmission occurs whenever data is ready to be sent. As data is gathered into a frame for transmission, start and stop bits, or frame delimiters are used to inform the receiving demultiplexer where the frame begins and ends. ATDM uses tags, or circuit IDs to provide flexibility and to accommodate "bursty" traffic. Asynchronous Transfer Mode (ATM) is the prevalent protocol used to take advantage of ATDM's flexibility. ATM will be covered in Chapter 2.

- Statistical Time Division Multiplexing (STDM). This method polls each node and skips those that have nothing to send. This is an overly simple explanation, however. This method requires additional overhead to allow reporting of which channels were actually sent. The DEMUX uses this information to properly reconstruct signal fragments in the correct order. Much data must be buffered, or stored, until the entire signal is received in order for proper demultiplexing.
- Synchronous Time Division Multiplexing. This method uses clock pulses to parse and arrange signal fragments. Each clock pulse determines the length and arrangement of each fragment. Interleaving of fragments can be at the bit, word, or block level of data. Transmissions are based upon regular timing pulses that synchronize the multiplexer and demultiplexer. Large amounts of data can be moved very reliably using this technique.

Frequency Division Multiplexing (FDM) divides the transmitting channels into smaller channels and uses these smaller channels to transmit data. Each of these channels is defined by a unique frequency within the transmitting carrier band. Between each channel are "guard bands," which separate each channel in order to avoid "bleeding" of data between channels. The original implementation of this technique was using "modem stacking." Each transmitting channel used a modem that transmitted at a unique frequency. Each modem's output was combined at the system output point and then separated at the receiving end by identical modems. FDM systems suffer from problems with cross talk if the guard bands are not sufficient. Intermodulation noise is also a problem, producing intermediate frequencies that produce audible interference in the transmission. Figure 10 illustrates FDM.

Wireless Communication Systems

This section covers the common wireless data and voice technologies available. Wireless telephony is a revolutionary advance in telephone technology. Since voice communication over wire was introduced by Bell, inventors sought to provide a means by which anyone could be reached, anywhere. T. A. Edison, of light bulb fame, developed a crude wireless system that allowed the transmission of telegraph pulses to a moving train without physical contact. American and British wireless

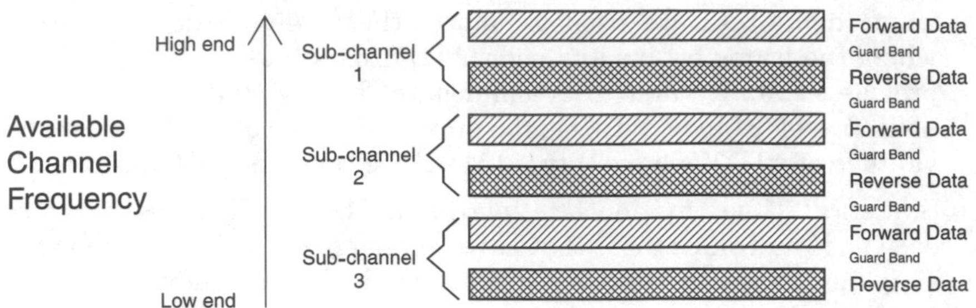

Figure 10 Frequency Division Multiplexing

companies operated a shortwave telephone link between Great Britain and North America, later extending service to South America and Africa. These services were quite expensive and their use was limited.

Wireless data communication is steadily coming to the fore, offering such services as wireless modems and Personal Communications Services (PCS). These technologies will become commonplace; the integration of telephones into small personal computers will literally bring the world to your fingertips. The following technologies will be explored:

- Improved Mobile Telephone Service (IMTS)
- Advanced Mobile Phone Service (AMPS)
- Narrowband Advanced Mobile Phone Service (NAMPS)
- Digital Advanced Mobile Phone Service (DAMPS)
- Personal Communication Services (PCS)

Improved mobile telephone service

Originally, wireless telephony was simply two-way radio with car mounted telephones connected to a human operator. The first use of mobile radio telephone was by the Detroit police department in 1921. The system operated at 2 MHz and quickly became overcrowded. By 1940, the frequency band between 30 MHz and 40 MHz was made available for two-way mobile radio. These early systems were not duplex systems however, requiring the press and release of a switch, or key, to allow each party to speak across the same channel. The mode of transmission is known as "Half-Duplex." The system is also called "Push to Talk," or PTT. Eventually, these frequencies also became overcrowded.

In 1946, the Federal Communications Commission (FCC) announced what it called a "Public Correspondence System," giving it the official name: Domestic Public Land Mobile Radio Service (DPLMRS).

The DPLMRS operated between 35 MHz and 44 MHz and served the area along the highway that ran between New York City and Boston. These systems were limited to commercial users and covered large areas served by a single base station. Repeaters were often used to extend the range of the system.

Overcrowding continued and forced the allocation of new channels by decreasing the channel spacing of existing channels. In 1956, eleven additional channels were assigned in the 150 MHz band and twelve more were allocated in the 450 MHz band. These early systems did not provide a dial tone or dialing capability. For a subscriber to make a telephone call, the system was keyed and service was requested. A mobile operator came on the frequency and completed the call. Several attempts were made to develop a system which would work without operator intervention.

In rural areas, it was not feasible to have a mobile operator on duty at all times. The Rural Electrification Administration issued a specification that outlined the performance criteria of a dial-based mobile telephone system. Several different firms attempted to implement the system, but it was not until 1958 that a workable system was developed and put into regular use. This early trial proved many designs and techniques and paved the way for improved systems. The development of two-way unattended mobile telephone service evolved into what became the "IMTS."

In 1969, the Improved Mobile Telephone Service, or IMTS, was introduced. IMTS was the first true mobile telephone system, allowing dial-up, duplex telephone conversations. This system operated in the band 150 MHz through 450 MHz and incorporated automatic frequency or channel selection. This was the mobile telephone service that was predominantly used by commercial subscribers through the early 1980s. Due to the limited number of channels, which allowed only one conversation at a time, delays in obtaining a free channel could be extensive. Some systems of this type are still in use today, but are being steadily replaced by modern cellular technology. Other mobile systems are still in use and operate as PTT systems. These types of systems are the staple of taxi fleets and some small commercial operations.

IMTS operated from powerful base stations that were as centrally located to a service area as was practical. In large urban areas, the system worked extremely well, given the limitations described. Once a subscriber acquired a channel, the connection was maintained until the call was completed or the mobile telephone went out of range. The frequency was "locked" by the subscriber using it. This proved to be the major impetus for development of an improved mobile communication system.

Advanced mobile phone service

We know wireless as "Cell Phones" because of the technology that makes their use possible. As has been said by many whom were consulted during this writing, the theory behind cellular telephones is simple, whereas the actual implementation is extremely difficult. The problems of the efficient use of available channels, power and size of mobile units, and connection to the wired telephone network had to be addressed and dealt with before any attempt could be made to implement mobile telephone service for the masses. The large-scale integrated circuits made the size of the telephone manageable, and almost constant research and development by Bell Labs solved most of the problems encountered.

Early in 1981, a new mobile telephone standard was introduced. Called Advanced Mobile Phone Service, it was the first system to use the concepts of Cells and Frequency Reuse. Cells are defined as the area of coverage of a given base station and can cover from two to ten miles depending on location. Cells are represented by hexagons. This is purely convention, however, and is used for planning and statistical purposes. The true shape of the covered area may be completely different given the physical terrain of the cell. The optimum shape for a cell is a circular pattern of transmission coverage. These comparisons are illustrated in Figure 11.

Without the technique of frequency reuse, cellular telephones would not be possible. Each cell is assigned a specific number of channels (frequencies), which cover users in that cell area. As a user travels from cell to cell, the user is switched into another channel, freeing up the previously used channel. This is a departure from the older IMTS system where a user remained on the assigned channel until the call was terminated or the user moved out of range of the transmitter. Sophisticated switches called Mobile Telephone Switching Offices (MTSO) connect the cell site to the wired telephone network and to other cell sites. The terms cell site and base station will be used interchangeably. Both terms describe the radio transmitter that serves the cell.

In addition to frequency reuse, multiplexing techniques are used to allow a higher concentration of calls over a given channel. The multiplexing techniques used are similar to those used in wired telephone technology. One variant of the AMPS system is NAMPS, which narrows the channel bandwidth to enable more voice channels. Fiber optics has also allowed a larger volume of calls to be completed reliably. Most cell sites are connected to the MTSOs using fiber optics. All connections between MTSOs and the wired system's switches are T1 and run over a fiber-optic trunk.

Channels and clusters. The original AMPS standard had 666 channels available. Of these, 42 channels were used for control purposes. We

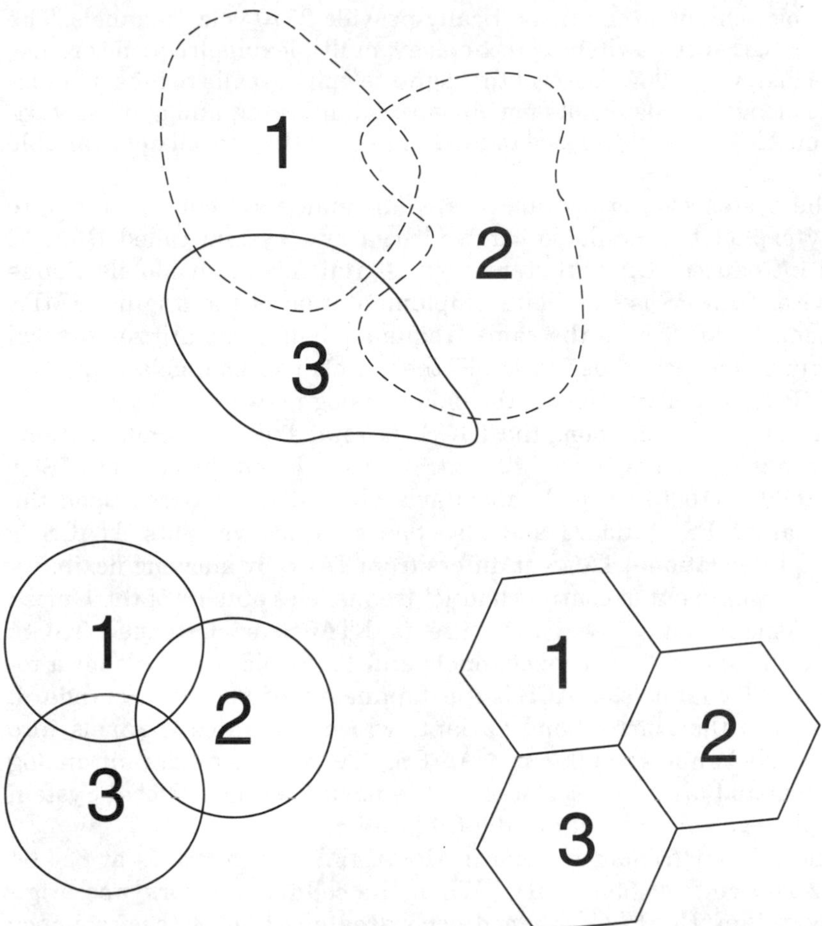

Figure 11 Actual Vs. Idealized Cell Coverage

will explore these signals below. 624 channels were available for voice traffic. It became apparent shortly after introduction that the system would need to be expanded. This led to the expansion of the system to a total of 832 channels. This system also has 42 control channels, and an additional 166 voice channels. The newer system is occasionally referred to as EAMPS, or Extended Advanced Mobile Phone Service.

A further variant to the AMPS standard is called NAMPS, or Narrowband Advanced Mobile Phone Service. In NAMPS, each channel is split into three separate channels, allowing for a tripling of capacity within the same allocated bandwidth. This system is used in large urban areas utilizing much smaller cells than would be used for an extended AMPS system. This system uses 42 control channels also. The

790 voice channels can theoretically provide 2370 voice channels. The use of high-speed switching technology, multiplexing, frequency reuse, and clustering allows several thousand telephone calls to take place simultaneously. This is not enough capacity, however, and another variant on AMPS was developed to further extend the capacity of available channels.

The systems described above are all analog systems. In order to truly exploit the available bandwidth, a new system called DAMPS was introduced. DAMPS stands for Digital Advanced Mobile Phone Service. DAMPS is the digital implementation of the original AMPS system. DAMPS uses the same frequency band, but utilizes several multiplexing techniques to triple the number of channels available. DAMPS is rapidly replacing the older analog network.

In the United Kingdom, the TACS, or Total Access Communications System is used. TACS uses 42 control channels, but has either 558 or 958 (600 or 1000) voice channels available. TACS is based upon the original AMPS standard and also has spawned variants. ITACS is called International TACS, it differs from TACS by allowing flexibility in the assignment of control channel frequencies outside of the United Kingdom. A newer system in use is ETACS, or Extended TACS. ETACS also has 42 control channels and 1278 voice channels for a total of 1320 channels. JTACS is the Japanese variant of this standard. NTACS is the Narrowband variant, which separates channels into three subchannels similar to NAMPS. TACS systems are all analog systems and are being replaced by the newer GSM, or Global System for Mobile communication, a digital system.

The cellular frequency band in the AMPS system starts at 824.04 MHz and ends at 893.7 MHz. When the cellular network was originally set up, the FCC assigned each area one half of the frequency band. These halves are called the "A" band and the "B" band. This allows two providers in each coverage area. Coverage areas are referred to as Metropolitan Statistical Areas (MSAs), or Rural Service Areas (RSAs). The A band automatically went to the local exchange carrier (LEC), the B band to any group or individual who could meet the rigorous financial and professional requirements for ownership of the frequency band. No advantage presents itself to having either the A or B bands. One carrier may, with the proper application of technology, provide up to three times the capacity with the same frequency spectrum.

Within the approximately 70 MHz of bandwidth available for cellular telephones, three subbands exist. Cellular (mobile) telephones transmit within the band between 824.04 MHz and 848.97 MHz, roughly 25 MHz. Base stations, or cell sites, operate within the frequency band between 869.04 MHz and 893.97 MHz, again roughly 25 MHz. A 20 MHz guard band separates the transmit and receive frequency subbands. In addition, 45 MHz separates each transmit frequency from the corresponding receive frequency to prevent interference. For instance, 824.04 MHz is 45 MHz below 869.04 MHz. Remember that the transmit and receive frequencies are "paired" and fixed in order to maintain the 45 MHz separation. These frequency pairs constitute single full-duplex channels. Each channel in each band is 30 KHz wide. This allows a total of 832 channels (832 × 30 KHz = 24.96 MHz).

Of the above channels, 790 are assigned to carry voice traffic. The remaining 42 channels are used for control signals between the mobile unit and the base station. The control channels are also full duplex. The control channels carry information that controls the operation of the telephone and are integral to the efficient use of available frequencies. Each cell has 2 control channels, one for each system. The mobile unit monitors one of the two frequencies for information needed to initiate or complete a call. Control channels are used to set up the call; the actual conversation takes place on a voice channel. The control channel in each cell is usually the first channel assigned to the cell.

Frequency reuse is the underlying principle of cellular communications. The same bands of frequencies can be reused in different cells, allowing the most efficient use of the frequency spectrum. Each cell may have up to 45 voice channels. Each cell has one control channel assigned to it. Channels are assigned to cells to arrange the maximum geographical distance and assure channel separation. As a mobile telephone travels from one cell to another, the circuit is automatically switched to another channel in the adjacent cell. Continuing through several cells, a mobile telephone may be switched to many different vocie channels without any perceptible interruption in service.

Cells are generally grouped into clusters. In less congested areas, a 12-cell cluster, or Frequency Reuse Pattern, is used to handle all traffic. A typical cluster is 7 cells. Figure 12 depicts an idealized 7-cell cluster. A cluster utilizes 42 channel sets, 21 for each system. We will not consider the separation of cells into A and B systems for the remainder of this discussion. All data will pertain to the total system channel capacity of 832 channels. Each cell is assigned 6 channel sets, each channel separated from another by 7 channels. An example would be: Cell #1 is assigned channels 1, 8, 15, 22, 29, 36; Cell #2 is assigned channels 2, 9, 16, 23, 30, 37; Cell #3 is assigned channels 3, 10, 17, 24, 31, 38; and so on through the assigned channels. No frequency pair is reused within the cluster. In an adjoining cluster, channels are similarly assigned. This method of channel assignment and physical separation of cell sites is employed to avoid any occurrence of cross-channel interference.

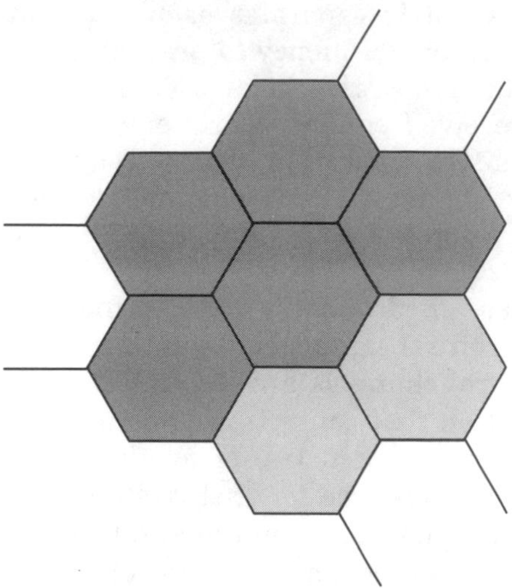

Figure 12 A Cell Cluster

Cells can be further divided into sectors. In the example above, an ideal cell was illustrated. An omnidirectional antenna is used to transmit in all directions outward from the base station. Within the cell, a mobile unit uses the assigned channel until it is handed off to another cell and assigned another channel. Using directional antennas, cells may be separated into 3 or 6 sectors depending on the number of antennas used. The effect of sectorization is to provide much better coverage in the cell. Theoretically, the amount of traffic handled is also increased. One disadvantage is a greater chance of cross-channel interference.

Frequency reuse is only possible with the careful assignment of frequencies and channels to cells that are separated by adequate distances. Figure 13 illustrates typical cell frequency reuse patterns based upon the 7-cell cluster. Ideally, this is the smallest possible distance. The larger the number of cell sites in a given geographical area, the greater the chance for interference. Sophisticated analysis of traffic needs and trends allows designers to manipulate coverage areas and channel assignments. The specifics of this analysis is well beyond the scope and level of this book; through this analysis it was discovered that cellular telephone systems can grow without geographical constraints. The minimum practical size for a cell is a coverage area of about one mile. Tunnels, bridges, and highly congested urban areas are often served by these "microcells," which may be up to 300 miles in diameter. Picocells are even smaller, covering an area of only 30 miles. Picocells are used in the office or on college campuses in wireless networking applications. At the high end, a cell could be as much as 35 miles in diameter, the major constraint being the transmitting

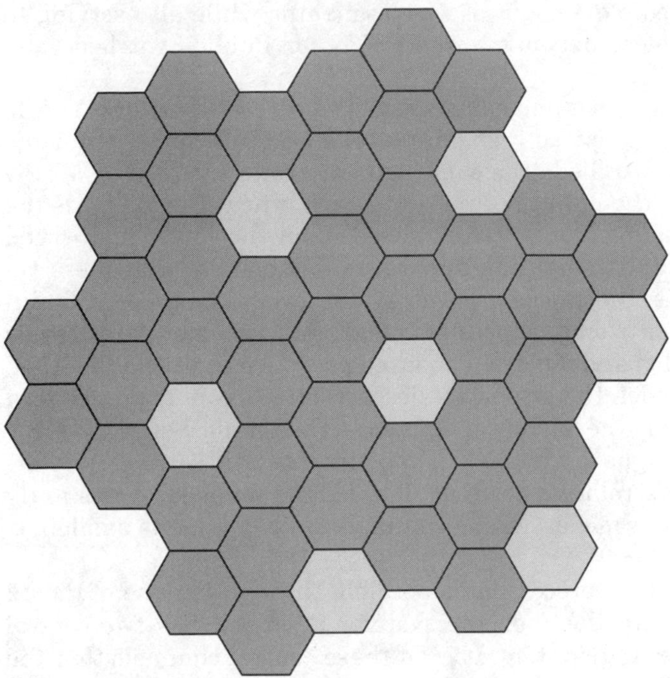

Figure 13 Cell Frequency Reuse Patterns

strength of the mobile unit. Other factors affecting propagation are trees, hills and valleys, mountain ranges, and buildings.

Presently, cellular telephones are limited to a maximum radiated power of 7 watts. This power is referred to as ERP, or the Effective Radiated Power. Base stations in MSAs are limited to 100 watts. RSAs may transmit to a maximum of 500 watts if the base station is located more than 25 miles from the nearest MSA. Larger cells do not depend as much on frequency reuse as do smaller cells. This is in direct consequence of the amount of available traffic in the cell's coverage area. As traffic increases in a given area, cells are split and clustered to take advantage of frequency reuse.

Call switching. Before we discuss the mechanics of making and receiving a call, call switching must be understood. The central control for all of the activity in a cellular system is the Mobile Telephone Switching Office, or MTSO. The MTSO monitors every call on the cells to which it is connected. The MTSO tracks every mobile unit within the range of its controlled transmitters. The MTSO measures each mobile's signal strength and controls the hand off of that mobile to an adjoining cell while dynamically reassigning channels. The MTSO handles an incredible amount of wireless traffic while also serving to connect the wireless network to the PSTN, or Public Switched Telephone Network.

The MTSO supervises and administers the wireless network. This is a fantastically complicated task. Many cells and cell clusters must be monitored and controlled. The same cells are connected to the MTSO for connection to the public network and each other. The MTSO is the brain of the system. The switch is based upon the latest high-speed communication and computing apparatus. Without having these resources available, the capacity of the system would be greatly diminished, if operation would be possible at all. All customer data travels through the switch and must be validated prior to handling a call. Customer data includes the status of the customer account, the class of service requested, the mobile telephone's Electronic Serial Number (ESN), and the signal strength. The signal strength is used to determine the mobile's relative position. The MTSO must also constantly monitor channel usage, assigning channels as they become available.

Call processing. Call processing is possible through the use of the 42 control channels available on the system. Each cell has two control channels, one for each system. It is on these control channels that the mobile and MTSO negotiate for the system's resources. Each mobile telephone monitors the control channel for the system to which it is subscribed. This can lead to delays when several thousand subscribers are trying to access the system. We will walk through the process of making and receiving a call and the various signals required to make it happen.

Of the two frequency pairs required for a voice channel, one frequency is dedicated to transmission from the mobile unit to the cell site. This is the lower of the two frequencies and is called the reverse channel. Communication from the cell site to the mobile unit is on the higher frequency and is referred to as the forward channel. Each call uses both forward and reverse channels. Both voice and control channels are divided into forward and reverse channels. Calls get set up on control channels and then are assigned a voice channel by the MTSO.

The first step in the process is registration. Registration begins when the cell phone is turned on. The unit immediately begins a sweep of the available frequency band, looking for the strongest forward control signal. This signal is typically supplied by the nearest base station and cell site. The phone then locks on to the signal and begins transmission of data on the corresponding reverse control channel. The data is sent to tell the system that the unit is operating and active and ready to take calls. This data includes the phone's ESN, the telephone number assigned to the phone, and the home system ID. The cell site then transmits the information to the MTSO, which then tells the world. The entire process takes only a few hundred milliseconds. Registration occurs every fifteen minutes or so and whenever the phone travels into another cell. The local exchange carrier first checks its database; if the unit's information does not match with valid data, the call will not be put through. The local database is regularly updated by larger databases. Databases maintained by organizations such as Electronic Data Systems and GTE keep track of valid ESNs and telephone numbers. The MTSO can also request registration data from the mobile by transmitting the request on the forward control channel. The phone responds by transmitting registration data to the cell site on the reverse control channel.

After the phone is successfully registered, it is ready to accept calls. The unit idles while monitoring a forward control channel for a paging signal. This paging channel is called an Initial Paging Channel, or IPCH. Large systems typically have multiple paging channels. The paging channel is unique in that it is transmitted across the entire cellular coverage area, from every cell site in every cluster. A paging signal is sent whenever a call is being made to the mobile's number. The public switched network sends the call to the MTSO, which sends the signal to the cell site nearest the mobile's location, which the MTSO has been monitoring. The paging signal contains the mobile's ESN and telephone number. Note that the forward control channel is used to complete calls, while the reverse control channel is used to make, or initiate, calls.

The mobile unit has been monitoring the forward control channel and recognizes its telephone number. The mobile responds by sending registration data and an acknowledgement of the page on the reverse channel to the cell site. The cell site sends this message back to the MTSO, which then assigns an available voice channel in the cell the mobile is in. This all happens within a few milliseconds. The MTSO then sends a Supervisory Audio Tone (SAT) over the assigned forward voice channel. The mobile, if it is tuned to the correct channel, will receive this tone and send it back on the reverse voice channel. This does two things. First, it completes the loop between the cell site and the mobile telephone, verifying that the phone is on the correct frequency. If the SAT is not returned, the connection is not made and the call is not completed. Second, it gives the MTSO a better idea of where the mobile is, owing to the time it took for the SAT to be returned from the phone. The MTSO could then find a cell site closer, if possible, to improve reception.

When the SAT is returned to the MTSO, the forward voice channel (FVC) is released, or unmuted, and a digital signal is sent alerting the phone to an incoming call. An acknowledgement is sent over the reverse voice channel (RVC) to the MTSO after the RVC has been unmuted. This signal is called the Signaling Tone (ST) and continues until the call is answered. The ring generator in the LEC sends a ringing tone to the caller while the phone rings. The signaling tone is a 10 kHz tone which is used to communicate with the cell site over the reverse voice channel. The signaling tone is used to acknowledge commands and in the hand-off procedure between cells.

When a call is about to be passed or handed-off to another cell, the ST is sent 50 milliseconds before the hand-off occurs. The SAT is also sent. The MTSO assigns a new channel, confirms the correct SAT and then listens for the ST after the handover from the previous cell. The whole procedure happens in less than a second and usually without perceptible interference. The ST is also used for signaling status and activity back to the MTSO over the reverse voice channel.

Making a call from the mobile unit is slightly more complicated. After the telephone number is punched into the phone and the Send button is pushed, the mobile sends the number and a request for service over the strongest reverse control channel available. After the MTSO validates this information, it assigns a voice channel and transmits the assignment to the mobile, which tunes to that channel. The MTSO transmits a SAT over the forward voice channel assigned. The mobile, if it detects the SAT at the assigned channel, transmits the SAT back to the cell site on the RVC. The MTSO receives the SAT and switches the call through from the mobile.

While it seems that this process is just the reverse of the process of receiving a call, the similarity ends there. Before the MTSO even assigns a voice channel, it must determine if the caller requesting service is, in fact, a valid user. This “Pre-Call Validation” is necessary because of fraud and the possibility of enormous loss to cellular carriers, who are responsible for the cost of most fraudulent calls.

The mobile unit requests service when the send button is pressed on the mobile telephone. This request is sent over the RVC to the cell site along with information about the telephone. This information allows the MTSO to check it against the databases maintained by GTE and others. These databases are regularly updated and allow carriers to keep track of subscribers. This information is sent repeatedly to ensure that it is received. Four separate identification numbers identify the telephone:

- The home system identification number (SID)
- The mobile identification number (MIN)
- The Electronic Serial Number (ESN)
- The station class mark (SCM)

The SID identifies the cellular telephone carrier to which the user is subscribed. This number tells the MTSO if the mobile is in its home area. If it is not, the roaming indicator is shown. The MIN is the number assigned to the mobile phone. The ESN is a unique number that is burned into the mobile unit's read-only memory (ROM) and cannot be changed. Every mobile telephone in existence possesses a unique ESN. The SCM informs the MTSO at which power level and frequencies the telephone is operating. The cell site can instruct the mobile to turn down or lower its transmitting power level if a lower power level can do the job. This avoids any interference within the cell coverage area.

Multiplexing. A brief description of three common multiplexing techniques close out this discussion of cellular telephony. Three methods are currently used:

- FDMA—Frequency Division Multiple Access
- TDMA—Time Division Multiple Access
- CDMA—Code Division Multiple Access

The first, FDMA, was widely used in analog systems, and has been largely replaced by the following two techniques, which are more suitable for use in digital systems. Aside from providing advanced features such as caller ID, digital telephony does not provide any great advances in the technology. Coupled with advanced multiplexing techniques, though, it allows marked increases in cellular frequency-use efficiencies. TDMA is used with the remaining analog systems as well as the newer digital systems. CDMA is, by definition, a purely digital system.

FDMA. Frequency Division Multiple Access was the first technique used for increasing the capacity of available cellular channels. It was the basis for the NAMPS system, the "N" designating Narrowband. Each 30 kHz cellular frequency was broken into three 10 kHz subchannels over which a conversation was transmitted. This allowed three times as many conversations on a given channel.

TDMA. Time Division Multiple Access, is another multiplexing technique used to increase frequency-use efficiency. In TDMA, each narrowband subfrequency is divided into six repeating time slots, or frames. Two slots in each frame get assigned to each call. TDMA suffers from "multipath fading" and distortion. These phenomena are caused by reflections of the signal from any number of physical interferences. These reflections cause interference and distortion due to time differences in the arrival of the signal from different paths. Also, fading is caused by the same physical interferences. A way to work around these problems is to put a time limit, a "window," during which a signal transaction can occur. This will effectively eliminate multipath signal interferences. Fading can be cured by better cellular coverage. Fading usually occurs at the fringes of the coverage areas. Lack of available time slots poses another problem to the use of TDMA. If all time slots are being used, then a subscriber will not get a dial tone. If a subscriber travels from one cell to another cell that has no available time slots, the call will be dropped.

ETDMA, or Enhanced TDMA makes even more efficient use of the frequency spectra. ETDMA actively monitors a conversation, sensing pauses or dead spots in the communication. The system utilizes these holes in the transmission and stuffs them with bits from other conversations. This system is not universally used. This technique makes the most effective use of the available time slots.

CDMA. Code Division Multiple Access is the latest method for improving spectral efficiency. CDMA works by assigning a unique code to each message and then transmitting the encoded message over single or multiple frequencies. Using this method, channel efficiency can be increased by a factor of 20. This is a vast improvement over FDMA and TDMA, which allow up to a maximum of 6 to 1 improvement in channel efficiency. The principles of CDMA are well known, having been used in military communication for the past 35 years. Chief among the advantages of CDMA are its excellent immunity to noise and improved security from interception (improved privacy). Call quality is also greatly improved.

CDMA utilizes a method of Spread Spectrum technology called Direct Sequence. Spread spectrum is a technique by which the transmitted information is spread over a bandwidth greater than that of the transmitted information. Spread spectrum will be discussed in a later section of this chapter.

In CDMA, a conversation is encoded with a "pseudonoise" code which spreads the information out over a much larger frequency than is used in the previous methods discussed. This has the effect of lowering the "Signal to Noise Ratio," or SNR. Being as each conversation is assigned a unique code, the entire frequency spectra can be used for the call. Multiple calls can exist simultaneously on the same frequency. An analogy has been made of a party where persons of several different nationalities are speaking to each other from across the room. The pairs speaking the same language can understand each other from across the room, without causing verbal interference to other pairs speaking in another language. This is an oversimplification, of course, but the principle is the same. At the receiving end of the transmission, the call is decoded and assigned to the proper receiving party. Further to the analogy, it can be said that the room is analogous to the frequency channel. There is a limit to how many people can be speaking before the cacophony causes everyone to become unintelligible. In other words, there are still limits to how many calls can be carried on a given channel.

Systems utilizing CDMA can also regulate the power output of the mobile units communicating with the cell site. This is done to maintain the lowest possible and usable signal needed to effect communication. This also avoids the "drowning out" of a mobile unit that is farther from the cell tower than a mobile unit in close proximity. Multipath fading is not an issue using this system of multiplexing because of the spreading and encoding of the signal. The details of CDMA operation were developed and implemented over a relatively short time. The benefits of this technology were easily recognized by those entities providing cellular service. CDMA is a purely digital system, and it is used in all of the major service areas. CDMA will completely replace TDMA as the preferred method of wireless data communication when the last analog systems are eventually replaced by digital systems.

Spread spectrum. Spread spectrum (SS) is a modulation technique which spreads the transmitted signal over a much larger frequency bandwidth. Fundamentally, spreading a signal over many narrowband (read "noisy") channels increases the probability that the signal will be received with little or no interference. By spreading the signal, the SNR is decreased. Lower power may be used over a larger range of frequencies as opposed to a high-power signal over a narrow range of frequencies. This has a direct effect on noise. Noise is an integral component of the signal in the form of other frequencies that are multiples of the base, or carrier, frequency. By using lower power, the effect of these multiples is also decreased. The effects of other noise components due to atmospheric factors are also similarly decreased. Also, the information is spread over a large number of frequencies, allowing the system to ignore noise specific to each narrowband frequency channel. The effect of this technology is to produce secure and reliable wireless communication while dramatically increasing system capacity over previous systems.

Three main types of spread spectrum are in use:

- Direct Sequence Spread Spectrum (DSSS)
- Frequency Hopped Spread Spectrum (FHSS)
- Chirp

We will discuss the first two techniques, which are widely used. The third, Chirp, is used in a very narrow range of applications, including military radar. In the Chirp system, the frequency carrier is "swept" over a broad range of frequencies. Another type of SS called Time Hopped Spread Spectrum is not currently used, but holds much promise for the future of wireless communication. Hybrids of the first two systems are common, taking the best features of both systems to provide a subscriber superior performance at low cost.

Direct sequence spread spectrum. DSSS is the most widely used form of spread spectrum. It is an all digital system using "pseudorandom" or "pseudonoise" codes to control frequency spreading. These codes are called "pseudo" because they are not actually random or noise. Information to be transmitted is modulated with the pseudonoise (PN) code. Modulation techniques in use are BPSK, Binary Phase Shift Keying, and QPSK, Quadrature Phase Shift Keying. BPSK uses a single data stream to modulate the information, transitions from a binary 1 to 0 or 0 to 1 causes two phase shifts of 180° in the transmitted data. BPSK is used as the method of modulation in the CDMA system described above. QPSK has many variants, including Staggered and Differential QPSK. QPSK typically uses two data sources to modulate the information. Transitions in data cause the carrier to shift by either 90° or 180°. Because QPSK causes four phase shifts in the transmitted data, it has twice the efficiency of BPSK. These concepts are somewhat beyond the scope of this discussion; several references are included in the bibliography for those who wish to explore the concepts further.

Frequency hopped spread spectrum. FHSS also utilizes PN codes to spread the transmitted data. FHSS is the easier of the two main SS methods to implement. The technique employs precision frequency synthesizers, which continously shift the center carrier frequency based upon the impressed PN code. As the signal is "hopped" from frequency to frequency, care must be taken to avoid interference with other users on the same channel who may still be using older technology. The period of time a signal remains on a specific channel is called the Dwell Time. The dwell time is typically very brief, less than 10 milliseconds. The manipulation of dwell time is very effective in eliminating interference to other users. There are two methods within this technique: Slow Frequency Hopping and Fast Frequency Hopping. SFH transmits very few (< 2) data bits. This allows improved data detection at the receiving end. If a frequency is blocked or jammed, however, data may be lost. This requires the use of error correcting codes. FFH divides the same bit over several frequencies. Error correcting codes are not needed, but require additional processing overhead at the receiving end to improve data detection.

Modulation of the signals used in FHSS is done with Frequency Shift Keying (FSK). FSK modulation is achieved by using two distinct frequencies to represent binary 1's (mark) and 0's (space). FSK uses tones to distinguish between the two data states. These are the tones that are then "mixed" with PN codes to create the spreading sequence used in FHSS.

Wireless data

Cellular telephone systems also support wireless data communications. This technology lends itself to such applications as:

- Remote information retrieval
- Remote inventory checking
- Warehouse management systems
- Wireless credit card authorization and point of sale
- Remote utility meter reading
- Wireless fax and email

Three systems are described here, showing the evolution of this technology: Specialized Mobile Radio (SMR), Cellular Digital Packet Data (CDPD), and Personal Communication Services (PCS).

SMR. Specialized Mobile Radio is a half-duplex method of communication designed for the trucking industry in order to aid in the dispatching of vehicles. SMR was capable of sending small amounts of data over the existing analog channels. Transmission rates of only 4.8 kbps are possible on this system and interface software usually is custom and therefore costly. SMR has been rapidly expanding into the digital realm in recent years, offering paging, telephone, and data services to their subscribers. Many large trucking and package handling firms still use these cell-type networks to track packages and dispatch vehicles.

CDPD. Cellular Digital Packet Data is the preeminent method of transmitting data over wireless networks. Originally developed as a technique for transmitting data over existing analog cellular networks, CDPD has made the transition to modern CDMA-based cellular networks. CDPD supports many different data transfer protocols, allowing interoperability with many different applications and networks. Also,

CDPD supports a wide range of equipment options and standard interface applications. There are two types of CDPD networks:

- Dedicated channel
- Channel hopping

In a dedicated channel network, certain channels or a range of channels are set aside exclusively for the use of the CDPD network. This is similar to the channel–frequency allocation to cellular networks. These channels are outside the range of channels specified for cellular telephone service. In a channel-hopping network, cellular telephone and CDPD devices share all channels within the cellular band. Data transfers are sent over vacant channels until the CDPD software senses voice traffic, at which time the data transfer "hops" to another vacant channel and continues the transfer. The cellular call has preemptive priority over the CDPD transaction and if there are no vacant channels available for the data transfer, the transfer will be dropped. The CDPD standard specifies that the transfer must occur within 40 ms of the detection of voice traffic; the cellular caller will never know that a CDPD transfer is taking place. Each channel in the network can handle up to 30 calls at a time. CDPD transfers operate in a manner similar to the Ethernet computer network (described later), with its many users. Each device listens to the channel to make sure that no other device is transmitting; it then places data on the air. If two devices transmit at the same time, a "collision" occurs, and each device resets and retransmits after a random delay.

A CDPD network consists of five components as shown in Figure 14. These are

- The Mobile End Station (MES)
- The Mobile Database Station (MDBS)
- The Mobile Data Intermediate System (MDIS)
- The Intermediate System (IS)
- The Fixed End System (FES)

The MES is the mobile device and wireless modem. This can be a laptop computer, a point-of-sale (POS) terminal, or a remote measurement device; each of these devices are typically connected to a wireless (cellular) data modem. The MES operates as would any other terminal, accessing email or downloading files, for instance. POS or credit card verification are other functions the MES might perform. Anyone

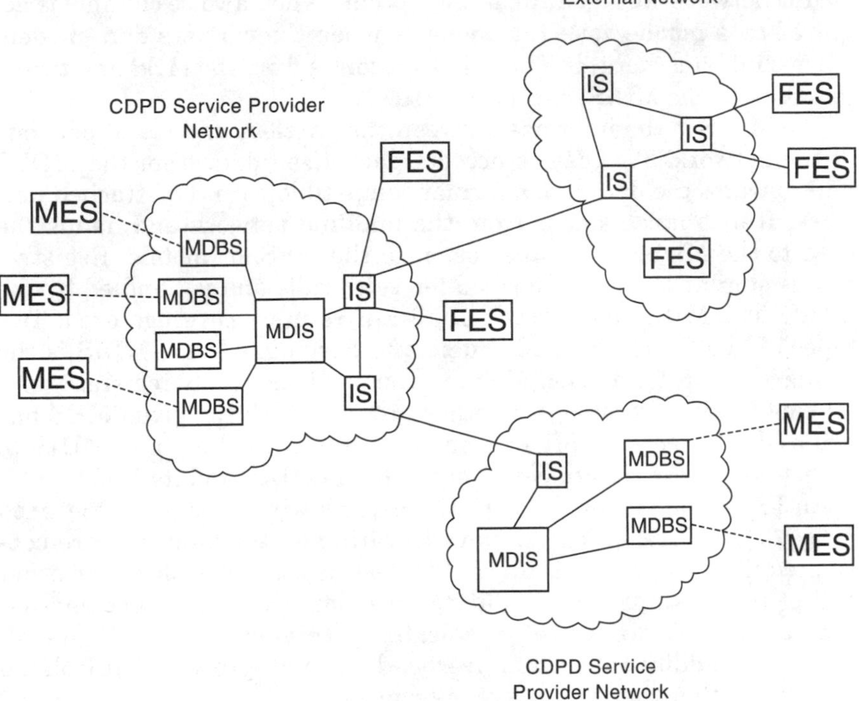

Figure 14 A CDPD Network

who has rented a car has seen an application of wireless data technology when turning the car back in. The attendant punches in the vehicle identification number (VIN); this is communicated to the server which handles these transactions. The data pertinent to your rental is tabulated and transmitted back to the hip-mounted MES, which then prints out your receipt. This is a limited application, but illustrative of a picocell as described previously. The MES sends data to the network via the MDBS at a maximum rate of 19.2 kbps. This rate is effectively less because of processing and other system overhead. Finally, the MES packetizes the data and may encrypt it for security before sending it out over the airwaves.

The MDBS is installed at the cell site, or base station. It uses the same antenna as the cellular voice network and controls the activities of the CDPD network. The MDBS actively monitors all voice channels in the cell whenever a CDPD transfer is taking place. The MDBS controls the "hopping" of transfers whenever it is necessary. The MDBS is responsible for monitoring all unused voice channels for use by the

CDPD network. To reiterate, a "hop" occurs when a voice channel is acquired by a mobile voice telephone or when the wireless data modem travels into another cell. The MDBS receives from the MES and transfers data to the MDIS, and vice versa.

The MDIS is the interface between the wireless and wired portions of the network. This device accepts packetized data from the MDBS and converts the data into a format compatible with the attached network. It also accepts data from the landline network and hands the data to the MDBS for transmission to the MES, or mobile. Registration is similar to that performed for voice calls and is handled by the MDIS. The MDIS keeps track of all MES's in its coverage area. Two types of MDIS's are defined: Home and Serving. A home MDIS is the primary MDIS for a given MES, or the one it normally registers to. A serving MDIS is located in another area to which the given MES has traveled. The serving MDIS communicates with the home MDIS to transfer packetized data between the home MDIS and its MES.

An IS can be any of the various systems which comprise the fixed network. Routers, switches, and the wiring on the landlines are considered the IS. The FES is any of the destinations possible at the other end of the landline. This could be an office network or just another computer. FES's are typically dedicated to serving wireless clients, allowing the additional system overhead to be absorbed by an isolated machine rather than the entire enterprise.

CDPD is very cost effective because of the existing cellular infrastructure. Charges for service are usually on a per-packet basis, whereas cellular subscribers pay for time whether it is used or not. Wireless data systems are often equipped with voice modems (or vice versa) allowing the most effective use of the service. CDPD does not overly impact the existing systems and can be a genuine enhancement to current modes of customer service. CDPD allows users to engage in multimodal forms of communication allowing voice, fax, and data from the same equipment.

PCS. Personal Communication Services is the newest wireless data technology to hit the market. PCS offers enhanced messaging capability, allowing services such as paging, voice, data, and Internet access to subscribers through the use of a single integrated mobile unit. Some enhanced features of PCS include:

- Voice and Data privacy with increased protection against eavesdropping
- Sleep Mode allowing extended battery performance
- Enhanced data services and accessibility
- Increased use of private and residential cellular systems
- Improved radio spectrum efficiency

While present units offer simple text messaging and email retrieval in addition to voice, equipment has been developed, and will be deployed, which are in actuality palmtop computers with integrated digital communication capability. PCS utilizes both cellular frequency allocations and has also been assigned to a new frequency band for use of PCS technology exclusively.

PCS has been assigned the frequency band between 1850 MHz and 1990 MHz, allowing approximately 1840 channels for each of the Transmit and Receive bands. Both TDMA and CDMA technologies are used to achieve spectral efficiency, though the use of TDMA is restricted to those mobile units that are capable of transmitting in the cellular bands. PCS makes use of CDPD and spread spectrum technologies to further enhance this feature-rich service. While cellular licenses were awarded based upon Metropolitan Service Areas (MSAs) and Rural Service Areas (RSAs), PCS licensing is based upon Major Trading Areas (MTAs) and Basic Trading Areas (BTAs). These designations are defined and copyrighted by the Rand McNally Corporation. The Federal Communications Commission (FCC) has modified these definitions for the purpose of more equitable awards of PCS licenses. MTAs are defined by the FCC as a collection of BTAs, while each BTA may cover several counties. MTAs are grouped into five regions, which theoretically cover the entire country.

The following is a brief overview of the technology defined under two standards published by the Telecommunications Industry Association (TIA). The first, IS-136, is defined for use in a TDMA-based network; the second, IS-95, is defined for use on a CDMA-based network. The differences between the two technologies have been adequately discussed. The handling of data, regardless of the method of modulation and transmission remains essentially the same. IS-136 was developed for use with dual band and analog–digital systems, while IS-95 was developed exclusively for use on digital systems.

At the core of the specifications is the Digital Control Channel (DCCH), which is used to control the function of the mobile units with which it is communicating. When a PCS capable mobile powers on, it will register with whatever cell site is broadcasting the strongest signal. It scans and "camps" on the strongest DCCH signal it finds. This may not necessarily be the serving cell. In the event the mobile is registered to a private or residential system, the strongest signal providing the required identification information will be the preferred cell onto which the mobile locks. The DCCH is a collection of logical channels. Each logical channel performs a specific function depending upon the direction of travel (to the mobile or reverse), and the type of information (control or data). The Forward DCCH contains nine separate logical channels while the Reverse DCCH contains only one. The FDCCH logical channels are

- Broadcast Channel (BCCH). This channel contains three primary logical channels: The Fast Broadcast channel (FBCCH), the Extended Broadcast channel (EBCCH), and the Broadcast messaging service (SBCCH).
- The SMS point-to-point messaging, paging, and access response channel (SPACH). This channel contains three logical channels: The Short Message Service channel (SMSCH), the Paging channel (PCH), and the Access Response channel (ARCH).
- Shared Feedback Channel (SCF).

The BCCH contains three channels, two of which are used. The third, SBCCH is about to be deployed for used with enhanced systems. The FBCCH is used for information that phones need immediately, such as system identification and registration information. The EBCCH is used to carry information of a noncritical nature, such as listings of neighboring cell locations and available frequencies. The SPACH contains three channels also, all of which are used. The SMSCH carries PCS messaging and programming information. The programming information includes over-the-air activation and programming (OAA/P) which controls the PCS mobile functions. The PCH carries system pages to the phone. The ARCH provides system responses to queries from the mobile and provides administrative information and commands. The RDCCH contains only one logical channel, the Random Access Channel (RACH). This channel is used for communication on the reverse channel from the mobile unit to the base station.

Each message sent to the mobile consists of three elements:

- Addressing information
- Alphanumeric messaging
- Message attributes

Addressing information tells the system which phone to send the information to. This information includes the MIN and ESN of the mobile. Alphanumeric messaging is the actual message itself. Message attributes are instructions to the mobile that instruct it on how to handle and display the message. Messages can be generated using any one of several methods: email, voice terminals, voice-mail systems, or live-operator message dispatching. Newer systems will allow direct access to the Internet.

A significant enhancement of digital communication brought by PCS is the ability of a mobile to exhibit different characteristics and offer expanded features depending upon the system to which it registers. Systems may be public, private, or residential. Other system configurations are available and will be mentioned below. Cells are defined as macro- and microcells and can cover geographical areas as small as 30 miles covering offices and homes. Each system may have different features available based upon the system constituent's unique needs.

While public systems provide the same basic service to all subscribers, private systems may provide specialized services to a predefined group of subscribers within a small geographical area and typically do not support public use. Private systems are used within buildings and are specific to one organization. A subtype to the private system is a semiprivate system. These systems may offer basic service to all customers, but still provide special services to a subgroup of employees or private customers. This type of system may be found within an office or college campus, for instance. Residential systems do not support public use of the cell, and provide specialized services to a small group of customers. These cells use a personal base station and behave as a cordless home telephone system. It follows that this type of system would be used within the bounds of a private home or apartment complex. Semiresidential systems are a subtype of the residential system and provide basic service to all subscribers while still providing specialized services to a predefined group of residential customers. This type of system can be found in a neighborhood which is served by a macrocell.

In order for PCS mobiles to determine the system ID and features available on the system, System Identities are used. SIDs allow mobiles to distinguish between public, private and residential systems. SIDs make these distinctions possible. Two SIDs are used: Private System Identities (PSIDs), and Residential System Identities (RSIDs). PSIDs are assigned to specific private systems by the system operators to identify them to mobiles in the coverage area. PSIDs can be broadcast by cell sector, allowing very small, tight coverage areas; or PSIDs can be broadcast over large geographical areas to create a large virtual system. RSIDs operate similarly to PSIDs in that the system operator broadcasts IDs that identify the system to mobiles as a private network. The system coverages are usually very small; there is no advantage to broadcasting IDs to a large geographical area. By definition, a Residential system is implemented through the use of micro and picocells.

Location IDs are used to identify the cellular carrier name and also to inform the subscriber when they have entered a private or residential network. A subscriber without rights to a private system would not see the company or residence name displayed on the mobile, rather, the carrier name would continue to be displayed. This display is important in that it is necessary to allow use of special features available to a private system subscriber.

PCS technology will continue to grow exponentially throughout the next few years. It is the "cutting edge" of wireless data technology and will continue to facilitate expansion in the industry while providing superior wireless information services to the masses. Access to information and people from anywhere on the globe will revolutionize the way we live and do business.

ISDN

Integrated Services Digital Network is a conglomeration of standards designed to provide an all inclusive digital data service for use with existing telephone systems. ISDN was developed primarily for use with voice circuits, but has become a major vehicle toward supplying a wide range of communication services such as Internet access, data communication, video and music on demand, and image retrieval services. Two types of ISDN currently available are Narrowband ISDN and B-ISDN, or Broadband ISDN. ISDN was also designed with fiber optics in mind, but standards have been revised to allow the use of this technology over copper wiring. This has made ISDN services available to the average consumer possessing Plain Old Telephone Service (POTS).

Two well-known ISDN interfaces are available:

- Basic Rate Interface (BRI)
- Primary Rate Interface (PRI)

BRI is the most familiar version of ISDN service. BRI is also referred to as 2B + D for reasons that will be described below. BRI consists of three independent data channels. Two Bearer (B) channels are used for the transfer of data and operate at 64 kbps each. These channels may be used independently, or may be combined for an effective throughput of 128 kbps. A third channel, called the Data or "D" channel, is used for the set up and control of the connection and runs at 16 kbps.

The second interface available, the Primary Rate Interface (PRI), is a higher level interface which runs at 1.544 Mbps, the speed of a T1 line. This is no mere coincidence. The PRI standard was designed to be compatible with the T1 line, which is a vital constituent of modern communication. The PRI consists of 23 B channels, each running at 64 kbps, and one D channel, also running at 64 kbps. B channels used in this interface can connect with a BRI if necessary, over POTS lines.

While BRI is used primarily in residences and small businesses, a PRI is typically used in a larger enterprise with multiple locations. The numerous B channels can be used to connect directly with remote offices or sites. BRI allows the use of up to eight devices on a bus-type configuration, including modems, telephones, fax machines, etc. PRI allows only one device connection per B channel. The use of switching equipment, such as a PBX (Private Branch Exchange) or server capable of interfacing and distributing the T1 line, may mitigate this restriction.

The D channel is used for call setup and termination, to ensure that the channel is available to receive a call, or to provide enhanced features such as caller ID. D channels operate at higher software levels than do the B channels, which are optimized for data transfer. The B channels are, in effect, "pipes" that carry data from point to point.

A new standard allows more effective use of channel bandwidth while allowing a subscriber to be "always on" as opposed to a dial-up connection. If a subscriber is always connected, or "nailed up," there is a tremendous waste of available bandwidth, which occurs during the times the subscriber is not using the connection, while sleeping or working, for example. A technique defined as Always On/Dynamic ISDN allows a subscriber to be continuously connected while not presenting the disadvantage of wasted bandwidth. Using the lower bandwidth D channel to process tasks that require low bandwidth is the key to this technique. Email, notification, and similar services are handled adequately by the D channel, thus freeing up the unused B channels. The B channels are still available for use by telephone and fax devices, but remain unused until the capacity of the D channel is approached. At that time, B channels are switched in to accommodate the increased load. When the file transfer or download is completed, the B channel will be switched out for use by another subscriber. Vast improvements in bandwidth efficiency are realized by using this method. The system at large is now much less burdened because of the increase in available bandwidth. This also results in a faster connection due to less congestion.

ISDN is available in most areas as of this writing. ISDN suffers from some restrictions, though, such as the requirement that a subscriber must be within 18,000 feet of a switching office capable of providing ISDN service. There is usually a one-time installation charge for system configuration and subscriber equipment. There are some packages that essentially waive this fee if a subscriber signs up for an extended period of service. Monthly and Use charges vary with region and provider. ISDN offers many different configurations based upon features desired. These features also vary according to region and supplier. Older telephone and fax equipment may be connected to an ISDN line using a Terminal Adapter (TA). This item is usually available at an additional cost. The adapter performs the analog to digital conversion (and vice versa). Two physical interfaces are used to connect equipment to the ISDN line. They are the "U" interface and the Subscriber/Termination (S/T) interface. The U interface carries ISDN signals over a single pair of wires and is designed for connection over the distance between a subscriber location and the Central Office (CO). The S/T interface is a four-wire interface used to connect multiple devices to the ISDN line. If your PC is the only device to be connected to the ISDN line, then the U interface is adequate. If more than one device is to be connected, then the S/T interface is required. If the device supports the S/T interface, then an adapter called a Network Termination 1 (NT-1) is required to allow connection of the device to the U interface. Devices with a built-in NT-1 adapter are available and allow direct connection with the U interface. Most devices available do support a S/T interface and will require an adapter. NT-1 adapters require an external source of power. Future standards may allow the powering of ISDN devices utilizing the additional four wires in the ISDN plug, however.

Further to the above, a discussion of reference connections may be in order. Reference configurations allow for a common vocabulary when discussing ISDN connections. Only one reference drawing is presented out of many different possible types. The actual network diagrams are much more complicated than the one shown in Figure 15.

Referring to the diagram, the following definitions apply:

- TE1: Terminal Equipment 1. This is the subscriber's equipment. It may be a computer, fax machine, or telephone.
- TE2: Terminal Equipment 2. This is an older analog telephone, fax machine, or modem.
- TA: Terminal Adapter. This device interfaces the older TE2 equipment with the ISDN line.

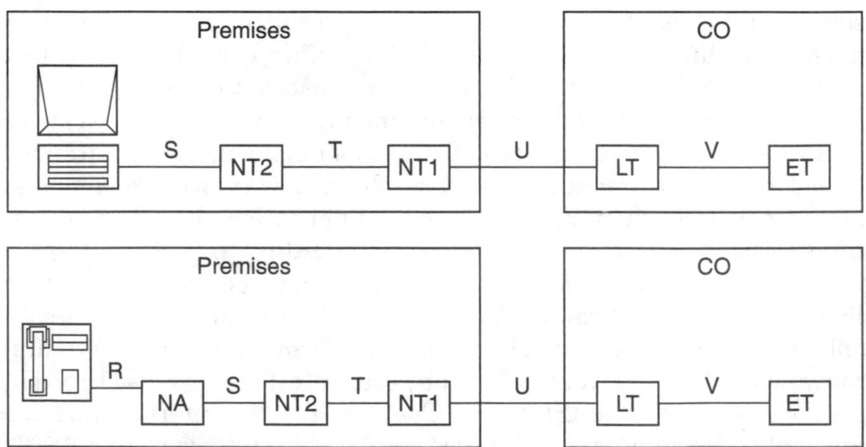

Figure 15 ISDN Reference Drawings

- NT1: Network Termination type 1. This is the "demarc" or line of demarcation between the telco's lines and the wiring inside your home or facility.
- NT2: Network Termination type 2. This is typically present in large organizations and could be the PBX or network server.
- LT: Line Termination. This is the physical connection to the ISDN line.
- ET: Exchange Termination. This is the telco switch connecting your ISDN line to the rest of the telephone network.

Attenuation and cross talk

Before we start a discussion of one of the newest methods of data communication, it is necessary to examine the reasons for the continued development of the technology. Several improvements to the existing telephone network were planned around providing increased consumer services into the home. Video on demand, interactive data services, on-line meeting facilities; these were some of the modern enhancements planned for the Public Switched Telephone Network. To make this possible, however, fiber-optic cabling was the medium of choice and would eventually be installed into every home in the nation. This was the impetus behind the development of ISDN and the technologies that followed. That was not where the market went, though; consumers did not enjoy waiting for six second video clips to interminably download. Video requires more bandwidth than currently available on the existing copper wiring present in virtually all homes with telephone service.

ISDN was the first technology developed to operate over existing copper lines. The maximum throughput is 128 kbps, even with newer variations. Real-time video is not practical at these speeds, and the answer was to increase bandwidth. The answer, as always, is obvious, though its implementation was somewhat less. Two obstacles had to be overcome before high-speed data communication over copper would be possible.

Copper wire possesses a property called resistance. As the length or thickness of the wire varies, so does the resistance. As the wire gets thinner or longer, the resistance becomes greater. Greater thickness or shorter length has the opposite effect. Also, resistance is frequency dependent. Attenuation is the term given to this resistance. Higher frequencies are weaker by nature than lower frequencies. It takes proportionately more power to transmit a high-frequency signal over the same distance as it would a lower frequency. Conversely, a high-frequency signal transmitted at the same power as the lower frequency will travel that much less distance on the same pair of wires.

This introduces a maximum loop length as one obstacle to be overcome while still maintaining the required high bandwidth. Given that the highest transfer rate is around 56 kbps due to bandwidth restrictions (the existing copper lines have an upper end of 3.4 kbps), methods had to be developed to allow higher frequencies and therefore higher throughput over existing lines.

ISDN achieves adequate loop length/bandwidth by splitting the high-speed data channels into two lower speed B channels. The lower frequencies allow a longer loop length and adequate bandwidth. But, as previously noted, 64 kbps or 128 kbps is not adequate for modern multimedia applications. In order to simultaneously raise the data transfer rate at the lowest possible frequency, and therefore length of the loop, it was necessary to develop newer forms of modulation, or Line Coding. Line coding refers to the methods by which data is impressed onto a base frequency. By manipulating the points at which the base frequency transmits individual bits of information, high throughput can be achieved while effectively lowering the transmitting frequency. ISDN originally required a carrier of 160 kHz. This frequency would have been attentuated to an unusable level on loop lengths under 18,000 feet. To mitigate this, each B channel is modulated using a coding scheme called 2B1Q, or 2 Binary, 1 Quaternary. Each analog cycle, or frequency cycle, transmits two bits on each cycle. This lowered the carrier frequency to a manageable 80 kHz, a frequency that would not be attenuated noticeably over the required 18,000 foot loop-length requirement. There are other methods of line coding, such as Carrierless Amplitude and Phase modulation (CAP), and Discrete Multitone. The former is an enhancement of the Earlier Quadrature Amplitude and Phase Modulation (QAM).

All of the techniques described are methods for "stuffing" more bits into a transmission cycle. Using a finite cycle time, a bit may be assigned to any part of that wave and then modulated at that position to indicate a digital 1 or 0. Using these techniques to move more data in a given amount of time allows for lower frequencies. This concept plays heavily into the future development of data communication using existing wiring plants. The future will bring a need for more bandwidth in already crowded frequency spectra. As frequency bands in the higher ranges are assigned, it will become imperative that the most effective use of these frequencies be made.

Another obstacle to the deployment of high speed data services over existing copper is the problem of cross talk. Two types of cross talk are typically encountered: Near End Cross Talk (NEXT), and Far End Cross Talk (FEXT). Cross Talk is the result of many wires being bundled together in a cable, each carrying high-speed traffic. The power required to transmit this information causes a magnetic field around each wire, which has the effect of "coupling" into the other circuit. This creates interference to both transmissions. NEXT is more of a problem than FEXT. NEXT occurs at the transmitting end of either system because of the higher power levels associated with transmission. NEXT is a major consideration of any system, and as will be discussed in a later chapter, can be controlled by proper design and installation of the physical wiring. FEXT is not usually a problem, the interfering signal having been attenuated to a power level that cannot impart any useful energy to a neighboring cable.

Performance of a data connection is usually quantified in terms of how many other data connections are operating on the same cable. A given number of ISDN connections may be able to function adequately over a given cable; after this number is reached, the performance of the system is derated. The loop length is the parameter most affected by this. If a cable is rated for a maximum of 25 data connections on a given cable in order to deliver the desired bandwidth at the required loop length, the addition of the 26th connection will shorten the loop length for all connections proportionately.

Several methods of controlling cross talk are in use. Echo Canceling circuitry cancels the duplicate frequency caused by the transmitter itself. This interference is called Self-NEXT. The offending frequency is simply subtracted out and eliminated. Another method is to use Frequency Division Multiplexing (FDM) to transmit and receive data. This as you will recall, separates the available channel into two subchannels, one for transmission and the other for reception of the data. The use of specially engineered wiring is also effective in controlling NEXT. Category 5 twisted-pair wiring allows bandwidths of up to 10 Mbps without major degradation due to cross talk; length is limited to 300 feet, however. The use of echo canceling is problematic due to interference from "like-type" echo canceling devices. In systems with high NEXT levels, multiple like-type echo canceling devices will actually act to degrade system performance. These devices should be considered for use in small systems with minimal cross talk. Other systems should consider the use of FDM systems to preserve system performance.

To summarize, the two major obstacles to high-speed data transmission are attenuation and cross talk. Attenuation can be made less of a factor by advanced modulation techniques and restricted loop lengths. Cross talk can be controlled by assessing the impact of NEXT and FEXT on the performance of the system and inplementing a suitable solution, such as Echo Canceling or Frequency Division Multiplexing. The following section details the newest advance in data communication, Digital Subscriber Lines (DSL). DSL utilizes the techniques we have just discussed in order to provide high speed and reliable data communication to the consumer for business user over existing copper wiring.

DSL. Digital Subscriber Lines is the newest technology to hit the field of data communication. DSL was originally under development for use by telcos to supply video and broadband data services to the consumer over existing wiring. Due to the explosion of the Internet and E-commerce, the standard was adopted to serve this market. This new technology, depending upon implementation, can attain transfer speeds in excess of 6 Mbps over standard copper telephone lines. This holds great promise for the future of on-line data and entertainment.

HDSL. Sometime in 1991, a movement among vendors of data communication equipment sparked development of methods to improve the available data services. A T1 line, running at 1.54 Mbps, was split into two separate lines, each carrying 784 kbps. Using advanced line-coding techniques, this allowed a lower frequency spectrum to be used and increased the usable loop length without the use of repeaters (amplifiers) on the line. The technique was referred to as High Bit Rate Digital Subscriber Line, or HDSL. The use of this technique allowed lengths of up to 12,000 feet at rates up to 784 kbps. (The loop length dropped to approximately 9000 feet using a smaller gauge wire, #26). Modulation was done using the CAP technique, or Carrierless Amplitude and Phase Modulation. This technique allows up to nine times the amount of data to be impressed upon the carrier frequency. Cap is also about twice as efficient as 2B1Q coding.

SDSL. Symmetric Digital Subscriber Lines are the single-line enhancement to HDSL. In HDSL, the spectrum was split between two lines, allowing lower frequencies and longer reach. SDSL allows the use of the full spectrum over a single pair of wires. While the coding techniques have been refined to allow reliable service, loop length has been sacrificed. HDSL systems running at 784 kbps have a reach of approximately 12,000 feet over 24 gauge wire; SDSL can support data transfer rates of 1.54 Mbps over an 11,000 foot run of #24 copper wire, a negligible difference in distance when considering the savings in wiring costs per consumer.

ADSL. Asymmetrical Digital Subscriber Lines maximize loop reach and data transfer due to the nature of the transmission and the wiring plan. It was found that data could be transferred more reliably and faster when transmitting from the central office (CO) to the subscriber than the reverse. Speeds of up to 6 Mbps could be achieved up to 18,000 feet from the CO. This is possible because of the fact that the loops are more susceptible to cross talk at the CO. At the subscriber end, cross talk is negligible because of attenuation. This allows the use of high speeds on the line or channel from the CO to the subscriber.

This has been the trend in data communication, particularly when accessing the Internet. Most traffic is from the CO to the subscriber during internet transactions. These take the form of file and media downloads. Video is also largely a one-way transaction. Commands to the CO from the consumer make up a small fraction of the overall traffic on the connection. If your Internet browser allows line status monitoring, it will illustrate the asymmetric nature of the data transfer. In an ADSL system, the transfer of data is asymmetric in that the bulk of the transfer is one-way, as opposed to symmetric, which supports two-way traffic at identical rates.

ADSL works very well with FDM. System operators can take advantage of this technique by assigning lower frequencies to traffic going to the CO. This reduces cross talk and attenuation problems. Lower frequencies are attenuated less than higher frequencies, but can still deliver a usable signal at the noisy CO end. Traffic going to the subscriber can be delivered at very high speeds without the concern of cross talk at the receiving end. Higher data throughput allows higher bandwidth and services such as real-time video and faster information access.

RADSL. Rate Adaptive DSL is a variant on ADSL, which allows dynamic control of line speed by the service provider based upon line condition and system utilization. The service provider will provide the fastest possible speed available to a given connection based upon the above criteria. Using RADSL, the service provider can allocate different speeds to different users based upon cost. That is, a subscriber can pay for desired throughput, the higher transfer rates costing more than slower connections. Also, this enhancement makes a reverse (to the CO) channel available at speeds up to 1 Mbps.

MSDSL. Multirate SDSL is another emerging technique that will allow reliable high-speed connections in both directions on a single-wire loop. MSDSL dynamically changes the line speeds and loop lengths to suit individual needs. Modulation techniques such as CAP allow up to eight distinct throughput/distance rates ranging from 64/128 kbps over 29,000 feet to 2 Mbps over 15,000 feet.

A comparison of the primary Digital Subscriber Lines is summarized here:

- **HDSL** High Bit Rate Digital Subscriber Line: Utilizes two wire pairs for symmetrical transmission of data at speeds to 1.544 Mbps.
- **SDSL** Symmetric DSL: Utilizes a single wire pair for symmetrical transmission of data at multiple speeds up to 1.54 Mbps.
- **MSDSL** Multirate Symmetric DSL: Utilizes one wire pair to provide symmetrical transmission of data through eight ranges of speed/distance to 2 Mbps at minimum distances.

- **ADSL** Asymmetric DSL: Utilizes one wire pair to transmit data asymmetrically at speeds up to 6 Mbps and includes conventional telephone service.
- **RADSL** Rate Adaptive DSL: Utilizes one wire pair to transmit data symmetrically or asymmetrically at speeds up to 12 Mbps and includes conventional telephone service.

Components of a DSL system

The area we are interested in lies between the CO and the subscriber's terminal. At the CO, which serves as the interface between the subscriber loop and the rest of the network, several specialized pieces of equipment are used to separate the DSL traffic from the voice traffic on the subscriber loop. The first piece of equipment used is called a Digital Subscriber Line Access Multiplexer, or DSLAM. The DSLAM is responsible for the separation described above, and is the key piece of equipment in the CO for providing DSL services. The DSLAM concentrates all DSL data from the multiple DSL subscriber lines and then combines the traffic into higher speed trunk lines for connection to the remote CO and DSLAM, which is connected to the desired data source.

At the subscriber's location, two devices are used. The first is a DSL transceiver unit that provides the interface to the DSL loop. These units will provide multiple output ports to provide the many services available such as video, voice, and data. The interface will also be capable of being upgraded and controlled by the service provider from a remote facility, allowing enhanced billing and maintenance services. Higher end transceivers will offer additional services such as routing and bridging for local networks.

Splitters are used at the subscriber's and the service provider's ends to allow simultaneous voice and DSL traffic over the high-speed link. These are also called POTS splitters. These devices can be active or passive. An active splitter requires external power to effect the interface, while a passive splitter does not. Passive splitters will support telephone service in the event of a power failure on either end of the loop. An active splitter must be equipped with a backup power supply to allow use of the voice portion of the link. Future equipment will offer "splitterless" technology that will also dynamically manage the subscriber's network. These devices will automatically allocate frequencies for voice and data traffic to avoid any possible interference between the two. This will allow the fastest data transfer speed while simultaneously completing a voice call.

DSL is slowly becoming available to the major urban areas. As the technology matures, it will no doubt be implemented universally.

ISDN is provided using a variant of DSL at the speeds indicated for this service. Equipment development and installation are a major hurdle, which is the primary reason for delay of this technology. Contact your local telco to determine if DSL is available in your area. If you choose to go with someone other than your telephone service provider, be aware that additional wiring will be needed and that substantial installation costs could be incurred.

Physical Interface Standards

Introduction

A very functional but overlooked mode of sharing information between computers and computing devices is that class of connections known as Physical Interfaces. These interfaces consist of actual wiring between devices, hence the designation, physical. These are the most basic of interfaces, allowing direct "hard wire" connection between devices as opposed to the networks described later in this book. In a physical connection, the connection is almost always dedicated to communication between two devices. The most familiar of these interfaces is the connection between a personal computer and a printer. Also, connections between older external modems and PCs were dedicated physical connections.

Network connections, by comparison, "broadcast" information to many different machines, allowing resources such as modems and printers to be shared. All machines on the network can receive information sent from any other machine. Any machine on the network can access information on any other machine. This method of connection is really the only effective way to communicate information between nodes on larger networks, both from a cost and time standpoint. On small, personal installations, the converse is true. There is no need to invest in equipment needed to install a network, though the cost of a small network is now comparable to a good laser printer. In most cases, however, the average user has no need for communicating between multiple machines.

Situations do occur where information must be transferred between a desktop and a laptop, for example. This is where a physical, dedicated interface becomes necessary. Even with the advent of large capacity floppy disks, the efficiency of a direct cable or wire connection is unparalleled for speed and convenience. The following section briefly describes the major physical interfaces in use and provides basic technical information for the more technically inclined. Pinouts are given for all of the physical interface standards described. It is important to realize that before the widespread implementation of computer networking, these interfaces were the primary means of communication between computers. Many enterprising and future network administrators coupled these standards with "home brew" utilities to effect primitive data transfer systems that served their enterprise adequately until networking had developed into a mature technology.

Defined

As stated previously, a physical interface is a direct connection between two data processing or communicating entities. These interfaces are also referred to as "hardware" interfaces because they rely on physical hardware to make the connection. These interfaces consist not only of a physical component, but also an electrical and functional component. These various components are described in the standards summarized in the following list. All of these interface standards are available from the agencies and organizations described on the following pages. These organizations charge a nominal fee for printing and handling. It is highly recommended reading for those who wish to pursue an interest in this information. Contact information is listed in the back of this book.

The standards describe three parts of the interface:

- **Electrical:** This section describes the electrical characteristics of the signals used to effect the interface. Among the parameters described are the magnitude, duration, and polarity of the signals. Also, the interface is described as synchronous or asynchronous, as balanced or unbalanced, terms that will be further explored later. Three basic types of signal are used in physical interfaces: contact closures (relay or switch contacts), voltage, or current loop.
- **Physical:** This section describes the physical features of the interface such as the size and shape of the connector, the number and type of wires required, the type and method of shielding the wiring, the number and arrangement of pins in the connector, and the manner in which the connection is made at one or both ends.
- **Functional:** This section describes how either party to the interface interprets the electrical signals. This area specifies which pins are transmitting and which are receiving data, or in general, which pins carry which signal and what function that signal performs. These functions allow the interface to work.

The standards described below use the three specification sections in various combinations to achieve reliable and fast communication over a wire attached to two, and in some cases more, machines.

EIA/TIA. Several standards described on the following pages have the designation EIA as a prefix to the standard number. EIA stands for Electronic Industries Association, a standards organization responsible for developing and publishing standards for the characteristics of electronic devices. EIA has teamed with the Telecommunications Industry Association (TIA) to enable development of a broader scope of standards dealing with devices required for the effective communication of voice and data. Before these organizations evolved, government agencies and academic institutions were the prime movers behind the development of these standards. A previous designation has survived the revised nomenclature and is used colloquially. This designation was RS, which stood for "Recommended Standard," and it precedes many of the standard numbers shown on the following pages.

IEEE. The Institute of Electrical and Electronics Engineers is a professional organization that has developed standards dealing with electrical equipment. For the purposes of this book, many IEEE networking standards have been referenced. IEEE has also developed standards for physical interface as will be shown later. IEEE standards are the product of technical and study groups that meet regularly to further develop and refine the various standards. Many standards are in a constant state of flux, while many are still in development. IEEE standards for data communication are primarily concerned with the physical aspect of the data connection.

Types of interfacing

Asynchronous. Asynchronous interfaces are those that transmit data without relying upon a regular timing signal. Each transmission is "framed" by a start and a stop bit, which signals the beginning and end of the transmission, respectively. Asynchronous transmission is less efficient than synchronous, but well suited to bursty traffic such as that associated with the physical link between a terminal (PC) and a modem. Asynchronous transmission occurs irregularly, with no set schedule; transmission occurs at a variable rate. Other examples of asynchronous transmission are voice and printer communication.

Synchronous. Synchronous transmission relies upon timing to effect communication. Both sending and receiving nodes synchronize through the use of an agreed upon sequence of bits sent before the transmission. The data or message is then sent at a constant rate across the connection. The transmission does not need any additional synchronization information as in asynchronous transmission. This makes this method more efficient through elimination of overhead caused by additional bits.

Balanced and unbalanced. The choice between balanced and unbalanced systems is an important one. In a balanced system, signals are transmitted over two wires. The signal at the receiving end is the difference between the voltages transmitted over the transmitting pair. In an unbalanced system, the signal is transmitted over only one wire. A signal common provides a reference for this signal at the receiving end on both systems. In a balanced system, also called a differential system, noise will appear on both conductors and cancel out. This is called "Common Mode Rejection." On an unbalanced system, no such CMR occurs. The importance of these phenomena is that the range of an unbalanced system is much less than that of a balanced system. Balanced systems also suffer from noise, but can tolerate much higher levels of interference. It is preferable to use a balanced system where possible. In many cases, though, use of an unbalanced connection is acceptable over very short (< 50 feet) distances.

Recommended standards

EIA/TIA-232. This standard is officially listed as "*Interface between Data Terminal Equipment and Data Circuit-Terminating Equipment employing serial binary data interchange.*" This standard was originally developed for connecting data terminals (PCs) to modems. In general, however, the standard specifies a method of communication between Data Terminal Equipment (DTE) and Data Circuit-Terminating Equipment, also called Data Communication Equipment (DCE).

DTEs communicate with DCEs. A DTE may be a computer or a dumb terminal; it may also be a printer or a data collection device used for remote telemetry. A DCE is usually a modem. Equipment adhering to this standard typically communicates point-to-point and asynchronously. RS-232 is suitable for use as a synchronous communication method, but is rarely used in this way. Communication is full duplex, that is, simultaneously in both directions, and can transmit up to 20,000 (20 K) bits per second (bps). Over 100 kbps has been achieved in some instances. RS-232 is an unbalanced system and because of susceptibility to noise, transmission is practically limited to 50 feet on either a dedicated or switched connection.

Signal characteristics as specified must fall within plus or minus 25 volts of zero. A logical "1," also referred to as a "mark," must be more negative than −3 volts DC. A logical "0" or "space" must be more positive than +3 volts DC. The standard defines two configurations, a 25-pin (DB-25) and a 9-pin (DB-9) connector. The DB-25 connector is shown in Figure 16. The DB-25 connector is the connector specified for use under normal circumstances and allows a wider range of features. The DB-9 connector is used in applications where these features are not required. The functions of the various lines are divided into four areas: data, control, timing, and special functions.

Referring to the diagram for the DB-25 connector, the following functions are defined as referenced from the DTE:

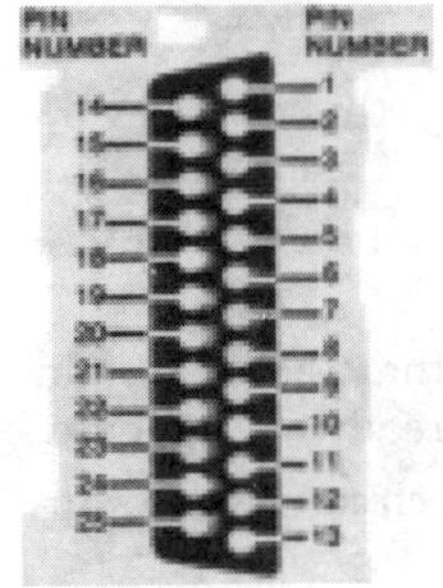

Figure 16 A DB-25 Connector

PIN	CIRCUIT	FUNCTION
1	AA	Shield, Protective ground.
2	BA	TX, Transmitted Data to the DCE.
3	BB	RX, Received Data from the DCE.
4	CA	RTS, Request To Send. A signal sent to the DCE to indicate the DTE is ready to send data.
5	CB	CTS, Clear To Send. A signal sent from the DCE to the DTE in response to the RTS indicating that it may begin transmission.
6	CC	DCE Ready. Indicates to the DTE that the DCE is in proper working configuration and ready to receive data.
7	AB	Signal ground.
8	CF	Received Line Signal Detector. Sent to the DTE from the DCE when a proper signal is being received from the receiving modem.
9	N/A	+ Voltage.
10	N/A	− Voltage.
11	N/A	Unassigned.
12	SCF	Secondary Received Line Signal Detector. Equivalent to Circuit CF in function, but for a secondary data channel.
13	SCB	Secondary Clear To Send. Equivalent to Circuit CB, for secondary channel.
14	SBA	Secondary Transmitted Data. Equivalent to Circuit BA, for secondary channel.
15	DB	Transmitter Signal Element Timing. Provides timing signals to DTE from DCE.
16	SBB	Secondary Received Data. Equivalent to circuit BB, for secondary channel.
17	DD	Receiver Signal Element Timing. Provides the DTE with timing information generated by DCE based upon received data.

PIN	CIRCUIT	FUNCTION
18	LL	Local Loopback. Used for testing the circuit and equipment.
19	SCA	Secondary Request To Send. Equivalent to circuit CA, for secondary channel.
20	CD	DTE Ready. Indicates to the DCE that the DTE is in proper operating configuration and ready to send data.
21	RL	Remote Loopback. Used in testing circuit and equipment.
22	CE	Ring Indicator. Sent from DCE to DTE to indicate that a ringing signal is being received on the communications channel.
23	CI/CH	Data Rate Selector. Used to select between one of two rate of data, signaling when dual or multirate DCEs are being used. Bidirectional.
24	DA	Transmitter Signal Element Timing. Provides timing signals to DCE from DTE.
25	TM	Test Indicator. Indicates whether the DCE is in test mode.

The below circuit and function descriptions are applicable to pin assignments in the DB-9 connector shown in Figure 17. Typically, the DTE utilizes a male connector, while the DCE uses a female connector. The pin functions for the DB-9 connector are described below:

PIN	CIRCUIT	DESCRIPTION
1	CF	Received Line Signal Detector
2	BB	RX, Received Data
3	BA	TX, Transmitted Data
4	CD	DTE Ready
5	AB	Signal/Common Ground
6	CC	DCE Ready
7	CA	Request To Send
8	CB	Clear To Send
9	CE	Ring Indicator

EIA-RS-422. Entitled "Electrical Characteristics of Balanced Voltage Digital Interface Circuits," this standard was developed to overcome the noise and range shortcomings of the RS-232 standard. RS-422 is a balanced system, using two wires to transmit and receive data. The improved standard allows transmission rates of up to 100 kbps over a range of 4000 feet and a maximum data rate of 10 Mbps over 50 feet. RS-422 is, as stated, a balanced or differential system. This means

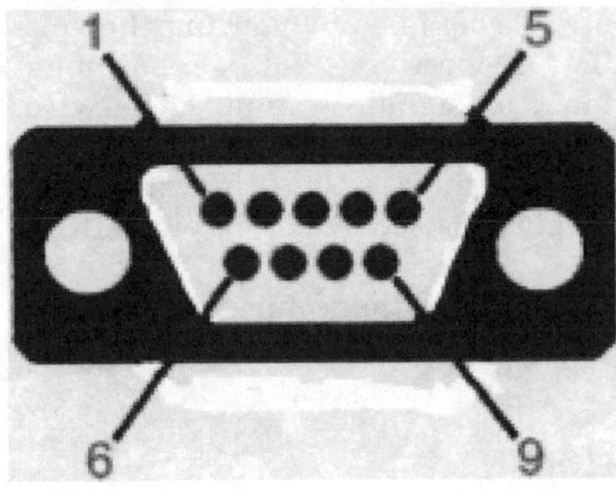

Figure 17 A DB-9 Connector

that the difference in the level of the signals received is used to define the logical state of the signal. Typical signal levels for this interface are −25 VDC to +6 VDC. The standard defines a logical 1, or mark, as a negative differential voltage between the signal wires with respect to the negative input. A logical 0, or space, is a positive differential voltage with respect to the negative input. Contrast this to a "single-ended" input, such as the RS-232 interface previously described, which references the input signal to a common ground. While the RS-422 standard does not specify a physical connector, EIA-449 does and is described in a following section. A DB-9 connector is also used with the following configuration.

PIN	NAME	DESCRIPTION
1	GND	Signal Ground
2	RTS+	Request To Send+
3	RTS−	Request To Send−
4	TXD+	Transmit Data+
5	TXD−	Transmit Data−
6	CTS+	Clear To Send+
7	CTS−	Clear To Send−
8	RXD+	Receive Data+
9	RXD−	Receive Data−

RS-422 allows multiple drops at half duplex. The system allows one master node and up to 32 slave nodes. The master node communicates with a selected node and then awaits that node's reply. EIA-485, described in a later section, enhances and improves this multidrop capability.

EIA-RS-423. RS-423 is the unbalanced version of the RS-422 standard. Voltages are not measured as a differential as in the previous standard, rather logical 1's and 0's are signaled by varying the voltage between −5 VDC and +5 VDC, respectively. Voltages are measured with respect to the common, or signal, ground. Otherwise, this standard is identical to the RS-232 standard previously described. No physical connector is specified under this standard, though a DB-9 connector is commonly used when implementing these interfaces.

Following is a description of the pinouts for that connector; the functions of the pins are the same as those described for use in an EIA-232 circuit.

PIN	CIRCUIT	DESCRIPTION
1	CF	Received Line Signal Detector
2	CD	Data Terminal Ready
3	CA	Request To Send
4	BB	Received Data
5	N/A	DCE Common Return
6	CB	Clear To Send
7	AB	Signal Ground
8	BA	Transmitted Data
9	N/A	Unassigned

EIA-RS-449. This standard was put into place to specify the physical and functional aspects of the EIA-422 and EIA-423 interfaces. It specifies a 37-pin connector (DB-37), the configuration of which is shown in Figure 18. Pin functions are described below.

PIN	CIRCUIT	DESCRIPTION
1	N/A	Shield
2	SI	Signal Rate Indicator (from DCE)
3	N/A	Unassigned
4	SD	Send Data (To DCE)

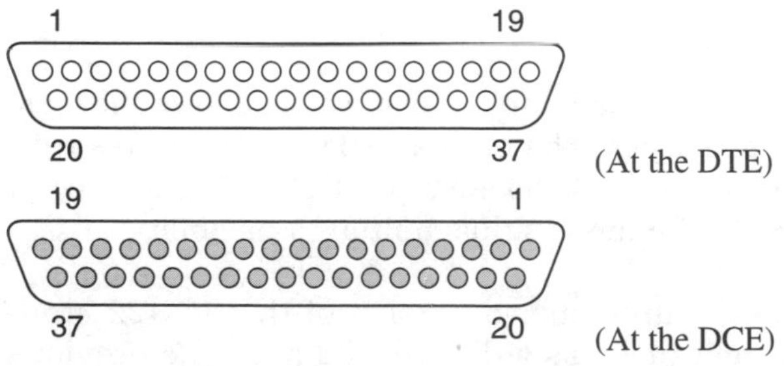

Figure 18 A DB-37 Connector

PIN	CIRCUIT	DESCRIPTION
5	ST	Send Timing
6	RD	Receive Data
7	RS	Request To Send
8	RT	Receive Timing
9	CS	Clear To Send
10	LL	Local Loopback
11	DM	Data Mode
12	TR	Terminal Ready
13	RR	Receiver Ready
14	RL	Remote Loopback
15	IC	Incoming Call
16	SR	Signaling Rate Indicator
17	TT	Terminal Timing
18	TM	Test Mode
19	SG	Signal Ground
20	RC	Receive Common
21	N/A	Unassigned
22	SD	Send Data Return
23	ST	Send Timing Return
24	RD	Receive Data Return
25	RS	Request To Send Return
26	RT	Receive Timing Return
27	CS	Clear To Send Return
28	IS	Terminal In Service
29	DM	Data Mode Return
30	TR	Terminal Ready Return
31	RR	Receiver Ready Return
32	SS	Select Standby
33	SQ	Signal Quality
34	NS	New Signal
35	TT	Terminal Timing Return
36	SB	Standby Indicator
37	SC	Send Common

The DB-9 connector shown in the section on RS-422 is widely used, however, and the 37-pin connector is not. A new standard, EIA-530 will largely supplant the EIA-449 standard utilizing a DB-25 connector. This standard is described further in the next section. The pin functions for a DB-9 connector as specified by this standard are shown on page 62.

PIN	CIRCUIT	DESCRIPTION
1		Shield
2	SRR	Secondary Receiver Ready
3	SSD	Secondary Send Data
4	SRD	Secondary Receive Data
5	SG	Signal Ground
6	RC	Receive Common
7	SRS	Secondary Request To Send
8	SCS	Secondary Clear To Send
9	SC	Send Common

Functions are identical to those described in the EIA-232 section. This standard is broken into two categories, I and II. Category I addresses interfaces that communicate at data rates below 20 kbps. These interfaces are typically done with equipment conforming to EIA-422, though it is possible, with proper interfacing equipment, to use EIA-423. Category II deals with interfaces running at a rate above 20 kbps, and are always EIA-423. RS-449 is interoperable with EIA-232 circuits at speeds below 20 kbps if the EIA-423 signaling standard is used. The circuits described in the above list function exactly as those defined in EIA-232. Voltages differ, however, and precautions must be taken to avoid damage to the interfacing equipment. EIA-232 specifies signaling voltages within the range of ±25 VDC. EIA-423 allows a range of ±5 VDC. Overvoltage can occur at the EIA-423 end if the proper interfacing equipment is not configured correctly.

As previously stated, the high-speed circuits (Category I) are always implemented with the EIA-422 interface; the low-speed circuits (Category II) are always interfaced with the EIA-423 interface.

EIA-RS-530. Partly because of the industry's lack of enthusiasm for the specified 37-pin connector, the EIA-530 standard was developed. EIA-530 has the formal designation of "High-Speed 25-Position Interface for Data Terminal Equipment and Data Terminating Equipment." EIA-530 is an attempt to codify the standards that have come before it. This standard specifies compatibility with the EIA-422 signal standards while specifying the physical and functional characteristics of a DB-25 connector. This standard is rapidly replacing the standards we have just discussed and is supplanting many older interfaces. The pinout for the DB-25 is shown in the following list. Pin functions are identical to those described in EIA-232. The pinout assignments are as follows:

PIN	FUNCTION
1	Shield, Protective ground
2	Transmitted Data

PIN	FUNCTION
3	Received Data
4	Request To Send
5	Clear To Send
6	DCE Ready
7	Signal ground
8	Received Line Signal Detector
9	Receiver Signal Element Timer
10	Received Line Signal Detector (B)
11	Transmitter Signal Element Timing
12	Transmitter Signal Element Timing
13	Clear To Send (B)
14	Transmitted Data Return
15	Transmitter Signal Element Timing
16	Received Data Return
17	Receiver Signal Element Timing
18	Local Loopback
19	Request To Send
20	DTE Ready
21	Remote Loopback
22	DCE Ready
23	DTE Ready
24	Transmitter Signal Element Timing
25	Test Mode

EIA-485. EIA-485 is actually an improvement to EIA-422. The standard allows up to 32 "Slave" nodes to be addressed and controlled by one "Master" node. Each slave node must have a unique address to avoid cross talk between nodes. The master node will poll each node, in effect asking each node if it has something to communicate. The system as implemented in a two-wire configuration is half duplex; allowances were made in this standard to allow four-wire communication. Four-wire circuits allow simultaneous communication between master and slave nodes. A unique feature of this standard, which is the chief difference between EIA-485 and EIA-422, is the ability of each slave to electrically isolate itself from the network. Otherwise, the two standards are virtually identical. Figure 19 illustrates two- and four-wire systems that are half and full duplex, respectively.

EIA specifies a maximum cable length of 4000 feet and a maximum transmission rate of 10 Mbps. The ability of a slave node to electrically isolate itself from the network was briefly mentioned. Signaling can be said to have three states: On (1), Off (0), and tristate (High Z). If the

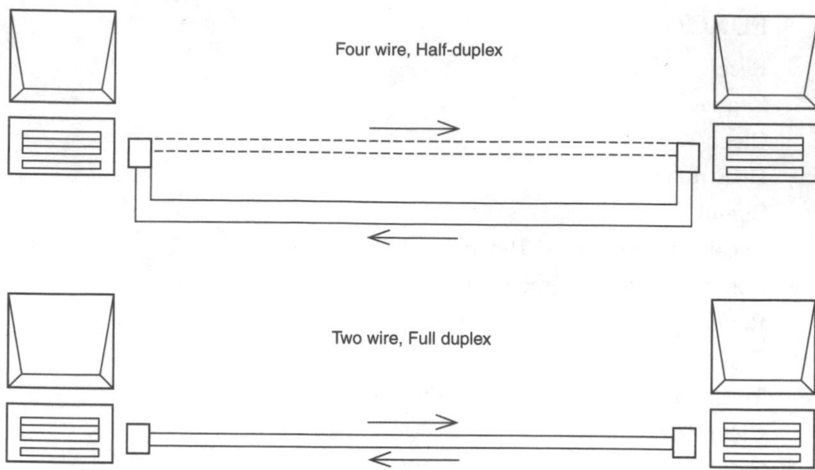

Figure 19 4 Wire Half-duplex & 2 Wire Full Duplex Transmission

slave is not actively communicating, it will present high impedance to the physical wiring attaching it to the network. The node will respond to polling, however, but will not initiate any communication or respond to any poll not addressed to that node. The system is differential, data being read as the voltage difference between the wiring as in the EIA-422 standard. Signal voltage ranges between +1.5 VDC and +5 VDC. A DB-9 connector is typically used and pinouts conform to that shown in the section on EIA-422, though a DB-37 as shown in the section on EIA-530 is also used. In practice, the EIA-485 interface will operate reliably using one twisted pair and a ground. It is recommended that a shielded cable must be used, but for short distances, under 100 feet, plain telephone wire will work.

IEEE-488. This interface is called the "General Purpose Interface Bus," or GPIB. The first standard was developed by the Hewlett-Packard Company and was accepted in 1975. GPIB is used to connect instrumentation devices together and allow centralized data acquisition. Up to 15 programmable instrumentation devices may be connected to this bus, which has a maximum length of 20 meters. Instruments must have a maximum separation of 4 meters and a minimum of 2 meters. Higher speed variants, such as National Instruments HS488 bus, allow higher speeds up to 8 Mbps depending upon how many connected devices are actually powered up and transmitting. The average speed of the standard bus is 1 Mbps. GPIB is configured in either a bus or star configuration.

When the bus is powered up, one device becomes the System Controller. The System Controller is capable of clearing the bus and then reinitializing all interfaced units and setting each to listen for data on the bus. Three types of devices are connected to the bus: Listeners, Talkers, and Controllers. Some devices may function as a combination of the three. A minimum system consists of at least one Controller and one Talker or Listener. A Talker sends data to one or more Listeners, which receive the data. A Controller manages the flow of information by sending out commands to all of the devices on the bus. When a Controller is notified that a Talker has data to transfer, it will connect the Talker to the intended Listener. On some configurations, it is possible to operate the bus without a controller. Talkers configured as "talk only" devices are connected to one or more "listen only" devices. Controllers may pass active control of the bus to other controllers at any time based upon programming.

GPIB uses 24 lines to control the bus and transfer data. Sixteen of these lines are signal lines, with the remaining 8 serving as ground lines. The 16 signal lines are divided into three groups: 8 Data lines, 3 Handshake lines, and 5 Interface Management lines. Data lines are designated DIO1 through DIO8 and are used to transfer address and control information and data. Data formats are undefined. Handshake lines are used to control the transfer and acknowledgement of message bytes among the connected devices. The three handshake lines are designated:

- NRFD—not ready for data. Indicates that a device is not ready to receive data.
- NDAC—not data accepted. Indicates when a device has or has not received data.
- DAV—data valid. Indicates when data is stable and valid and may be accepted by connected devices.

The NRFD is asserted, or set high (vernacular for setting the line to an agreed upon voltage), whenever a Listener is busy and not ready to receive data. If there is more than one device on the bus, the controller will not see the NRFD released until all devices release it. The NDAC is asserted by a Listener to indicate that it has not yet accepted the data sent by a controller. Again, all connected devices must release this line in order for the controller to see it. The DAV is asserted by the controller when it is sending commands and by a Talker when it is sending data.

Handshaking is the term for the functions these three lines perform, controlling the transfer of command or message data. A typical handshaking sequence is as follows. When a Controller or Talker wishes to transfer information, it asserts, the DAV, or sets it high. It then checks to see if the NRFD and NDAC lines are both low. This indicates that all devices are ready to receive data. When the last Listener takes its NRFD low, the line then goes high again, signaling the Controller, which then takes the DAV low, indicating that there is valid data on the bus. Each Listener responds by taking the NRFD low again to indicate it is busy, and sending the NDAC low when it has received the data. When the last Listener has received the data, the NDAC goes high again. This resets the bus by signaling the controller or talker to set the DAV high again, allowing the next transmission of data. If, after setting the DAV high, the controller or talker senses that both the NRFD and NDAC are high, an error will occur and no data will be transmitted. The slowest device on the bus controls the speed of the data transfer. Figure 20 illustrates the relative levels of the various lines during the handshaking process.

The five Interface Management lines manage the flow of control and data across the interface. These lines function as described below.

- ATN—Attention. This signal line is asserted by the Controller to indicate that control or address data is being put on the bus. The addressed Talker is then enabled to place data on the bus. The Controller can regain control of the bus by reasserting the ATN line.
- EOI—End or Identify. This line is asserted either by the Talker when the last byte of data is sent, or by the Controller when more than one device is being addressed.
- IFC—Interface Clear. This line is asserted by the System Controller to reinitialize all connected devices to a known state.

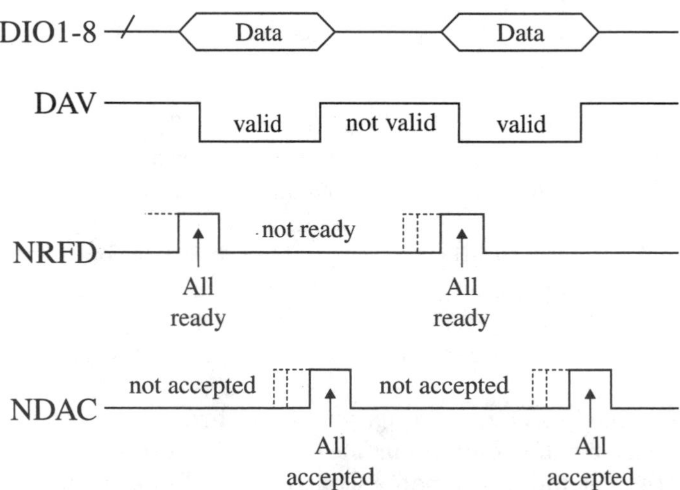

Figure 20 IEEE 488 Handshaking

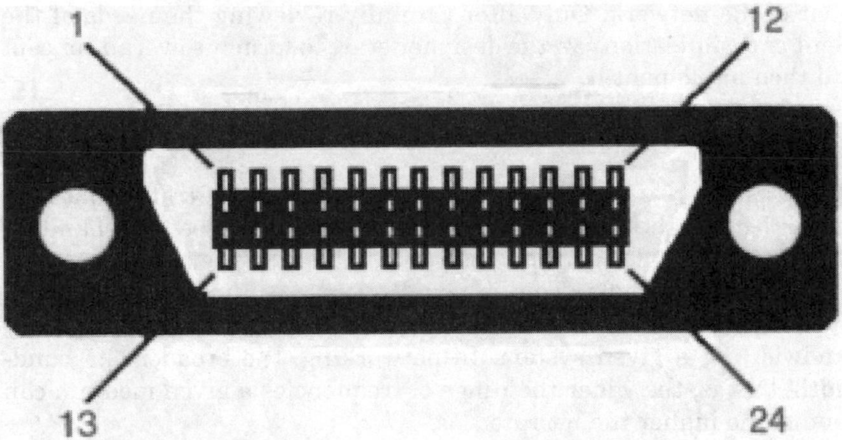

Figure 21 IEEE 488 Connector

- REN—Remote Enable. A System Controller can assert this line along with a device address to place that device into Remote mode, which causes the device to ignore any command from its front panel controls.
- SRQ—Service Request. Any connected device may assert this line when it requires the System Controller to perform some action. The System Controller must poll each device to determine which device has requested service. The line is released when the Controller polls the requesting device.

Each device is assigned a unique address by setting dipswitches on the device. Addresses range from 0 to 30. Typically, devices have only one address; some devices allow the setting of a secondary address, also in the range 0–30. The physical characteristics and pinouts of the interface are shown in Figure 21.

Network Types and Classifications

Every network, regardless of its size, is made up of communicating nodes arranged in a way that takes advantage of features unique to a particular installation. Many factors come into play, and the network designer must arrange the network so as to provide the highest quality service available. Described on the following pages are a number of items that must be considered before anything is "put down on paper." Each network is different, from the number of communicating nodes to the type of medium used to connect the nodes together. Factors include methods of signaling between nodes and the type of physical arrangement of the network. Only after carefully reviewing the needs of the client or organization can the designer start to form his overall concept and then implement it.

Capacity

Networks can be classified according to the amount of traffic that may be carried on the physical media. Networks may be Baseband or Broadband. This designation is directly related to the amount of data a network can handle. This is usually expressed in bits per second (bps), however, the following discussion will center on the frequency bandwidth of a given system. In networking, the broader the bandwidth, that is, the wider the range of frequencies a given medium can handle, the higher the bit rate.

Baseband. Baseband networks are digital networks that use a single, or base, frequency to transmit information over the medium. The entire capacity of the medium is used to transmit one frequency. This allows the maximum possible data throughput for a given medium. Baseband networks can use a variety of media, including twisted-pair cabling, coaxial cabling, and fiber-optic cable. The data is placed upon the media with no modulation, the individual bits being represented by changes from a base voltage. The base or reference voltage represents a logical zero; a logical one is represented by a change from the base voltage. A zero is typically represented by a voltage in the range of 0 to +2.5 VDC. A one is usually represented by a voltage in the range of +3 to +5 VDC. A baseband network is also called a carrierband network; these networks operate in the 10 Mhz range.

Broadband. A broadband network is an analog network that transmits many different frequencies simultaneously. The media used in broadband networks must have a high bandwidth in order to be useful. Coaxial cables are typically used in this type of network, though fiber-optic cable is steadily replacing coaxial cable in long-distance applications. In this type of network, data is used to modulate a frequency that is then transmitted over a band, or channel. Each channel has a unique frequency. Between each channel, a guard band is inserted to avoid cross talk between channels. Broadband networks operate in the 300 Mhz range and are capable of high-data throughput due to high media bandwidth. Many different types of data may be transmitted over the same cable using this method of transmission. A cable may carry data, voice traffic, and video (CATV) signals at the same time, making this type of network very cost-effective.

Range

Another way to classify networks is by the geographical range the network serves. The most common ranges are Local Area Networks (LANs), Wide Area Networks (WANs) and Metropolitan Area Networks (MANs). The designations sometimes run together forming hybrids, which are described later. Some designations are specialized or are not in common use and are presented for information only.

LAN. Local Area Networks, or LANs, are collections of computers and computing peripherals, which are connected together on a common medium. The computers communicate using a variety of techniques and share resources and information. LANs cover a small geographical area, typically a single building within the structure of an organization. Resources, to the user, appear to be seamlessly connected together on a LAN; that is, all available resources appear to be a part of the computer that the user accesses, though the resource may be located a considerable distance from the user's machine.

A variation of the LAN is the wireless LAN, which uses various wireless techniques to communicate data between machines. The first large-scale, nonmilitary implementation of Ethernet was a wireless LAN, or more properly, a WAN. The distinction is discussed in the following sections.

CAN. Campus Area Networks, CANs, are implementations of LANs over larger geographical areas. CANs may be several individual LANs connected together or may be several widely separated machines. The major difference between a CAN and a WAN is that the CAN does not rely on public communication systems to connect LANs together. A CAN is used where several different buildings within an organization, situated relatively near to each other geographically, must connect their individual LANs together.

DAN. Departmental Area Networks are specialized, highly secure Local Area Networks. These networks exist within organizations that are concerned about the proliferation of sensitive data. These networks are Local Area Networks in isolation, existing wholly within a single department within a larger organization.

WAN. Wide Area Networks are complex networks covering a large geographical area. WANs are used to connect widely dispersed LANs or even individual machines operated by a single organization. A WAN may cover a large manufacturing facility, a town or city, or offices in several different cities. WANs make use of dedicated, leased high-speed telephone lines. WANs allow users in one city to access data and resources in another part of the city or in another city. Many WANs utilize satellites to form links between offices in different cities.

MAN. A Metropolitan Area Network is a more strictly defined WAN, covering areas up to a radius of 75–100 miles. Typically, MANs have a higher bandwidth than WANs, using high-speed communication lines with bandwidths of 100 Mbps or more.

GAN. Global Area Networks are another variation on the WAN concept and are used by large multinational organizations for communication among LANs separated by large (global) distances, usually crossing national borders.

Enterprise. Enterprise Networks are large corporate WANs or GANs that connect any and all computing resources with which an organization must communicate. An Enterprise Network must communicate with different types of computers that may run on different operating systems and could use different networking protocols. The machines may be anywhere in the world.

Node type

This category breaks down into two basic types (Figure 22). The PC-based LAN consists of individual PCs that communicate using Network Interface Cards (NICs), which connect directly to the network; the Mainframe type consists of a large, centrally located computer using terminals that allow users to access data and resources. PCs are also used in Mainframe-based networks, but must run terminal emulation software in order to communicate with the mainframe.

Other than IBM-based mainframe networks, network architecture and protocols become proprietary and beyond the scope of this book. IBM developed Systems Network Architecture (SNA) for use on their networked systems. SNA is a communication model that allows the structuring of network functionality within a predefined framework. SNA and other communication models will be discussed in the next chapter.

Node relationship

How a node interacts with other nodes is an important way to classify a network. In any LAN, the object is to allow widespread access to information and services. How nodes provide the user with this access is defined by several unique arrangements. In all cases, some method of medium access arbitration is required to keep all of the nodes from trying to access resources at the same time. These methods are called Protocols, and are discussed in the next chapter.

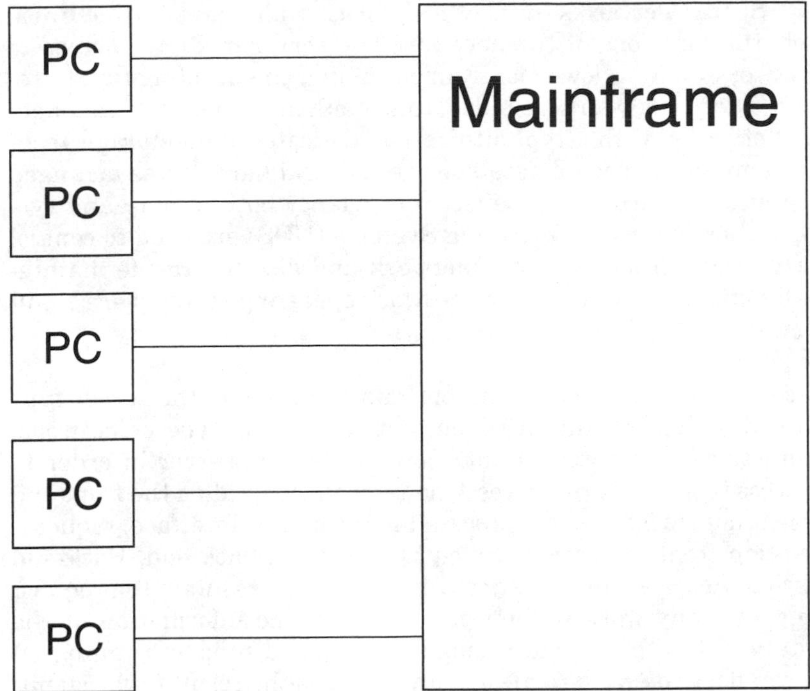

Figure 22 PC & Mainframe

Distributed. In this arrangement, any node may communicate with any other node. Information may be transferred freely without restriction between connected machines. There is no controlling entity in this type of network. Servers perform the function of providing distinct services to the requesting nodes and do not initiate communication with other nodes outside of normal functions. A server may be another machine or a program running on a peripheral device. A server performs no controlling action on this type of network, it merely behaves as a node with enhanced functionality as defined by the service it provides. All nodes communicate with all other nodes.

Peer-to-peer. In this type of network, also called a peer network, all nodes are equal, hence the designation Peer-to-Peer. Also, each node can perform the functions of a client or server (see following sections). These networks are simple and easy to implement for small networks of under ten or twelve machines. Each node must use a network interface of some sort to attach to the medium. Peer networks are useful for small offices or for small batch processing networks.

Server. Server networks employ a dedicated file server, which also controls the functions of the network. The server mediates access to the network, and allows centralized management of network resources. Larger networks employ this method of network management. Networks of this type utilize sophisticated data-storage techniques to maintain secure data storage. Several hard drives arranged to form an array are used to effect redundancy both in data and system operation. Network Operating Systems (NOS) are used to control interaction of the nodes on the network and also to provide maintenance functions. Several popular network operating systems are available, most notably Novell Netware.

Client-server. Client-Server is an optimized version of the server-type network described in the previous section. In this type of network, fully functional computer workstations access the server in order to gain access to network resources. The workstation, called the front end because it does the front-end processing at the user interface, requests information from the server, which is called the back end. Back-end processing occurs at the server, which returns the result of the request to the client; any work is then performed on the information at the workstation. In other applications, the back-end process running on the server does the work requested and returns the result to the client.

Network topology

The physical layout of the network, called the *topology,* is categorized by the method by which the nodes are connected together. Two broad categories have been defined. A network topology may be logical, physical, or both. Hybrid topologies take advantage of the features presented in both categories. Logical topologies are defined as topologies that correctly route data from one node to another, regardless of the physical layout. Physical topologies are concerned with the electrical routing of the signal, that is, the cabling between nodes.

The two major logical topologies are the Bus and Ring topologies. Both are described in following sections. In both, only those nodes to which the data is addressed accept the data. In the bus network, information is broadcast for all nodes to receive. The node to which it is addressed will accept and process the information, while the other nodes will ignore it. The bus network allows for all nodes to "hear" the information at the same time. This arrangement also allows multiple nodes to transmit simultaneously. In a ring topology, a token is used to give a node transmission rights. The ending node attaches the data to the token with an address and sends it off. Only the node it is addressed to receives it. The receiving node acknowledges receipt of the message and then sends it back to the sending node. In this manner, only one node has transmission rights, and message collisions and possible corruption are avoided. In a bus network, collisions do happen and are expected. Mechanisms built into the various applicable protocols manage this problem with timing and error-correcting schemes. Hybrids of the two types, such as the star-wired ring, are also possible.

Physical topologies can also be divided into two methods by which nodes are connected. In point-to-point connections, nodes are connected directly to one another. Mesh topologies are examples of this type of connection, and are typified by direct connections with no intervening switches or routers. Multipoint connections, also called multidrop connections, connect to a single node only, as in a star or star-wired ring. This distinction is not crucial, but provides further illustration of the different modes of transmission as affected by physical topology.

The following chapter explains the various protocols and the medium access and arbitration methods in use.

Bus. Bus is the most basic and simple network topology to implement. A bus network (Figure 23) is both a logical and a physical topology. It utilizes a single transmission line or cable called a trunk, or backbone, cable. This type of topology is sometimes referred to as a backbone network. In a bus network, communicating nodes are connected to the backbone either directly or by a small length of cable. The overall length of the backbone and connecting, or "drop" cables, are defined in physical layer standards. This type of topology requires the trunk cable to be terminated at both ends.

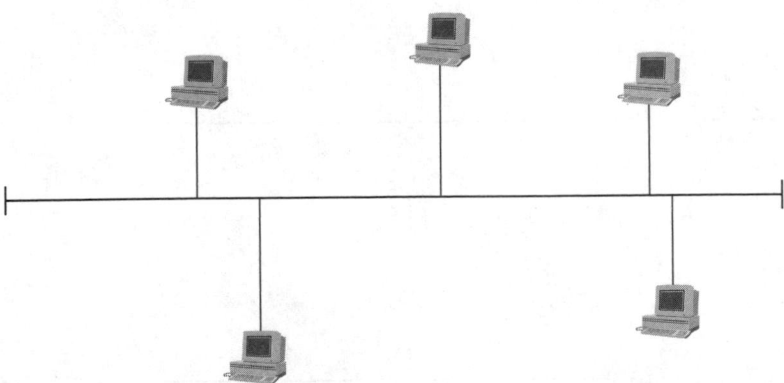

Figure 23 Bus Network

Advantages of the bus topology are low cost, ease of installation, and configuration. Nodes can be added and removed without major disruption to or reconfiguration of the system. Disadvantages include susceptibility to "bottlenecking" over the backbone during periods of high traffic. Also, fault diagnosis can be difficult. The single transmission path also leaves this topology vulnerable to complete or partial failure due to backbone failure, such as cut or damaged trunk cables.

Ethernet is an example of a bus network, and IEEE-802.3 describes the physical and electrical characteristics of this type of network.

Tree. The Tree (Figure 24) is a variant of the bus network topology. It is also called a distributed bus or branching-tree topology. This topology borrows from the star topology in that it combines bus networks into a tree structure using hubs at the branch points. Cable television systems are set up using this topology. As shown in the illustration, the main branch is called the "root" or "head end." This topology is somewhat more expensive to implement and configure because of the additional equipment required.

There is theoretically no limit to the number of branches from a given root, though management and fault diagnosis can become quite difficult on large installations. Other disadvantages include the loss of

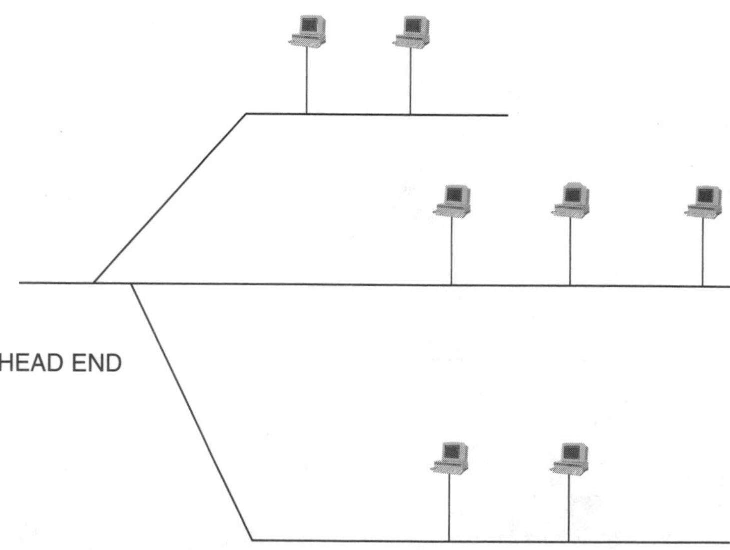

Figure 24 Tree Network

the entire enterprise should the root go down. Also, if any branch's hub is down, so is that branch.

Advantages of this topology are the ability to expand the network by adding branches. This also segregates network traffic, reducing the bottleneck phenomenon. Fault diagnosis is easier than in a single-backbone network because of the inherent isolation of branches.

Backbone bridge. A Backbone bridge topology (Figure 25) is another variation on the backbone network topology. In a backbone bridge network, several networks may be connected together using bridges. Bridges are devices that provide simple switching between the connected networks. A backbone bridge topology eases the load on the individual networks filtering out unnecessary traffic from a given network, and allowing only data meant for a node on a specific network to be switched in.

This type of network is more complex than a simple bus network, requiring configuration of the bridges, which also adds to the cost. The loss of Bridge 2 in the illustration would effectively isolate all three networks.

Cascaded bridge. The Cascaded Bridge topology (Figure 26) also allows multiple networks to be connected together, with some of the same advantages shown in the backbone bridge topology. The difference is that one network will carry an inordinate traffic load. This topology saves one bridge by forcing network 1 to go through network 2 in order to communicate with a node on network 3. Loss of network 2 would isolate two networks; loss of either bridge isolates one. Cost is

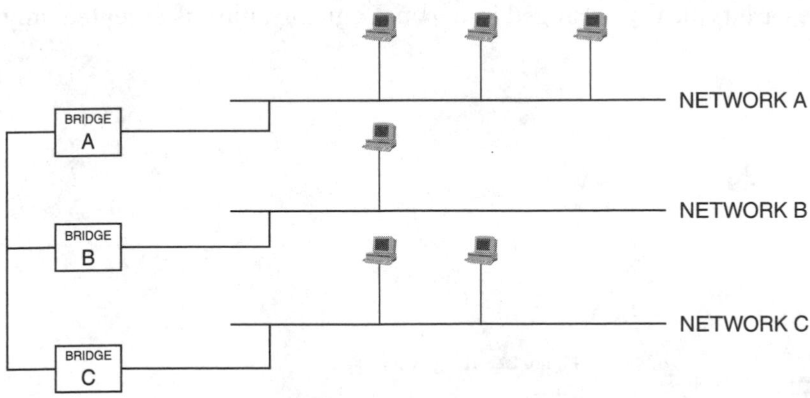

Figure 25 Backbone Bridge Network

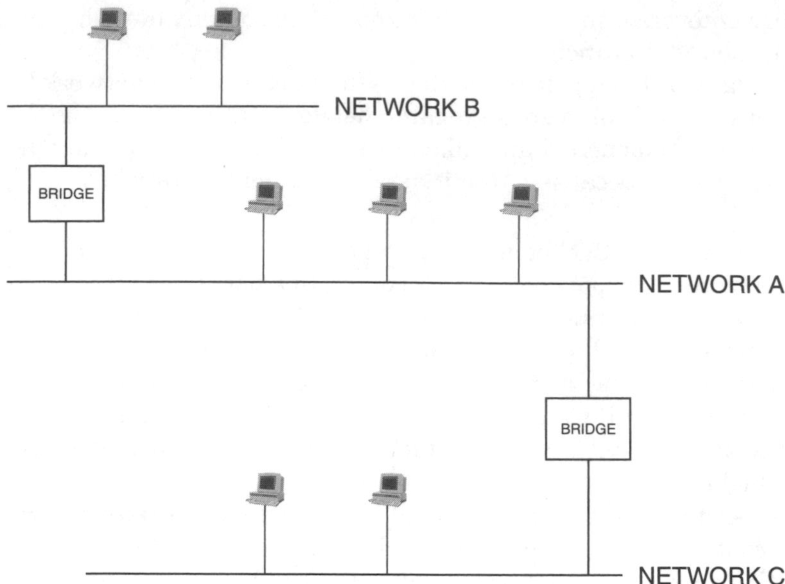

Figure 26 Cascaded Bridge

comparable to the backbone bridge; however, management and configuration overhead rises.

Ring. Ring topology (Figure 27) is also both a logical and a physical topology. In ring networks data is transmitted from node to node in a sequential fashion. This may happen by following a physical cabling scheme that connects one node to another, as is usually the case. Nodes are typically arranged in a loop. Logically, data is accepted only

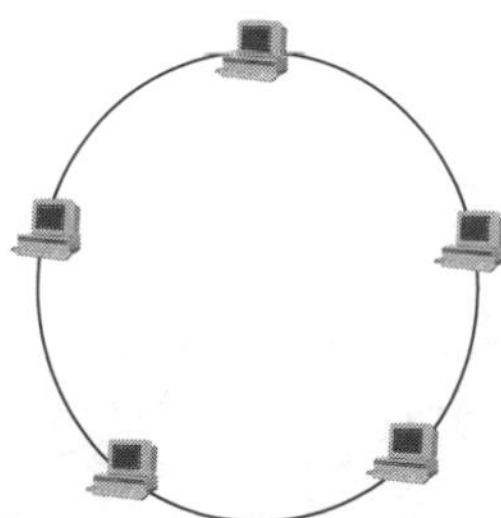

Figure 27 Ring Network

by the nodes to which the packet is addressed. Ring networks use token-passing schemes in order to transmit data. The first node to place data on the network is the last to receive data.

Ring networks are relatively simple to install and maintain. Wiring is minimal, and wiring closets or concentrators are not needed. Ring networks are faster than other networks used for comparably sized installations. A major disadvantage is the loss of the entire network with the loss of a single node. Also, adding or removing a node disrupts the whole network. Fault diagnosis becomes a problem because of the mode of transmission. An example of a ring topology is IBM's Token Ring technology. It is described in IEEE-802.5.

Star-wired ring. In the Star-Wired Ring (Figure 28) variation of the ring topology, which is actually an enhancement to the ring, a central

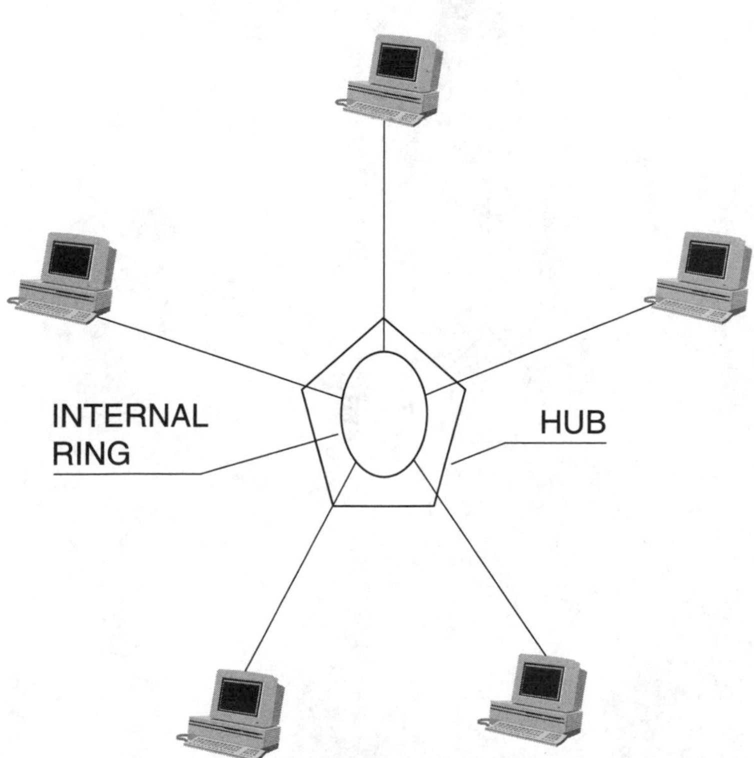

Figure 28 Star-wired Ring Network

ring of hubs or routers is used to feed star-wired nodes. This is an example of a physical hybrid. The major advantage of this type of topology is that it overcomes the major disadvantage of the ring, that being the loss of the network with the loss of one node. If a node in this network is not functioning properly, the hub, or Multistation Attachment Unit (MAU), may be set up to bypass that branch and the malfunctioning node(s).

MAUs may feed other MAUs on the star-wired branches, thus creating other star-wired rings. Other advantages include ease of expansion and a modular, flexible network. This flexibility, however, can lead to complex cabling and configuration. This type of network also tends to be more expensive, albeit reliable.

Star. The Star (Figure 29) physical topology actually operates as a logical bus topology. In this arrangement, all nodes are wired through

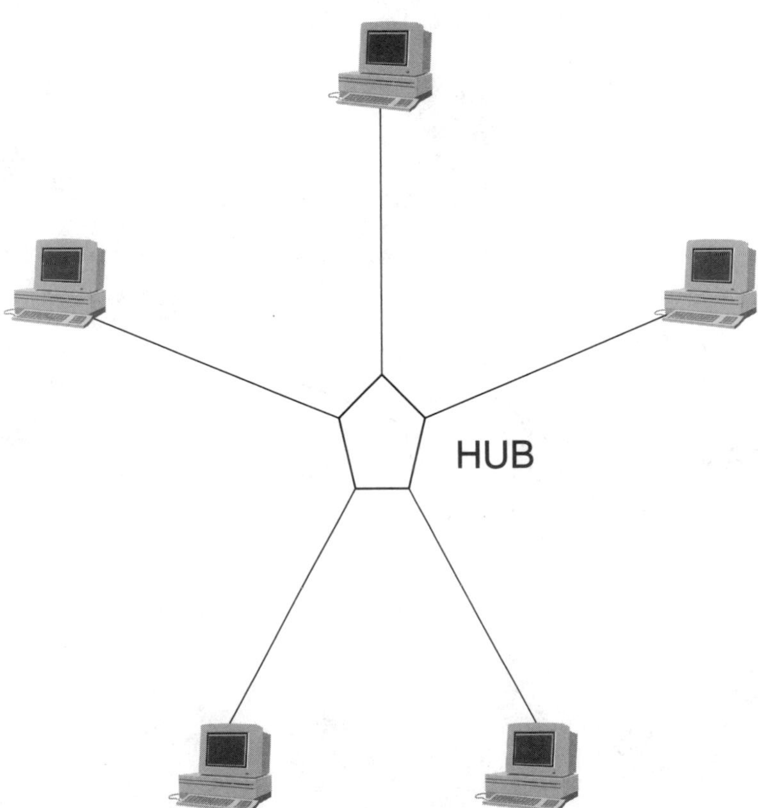

Figure 29 Star Network

a central hub or router. The hub may be a file server, or may be connected to the file server. This arrangement may also be connected to another hub, forming the distributed-star topology described next. A good example of a star topology is your local telephone service, which connects subscribers to a central switching location. All messages go from the subscriber (node) to the central office where it is routed. The message is then sent to the receiving subscriber.

An advantage of this type of network is the ease of fault isolation. If a node is malfunctioning, it can be removed from the hub without disrupting the network. The chief disadvantage is the loss of the entire network if the central hub fails. Also, star networks require the most cable of the networks discussed so far.

Distributed star. A Distributed Star (Figure 30) is a multiple implementation of the star arrangement. Star networks are connected together using wired connections between hubs. Many networks can be connected together in this manner creating quite large and complex networks. The advantages and disadvantages of this type of network

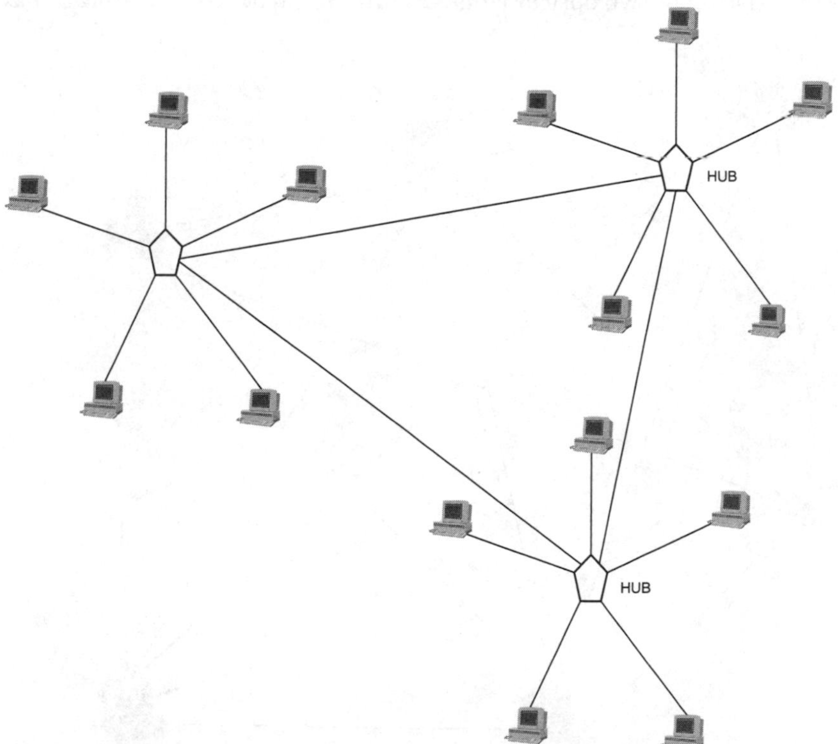

Figure 30 Distributed Star Network

are basically the same as the star network. In a distributed star, however, loss of a hub can take down larger portions of the distributed network, or at best, isolate them. Proper physical arrangement of the hubs would allow redundant paths that could minimize the impact of a hub failure, possibly isolating the failure to only the nodes connected to the failed hub. Another way is to provide backup hubs at each node, but this would effectively double the cost of the network.

Mesh. In the Mesh topology (Figure 31), there are at least two paths in and out of a node. The primary use for this type of topology is in environments where the physical connection could be subject to interruption. This physical arrangement requires the use of redundant data paths, often routed a considerable distance from each other. In a strictly considered mesh network, each node is directly wired to every other node. This topology requires extensive and complex cabling, larger than normal wiring closets, and is typically used for small, highly secure networks operating in isolation.

FDDI. Fiber Distributed Data Interface is an optical networking system consisting of two optical rings (Figure 32), with counter-rotational

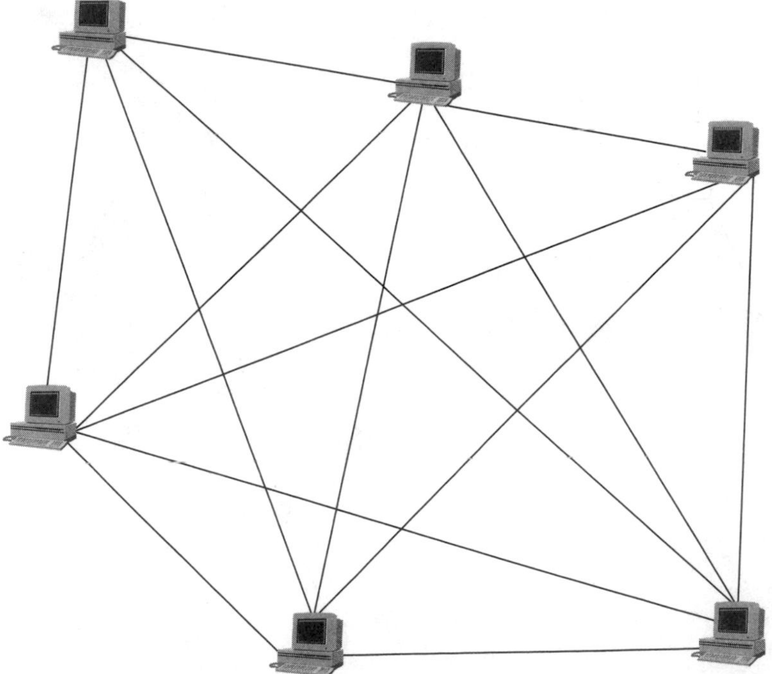

Figure 31 Mesh Network

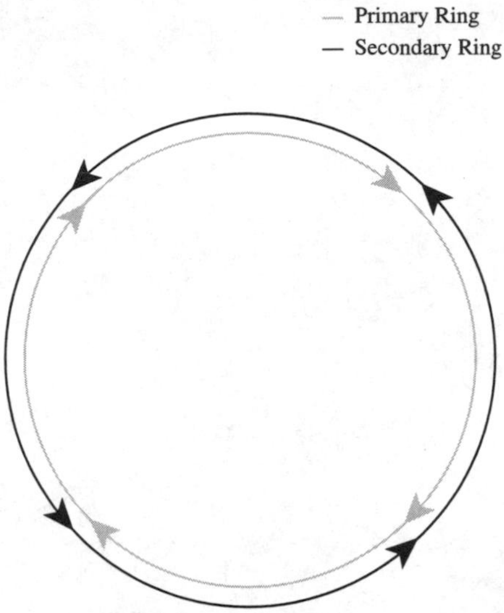

— Primary Ring
— Secondary Ring

Figure 32 FDDI Rings

token passing as the means of medium arbitration and data transmission (Table 2). The FDDI token differs somewhat with a single ring token-passing network previously described. The network can handle throughput up to 100 Mbit/sec. The rings are called "primary" and "secondary" rings. The secondary ring provides an alternate data path for use in the event of a primary ring failure. FDDI standards also define a method of medium access and control that are described in the following chapter.

TABLE 2 An FDDI Token Frame

64 BIT PREAMBLE	8 BIT **SD**	8 BIT **FC**	8 BIT **ED**

SD: Starting Delimiter
FC: Frame Control
ED: Ending Delimiter

Chapter

2

Protocols

What Is a Protocol?

The concept of protocols escapes many. A protocol, simply put, is an agreed upon series of steps that must be accomplished, in turn, by the sending and receiving parties. In other words, a protocol is a set of rules for the behavior of the two devices at either end of the line. Several protocols will be discussed, all of which are in common use. Protocols should not be confused with physical standards. While physical standards are typically part of the protocol standard, the emphasis is on the software manipulation of the data. Protocols are logical agreements between nodes that enable each one to communicate with another.

In order for two nodes to communicate, both nodes must follow an agreed upon method for reading and decoding the data sent between nodes. A network communication protocol forms an "envelope" around the transmitted data according to agreed upon conventions. The envelope will consist of leading and trailing data fields used to mark the beginning and the end of the transmission. Within the data fields, decoding and routing information is also provided. This information permits the receiving node to provide any special handling the transmission may require, such as transmission to another node or network.

Data communication systems rely upon arranging information into patterns. The raw data may be this text, a photograph, a sound, or a computer program. In order to get these items from one point to another via data communication, raw data is received and subsequently processed from a real, tangible form into electrical impulses. These electrical impulses are arranged into patterns that conform to known standards, such as the ASCII code described in the previous chapter. These codes enable any data to be encoded and transmitted anywhere in a form that can be recognized by any machine. The means by which the data is converted from the tangible to the electrical vary greatly in cost and complexity. Virtually every sensation and reaction can be converted into an electrical signal. Light, heat, and sound are the most common sensations that can be measured. Touch and movement are other common parameters that are converted into usable electrical impulses. Consider that this page was initially typed using the conversion of the sensation of touch into electrical impulses via the keys on the laptop.

In practice, all digital data is actually a small voltage variation from zero to five volts for a digital "one" and the converse for a digital "zero." These voltage variations are known collectively as "bits." Each character on this page is a collection of bits. In its most basic form, this page would consist of line after line of "0" and "1" in alternating sequences. It might look something like this (Figure 1):

1011110100011010010101010100001001

0001101010100001101011111010101010

1111110101010010010010000110101010

0000101001101010100100011001110101

1110111010110101010100110101010101

1110101011011001010100111010110111

0011010010101001010100100101000101

Figure 1 Random Bits

Looking at the bits arranged above, it becomes apparent that without some definition of a regular pattern, it would be difficult to tell one character from another. The above could be considered one long pattern, and this would correspond to one letter of the alphabet. Or it could be seven different letters of the alphabet if each line is taken as a representation of each. Typically, bits are arranged into patterns called "bytes." Bytes consist of eight bits. A byte is the smallest "packet" of information that can represent a character. Early computers used eight bit bytes and four bit "nibbles" to transfer data between components and other computers. If the above sequence were arranged into bytes, the appearance changes somewhat (Figure 2):

10 11110100 01101001 01010101 00001001

00 01101010 10000110 10111110 10101010

11 11110101 01001001 00100001 10101010

00 00101001 10101010 01000110 01110101

11 10111010 11010101 01001101 01010101

11 10101011 01100101 01001110 10110111

00 11010010 10100101 01001001 01000101

Figure 2 Formatted Bits

Each byte is "read" from the right. Each position in the pattern represents a value which, if added together, have a total value. The bit on the extreme right side of the byte is called the "Least Significant Bit" (LSB) and represents the lowest value, or weighting, in the pattern. The bit on the extreme left side is called the "Most Significant Bit" (MSB) and is the bit with the highest value, or weight. By this method, a "0" in any position equals no value or zero. A "1" indicates that position has a value equal to the position it occupies.

In this manner, each byte can have a unique value as a unit. This value can be converted into a higher level of code. The weighting of each position determines the value of that position. The weighting of a byte resembles the following arrangement (Figure 3):

MSB 128 64 32 16 8 4 2 1 LSB

Figure 3 Bit Weights

In this example, 128 would occupy the position of the MSB, while the 1 would occupy the position of the LSB. A bit pattern such as 11000001 equals 193 if the individual values of only those positions with "1" are added together. 11000001 could be viewed as 128 + 64 + 0 + 0 + 0 + 0 + 0 + 1 = 193. If we look up the value of 193 in the ASCII table, we find that it is the 193.

What does all this have to do with communicating data? All of the numbers shown in this section have been arranged to make it easy and efficient to package information in a digital format. In the digital format, information can be encapsulated in other information that in turn is itself encapsulated by other information. It all comes down to the arrangement of bits into bytes and then bytes into some higher form of code.

As the level of coding gets higher, more and more bytes are used to represent a given piece of information. Information can be thought of as a long string of bytes that are read and interpreted one by one.

Another important function of a protocol is the establishment of a standard communication path. When two nodes establish communications, services such as flow control and error handling must be provided. Flow control establishes a virtual communications path after assuring that there is no contention for the network resource. Error handling provides for the loss or corruption of data and its subsequent retransmission.

Data transmission protocols are the "language" the network speaks between nodes in order to establish connections, transfer data, regulate the use of the network, and avoid contention for resources. Without protocols, data communication would not be possible. Whether the protocol describes a physical device or a method of encoding data protocols enable the efficient, reliable transfer of data between computers.

Prior to the widespread development of LANs in the late 1980s, the quickest and cheapest method of transfer was the floppy disk. Even this method, while frustrating, provided impetus to the development of techniques designed to increase throughput using the available medium. Data compression became a popular early method for increasing the amount of data one could stuff onto a floppy disk. Reliable encoding schemes led to higher density floppy disks.

Another method, which is almost universal given the predominant operating systems, is the use of shared, or common, files that applications use to eliminate high operating system interface overhead. This means that an operating system carries the burden of providing system services, such as printing, communication, and memory management. This technique allows a program to save data in the smallest file size possible, including formatting commands which present the data to the application for presentation to the operator. The application uses the common services to access and act upon the data.

Classification of protocols

Applicable layer

Application layer protocols. Protocols can be classified by the layer at which they function. At the application layer, application programs reside and perform the functions that are directly controlled by the operator. Two protocols that operate at this layer are FTP and SNMP, or File Transfer Protocol and Simple Network Management Protocol, respectively.

FTP is a powerful file transfer utility. The purpose of FTP is to define and manage the transfer of files, or data, between two machines. FTP sets up two connections between the communicating machines: The first is used for login to the remote machine and uses Telnet, terminal emulation protocol. The other connection is used for the file transfer. The application provides the operator interface, which communicates with the lower levels to accomplish these connections.

SNMP also provides communication capability from the application layer. It uses a system consisting of a management node and agents, which communicate with the node. A node is located at the machine running the management program.

Session layer protocols. Protocols at this level are responsible for maintaining a network connection, properly synchronizing transmissions and sequencing the data being transmitted. Session layer protocols are often part of a larger protocol suite, which straddles multiple levels. NetBIOS actually spans this layer and the two above it.

Transport layer protocols. Transport layer protocols are responsible for the sequencing and formatting of the packets during reception or transmission, depending upon the flow of the packets.

TCP, or Transmission Control Protocol, operates at the transport level of the model. TCP provides full duplex, connection-oriented transmission control, with acknowledgement of receipt of transmission from the remote node. It also provides flow control.

An alternate protocol used at this layer is UDP, or User Datagram Protocol. This is a much simpler, connectionless protocol without acknowledgement of receipt of transmission. It is faster than TCP, but less reliable. Upper layer processes such as SNMP (simple network management protocol) use UDP.

Network layer protocols. Network layer protocols are responsible for establishing and maintaining a network connection. It is at this layer that routing of packets across an internetwork is controlled and performed.

At the network layer, the predominant protocol is called the Internet Protocol (IP). IP is a connectionless protocol, handling transmission without guarantee of delivery. It permits the exchange of data between computers without prior set up of the connection. IP receives data and commands from the higher levels, such as the transport layer, where delivery acknowledgement is performed. This protocol routes and defines data packets using the services provided by the data-link layer below it. IP hides the subnet from the user making it transparent.

Another widely used protocol is X.25. X.25 is a set of recommendations by the International Telegraph and Telephone Consultative Committee (CCITT) for transmission of data over a packet-switched network. Packet switching basics are discussed in the previous chapter. Briefly, packet switching is a transmission method by which data packets are sent using any available path. The packets may not arrive in the order they were sent and may not arrive using the same paths. Packet switching works because of the services provided by the next highest layer, the transport layer. Protocols such as TCP reassemble packets in their original order and manage the gathering of packets to format a complete message. Packet switching is used primarily on Wide Area Networks (WANs). Large organizations with large amounts of data to be transferred typically use the X.25 protocol.

Data-link layer protocols. Protocols at this layer are the interface between the physical media and the application programs through interaction with the layers above it.

Two familiar protocols are used at this level. This first Point-to-Point

Protocol (PPP), is used on the majority of the dial-up internet connections in use today. The second is Serial Line Interface Protocol (SLIP). This protocol is also used to access the Internet, providing access to the Internet Protocol over serial lines.

Asynchronous vs. synchronous. In the most general sense, data transmission protocols fall into two broad categories: synchronous and asynchronous.

Synchronous protocols operate based upon clearly defined steps initiated by a timing pulse or signal such as the pulse generated by the system clock. These protocols transmit the timing signals with the transmitted data, allowing the receiving node to synchronize with the transmitting node. Typically, the sending node will send a *preamble* before the data for the purpose of achieving synchrony. The effect of synchrony is higher throughput, reliable error detection, and correction and fault-tolerant network control. Examples of this type of protocol are Ethernet and Token Ring.

Asynchronous protocols transmit the data without establishing synchrony. The transmitting node sends the data through the medium and performs the next operation without regard to whether the message was received or lost. Asynchronous protocols do not establish a "connection" in the sense that the synchronous protocols do. Some error correction schemes are in use, such as CRC (Cyclical Redundancy Check). Transmission of data can happen at any time between the communicating nodes when an asynchronous protocol is used. This type of protocol is used on systems where the communicating devices are wired together directly and utilize a single communicating channel exclusively. An example of an asynchronous protocol is the EIA/TIA RS-232-D serial communications standard.

Connection-oriented or connectionless service. Protocols can further be separated into connection-oriented service and connectionless service. This is a fundamental distinction that has guided much of the research on and subsequent implementation of internetworking protocols.

In the first, a data path is established between the sending node and the receiving node. Data is transmitted only after a path has been established. That is, the transmission path is "agreed" upon by the transmitting and receiving nodes. This predetermined path, or "virtual circuit" is established for the duration of the transmission and all packets are routed through it. The packets do not have to carry any addressing information with them. Certain other aspects of the routing can be taken for granted after set up, such as the reliability and quality of the connection. This allows a connection-oriented protocol to shed all but the most necessary error-checking functions, greatly speeding up transmission.

In connection-oriented network protocols (CONP), such as TCP, packets that are sent do not have to be assigned sequence numbers because packets are sent sequentially. This means that the packets do not have to be reassembled at the receiving end, further increasing transfer speed. If a packet is lost, these protocols will retransmit the packet. CONPs are very "tidy" protocols, allowing for fast, reliable communication, the setup of which is transparent to the user. Since the transmission path is set up and established before transmission, no confirmation of reception is required from the receiving node; also, after the transmission is completed, the connection, or circuit, is broken.

X.25 is a connection-oriented protocol, as is ATM (Asynchronous Transfer Mode), and frame relay. X.25 is actually a "set of recommendations" defined by the CCITT for interface to WANs and packet-switched networks. This set encompasses the lower three layers of the OSI model (physical, data link, and network). X.25 is described further in the next section. ATM and Frame Relay will also be discussed later in the chapter.

Connectionless service is the other type of transmission available. In connectionless service, data packets are sent without first establishing a transmission path. Therefore, the data transmission path is not predetermined. Packets are sent nonsequentially and can possibly travel over different data paths. Consequently, the packets may not arrive in proper order at the receiving node; further, there is no guarantee that the packets will arrive at all. Data packets being transmitted in this manner must carry complete addressing information about the sending and receiving nodes. Connectionless protocols must also perform extensive error checking and must reassemble packets into the proper order upon arrival.

UDP, or User Datagram Protocol, is a connectionless service protocol that is part of the TCP/IP protocol suite. UDP operates on the transport layer of the OSI model. This protocol will be discussed later in this chapter during a discussion of the TCP/IP protocol suite.

Networking protocols

Ethernet. Ethernet, as first implemented by Abramson in 1969, was developed at the University of Hawaii's Computer Science Department to link the resources of the school's seven campuses on four of the many islands that make up the state. The physical medium used for transmission was radio, hence the term *Ether*net. Metcalf and Boggs further developed the technology between 1973 and 1976 at the Xerox Palo Alto Research Center (PARC).

IEEE-802.3 is a physical layer protocol standard and is the result of work by the committee of the same designation. Physical medium, primarily cable specifications, are defined, as are the data transmission format or "frame," and medium control technique. Several options are available for use as a transmission medium. Radio transmission, as mentioned, is available and with state-of-the-art cellular network technology is being used increasingly for wide area networks. The vast majority of networks are implemented over copper or optical cables. The original Ethernet used a coaxial cable of approximately .436" diameter with taps or terminations allowed every 3.3 meters. IEEE-802.3 is slightly different from Ethernet, using a standard coaxial cable with a diameter of .405" (Figure 4). Tap placement remains unchanged. Maximum allowable segment length is 1762' or 500 meters. Data throughput or bandwidth is limited to 10 Mbps.

Transmission of data is broadcast over the network to all nodes on a single-frequency modulated channel. In radio parlance, this is known as baseband transmission. Broadband transmission, by comparison, uses more than one frequency to transfer data. IEEE-802.3 defines the use of this technology. The reader is referred to this document for further information.

As previously mentioned, data transmission format is commonly known as a data frame, data packet, or packet frame (Figure 5). This fame is the agreed upon format that defines a standard data element. The original Ethernet frame differs from the 802.3 frame in that it had a smaller minimum frame size. IEEE-802.3 changes the minimum data field size by making it larger.

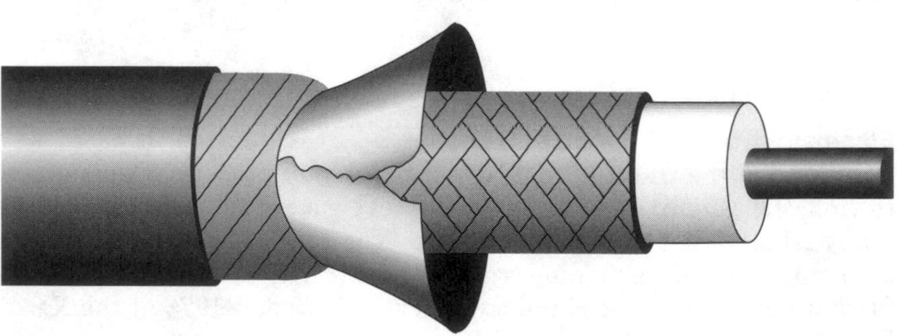

Figure 4 Thick Ethernet Cable

PREAMBLE	SFD	DESTINATION	SOURCE	LENGTH	DATA	PAD	CRC
7 OCTETS	1 OCTETS	2 6 OCTETS	2–6 OCTETS	2 OCTETS	VARIABLE		4 OCTETS

1 OCTET – 8 BITS = 1 BYTE

SFD = START FRAME DELIMITER
CRC = CYCLICAL REDUNDANCY CHECK

BITS WITHIN FRAME TRANSMITTED LEFT–TO–RIGHT

Figure 5 An Ethernet Frame

A data frame contains the following elements:

- The *preamble* is the first 48 bits of the frame and is used to synchronize the listening nodes to the new transmission.
- Following the preamble is the *start frame delimiter* (SFD) which consists of the bit sequence 10101011 and signals the beginning of the message.
- Following this is the *destination address field,* which contains the data necessary to route the message to the proper node.

This information is supplied by the network layer of the ISO stack using a protocol such as the IP (Internet Protocol) portion of the TCP/IP protocol suite. The network layer protocols handle establishment of a connection between nodes. This involves routing the message regardless of the actual physical location of the receiving node. The protocol establishes the connection, handles any acknowledgement from the receiving node(s), and then turns the data over to the link or transport layers, depending upon the direction of travel in the stack.

The remaining data-frame elements, are as follows:

- The next field is the *source address field* and is used during the establishment of a connection or "handshaking."
- The *data field* follows and is limited to a minimum of 48 bytes by convention thereby determining a minimum frame length.
- Finally, a *stop frame delimiter* is sent, signaling the end of the frame. The stop frame delimiter typically performs an error checking function, usually a CRC (Cyclical Redundancy Check) calculation.

IEEE-802.3 describes a method of medium access control designated as CSMA/CD. Carrier Sense Multiple Access with Collision Detection is used exclusively on backbone networks. Essentially, a node's network layer protocol receives data from a layer above it and signals the link and physical layers that it has a message to send. The message is encapsulated with routing information as previously described and is passed to the link layer. The link layer protocol has been monitoring the network for traffic and when a break in traffic is detected, the link layer instructs the physical layer to seize and assert a busy signal on the network. The other nodes see the busy signal and refrain from attempting transmission. Transmission of data then begins. If, at the same moment, another node has begun to send data by the above method, a BUS error or JAM signal, is generated by both nodes, aborting both transmission attempts. Both nodes then back off for a randomly selected delay time and retry the above sequence.

Once a message is successfully transmitted onto the network, the preamble signals the start of a new message and allows all listening nodes to get into synchronization with the sending node. The start frame delimiter and destination address are then transmitted, the nondestination nodes ignoring the message. The receiving node accepts the message and using the source address, acknowledges receipt and transmits the same to the sending (source) node. The network medium is then released for use by other nodes. The receiving node then decodes any special routing or control information included and passes the data up the stack to the appropriate level.

This method of transmission works very well with small traffic volumes. As traffic on the network increases, collisions become more frequent, adding an additional propagation time delay component in the form of cumulative transmission retry delay times. The above results in declining performance at high-traffic volumes. Performance is directly related to medium bandwidth, which is dependent upon the physical characteristics of the physical layer devices.

Token-Ring. The second major network type is known as *Token-Ring*. Token-ring was developed by the IBM Corporation from a network control system being used in the Netherlands in the mid-1980s. This class of network access and control is known as sequential control. Each node or station is allowed to transmit data over the network only when the node is given exclusive permission to do so. Any other nodes on the network are blocked from transmitting and must wait their turn to transmit. The device by which the nodes are given the right to transmit is known as the *token* and the technique used to transfer control is called *token passing*. IEEE-802.5 defines this protocol as a physical layer standard.

The token is a bit sequence as illustrated in Figure 6. A node is inactive until it receives the token. The node examines the token and based on the data it contains, takes action or passes on the token and any information it contains. If no data or data addressed to another

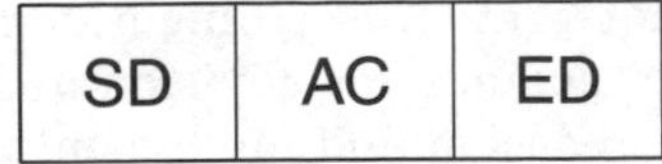

TOKEN FORMAT

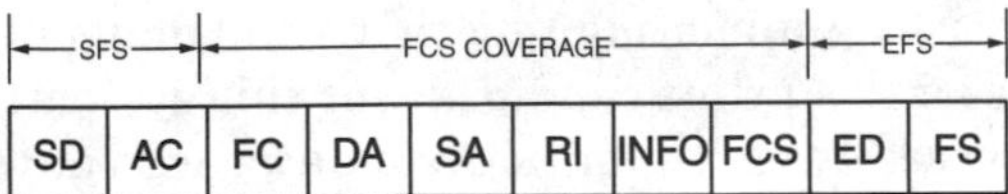

FRAME FORMAT

AC: ACCESS CONTROL (1 octets)
DA: DESTINATONADDRESS (2 or 6 octets)
ED: ENDING DELIMITER (1 octet)
EFS: END-OF-FRAME SEQUENCE
FC: FRAME CONTROL (1 octet)
FCS: FRAME CHECK SEQUENCE (4 octets)
FS: FRAME STATUS (1 octet)
RI: ROUTING INFORMATION (0 to 30 octets)
SA: SOURCE ADDRESS (2 or 6 octets)
SD: STARTING DELIMITER (1 octet)
INFO: INFORMATION (0 or more octets)

Figure 6 Token Format

node is present, the node retransmits the information to the next node. If the data is addressed to that node, the node begins to receive and pass the data to the appropriate level of the stack.

The node can use a token to pass data of its own if there is no data appended to the token. This is accomplished by appending a *data frame* to the token. The data travels over the network until the addressed node is reached. The receiving node acknowledges the receipt of the data and sends the token/frame back to the transmitting node, which checks for successful and error-free transmission and ultimately deletes it.

Token-ring networks operate at 16 Mbps and support both synchronous (real-time video) and asynchronous (real-time control) classes of service. Lack of collisions and subsequent recovery allow this type of network to perform at consistent transfer speed independent of traffic volume. Factors having an affect on network performance include server bus and hard disk access and transfer speeds, and to a lesser extent, data-frame length. A major drawback to this type of network is the loss of network on node failure. Token-ring networks require an active network management scheme, which adds to the cost of setting up and operating the network.

Token-Bus. The first such network to attempt this was a token-passing bus network known as MAP, or Manufacturing Automation Protocol. A consortium of manufacturers led by General Motors developed MAP. The goal was to integrate the coming automation of the assembly lines with the design engineering division and thereby eliminate the time delay between redesign and implementation. The vision was to eliminate costly and inefficient retooling and be able to reconfigure existing tooling on the fly.

Unfortunately, this vision was not practical. The best that could be hoped for after a manual retooling was to eliminate the inherent inefficiency of human labor by automating repetitive and dangerous tasks. Welding of auto bodies was a prime candidate for this system. The part or body could be processed twenty-four hours a day, seven days a week. The computer never took a holiday, went on strike, or asked for a raise—and it produced more parts per hour.

Programmable Logic Controllers (Figure 7) are highly specialized industrial computers. PLCs are typically used as stand-alone process control computers integrated into a larger system and control a local manufacturing process. The PLC was developed in the mid-1970s as a way of getting better control over industrial-process control instrumentation such as those that rely on a "recipe." The first PLCs were bulky and difficult to program effectively. Instructions are of the "ladder logic" variety as opposed to "coding." These instructions had to be entered individually through the use of an external keyboard. This often took days.

After the code was entered, the PLC was started in "run" mode, and the process logic tested with simulated inputs and outputs, or I/O. This step often took a few more days to debug and modify logic. The I/O was, and is, brought into separate and removable modules that fit into a PLC rack, or backplane. A separate instrument monitors each process parameter. This is an input. A control function, such as the starting of a motor or illuminating an indicator lamp, is an output. Inputs and outputs are wired discretely to separate modules, or cards.

Figure 7 A Programmable Logic Controller

Around 1985, the major PLC manufacturers developed the first PC interface. This greatly simplified the task of entering instructions. Instead of tediously inputting the instructions while sitting on the noisy, dirty factory floor, the program could be written on a desktop in another city and then loaded locally using a portable PC. The portables of the time resembled large pieces of hard luggage and used 8-inch CRTs (Cathode Ray Tubes) to display the program in bright green. This was very effective and saved much time and aggravation, despite the primitive nature of the equipment.

Laptops were further refined and miniaturized to the size of today's laptops. At the same time, PC networks were coming to the fore. It was a logical next step to attempt a link between the desktop and the factory floor. The subject had been discussed for many years; however, the technology was not available. This fact was not lost on organized labor.

General Motors used MAP almost exclusively for many years. It was the only system that had been proved in the factory environment. Unfortunately, Ethernet had begun to gain ground outside of the manufacturing environment. This would eventually spell doom for MAP. Being a highly specialized network protocol, it had little use outside of the factory. It required a large investment in equipment that could not be used anywhere else. The token-passing bus was inefficient as a means of passing the typically asynchronous data traffic of an office environment. This implied that the office would be using a better-suited protocol such as Ethernet or Token-ring to operate their network. This traffic would then have to be translated into a form that MAP could understand and act upon. This added another bottleneck in the transfer of data.

Most organizations found themselves with two separate data networks, one for office traffic, and one for factory traffic. A very temperamental and expensive piece of equipment called a Gateway made its appearance. Gateways are described in greater detail elsewhere. Briefly, a Gateway is a combination of software and translates protocols and allows data to be passed between disparate networks. Gateway are also widely used to communicate data between networks based upon computers with different operating systems.

It became apparent to all that the most cost-effective solution to the problem was to scrap the factory floor network and make it comply with the accepted office standard. This has come to be Ethernet. Most PLCs now have the capability of "talking" Ethernet. This gives the process engineer ample capability to monitor and modify the (Figure 8) manufacturing process as it happens. One thing remained to be done. The technology had matured to the point where a highly trained operator was no longer required to monitor the network. Once the code (ladder logic) had been entered into the PLC, now via the network from the desktop, all that had to be done was to monitor the various stages of the process in the same manner as had been done in the past

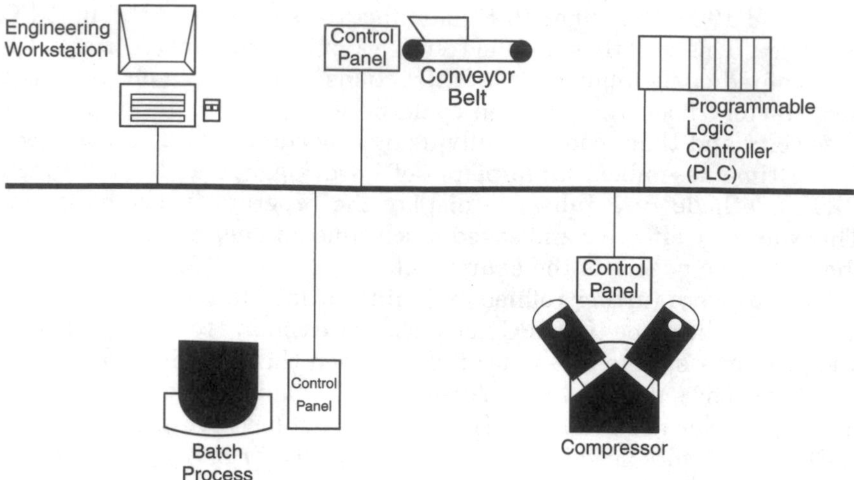

Figure 8 Factory Floor Network

simply by standing by. Using the new technology, however, enabled the operator to monitor several processes instead of just one. Where it had required ten men to weld one auto body together, now one man could monitor the building of ten auto bodies.

Other PLC networks emerged after the initial success of MAP. Several are still in general use. Space prohibits a thorough discussion of all protocols. MAP, as previously stated, stands for Manufacturing Automation Protocol. It is a token-passing bus type of a network that is similar in function to the token-passing ring described earlier. The IEEE-802.5 standard defines MAP.

Internet protocol. IP is a network layer protocol and performs routing of data packets from the sending node to the receiving node. IP is a connectionless protocol. An IP packet consists of a header and the data to be transmitted. Data may be up to 64 KB long, but must be at least 512 bytes. The IP header (Figure 9) consists of the following fields:

- Version: The version of IP being used.
- Internet Header Length (IHL): Contains the number of 32-bit words used in the header. This field must end on a 32-bit boundary to maintain field position within the header. Padding is used to assure this happens.
- Type of Service (ToS): The type of manipulation allowed for this packet in terms of handling and delays.
- Total Length: The number of bytes in the entire packet, including the header. The maximum is 64 KB (65536 bytes).

<table>
<tr><td>VERSION
(4 BITS)</td><td>HEADER LENGTH
(4 BITS)</td><td>TYPE OF SERVICE
(8 BITS)</td><td colspan="2">FRAGMENT LENGTH
(16 BITS)</td></tr>
<tr><td colspan="3">PACKET ID
(16 BITS)</td><td>FLAG
(3 BITS)</td><td>FRAGMENT OFFSET
(13 BITS)</td></tr>
<tr><td colspan="2">TTL (TIME TO LIVE)
(8 BITS)</td><td>PROTOCOL ID
(8 BITS)</td><td colspan="2">HEADER CHECKSUM
(16 BITS)</td></tr>
<tr><td colspan="5">SOURCE IP ADDRESS
(32 BITS)</td></tr>
<tr><td colspan="5">DESTINATION IP ADDRESS
(32 BITS)</td></tr>
<tr><td colspan="3">OPTIONS
(16 BITS)</td><td colspan="2">PADDING
(16 BITS)</td></tr>
</table>

Figure 9 The IP Header

- ID: A unique identifier for the packet, which is attached to all components of the packet. If the packet is fragmented, this information will allow the correct reassembly of the packet.
- Flags: This field consists of three bits that indicate whether the packet has been fragmented. The left-most (highest order) bit is always 0. The middle bit is 0 if the packet might be fragmented, 1 if not. The right-most (lowest order) bit is 0 if the fragment is the last, 1 if not.
- Fragment Offset: This value indicates the location of the fragment in the original packet and allows IP to properly reassemble the packet.
- Time To Live (TTL): Originally defined as the number of seconds a packet was in existence before being erased, it is now a hop-count value. This value is set to 32 by default and decremented at each router that passes it.
- Protocol: This field specifies the higher level protocol that is contained in the payload (data field) of the packet. TCP has a value of 5, UDP a value of 17.
- Checksum: This value is used in the error checking of the packet. It is updated at each router.
- Source Address (SA): The IP address of the source node. It is not a hardware address.
- Destination Address (DA): The IP address of the destination node. Again, it is not a hardware address.
- Options: Up to three user-defined option fields are allowed at this location.
- Padding: This field is used to make sure a header ends on a 32-bit boundary.
- Data: This is the "payload" or data, which has been passed from a higher level protocol for transmission to a remote node.

IPX/SPX. In Novelle Netware, these protocols are responsible for the addressing, routing, and delivery of packets to nodes on other networks. IPX (Internetwork Packet Exchange) is responsible for the addressing and routing responsibility. IPX works with network layer addresses as opposed to physical layer addresses, such as hardware addresses. IPX operates on the data-link layer, and works in tandem with SPX (Sequenced Packet Exchange), which operates on the network level. SPX is a connection-oriented protocol, assuring proper sequencing and delivery of packets.

X.25. X.25 is a collection of protocols that have been developed by the CCITT (International Telegraph and Telephone Consultative Committee), now known as the ITU–T, or International Telegraph and Telephone Union–Telecommunications. These protocols are used to connect different types of computer systems and networks to a Packet Switched Data Network (PSDN). A PSDN is a high-speed data communication network, which provides services to a single organization or to the public. Many organizations subscribe to PSDNs to allow remote or widely separated offices to communicate. A PSDN vendor charges an up-front connection fee, a monthly access charge, and a per-packet fee for packets sent over the network during a given time.

Each physical cable entering a premises may carry many messages concurrently. Therefore, a method is needed to route the data to the proper recipient. X.25 provides basic packet routing, or switching, by setting up each connection as a Virtual Circuit. Virtual circuits are data communication circuits between a sending and receiving node, and may have their route switched within the network to accomplish this. On each virtual circuit, a number of Logical Channels are provided. A logical channel is the circuit over which a specific Data Terminal Equipment/Data Communication Equipment (DTE/DCE) pair communicates (Figure 10). Logical channels allow many simultaneous connections over a single cable. Each DCE and DTE must share the same logical channel number when communicating directly. A DTE communicating across the network to another DTE need not have the same channel number.

Two types of virtual circuits are defined under the specification. Permanent Virtual Circuits (PVCs), are similar to leased telephone lines, but are virtual circuits set up by the network services provider. PVCs are simply an established, long-term link. A customer will pay for a PVC even though no data is being communicated. The link is permanently established and remains connected for the convenience of the subscriber.

The second type of virtual circuit is called a Switched Virtual Circuit (SVC). SVCs are virtual circuits that are set up and maintained for the

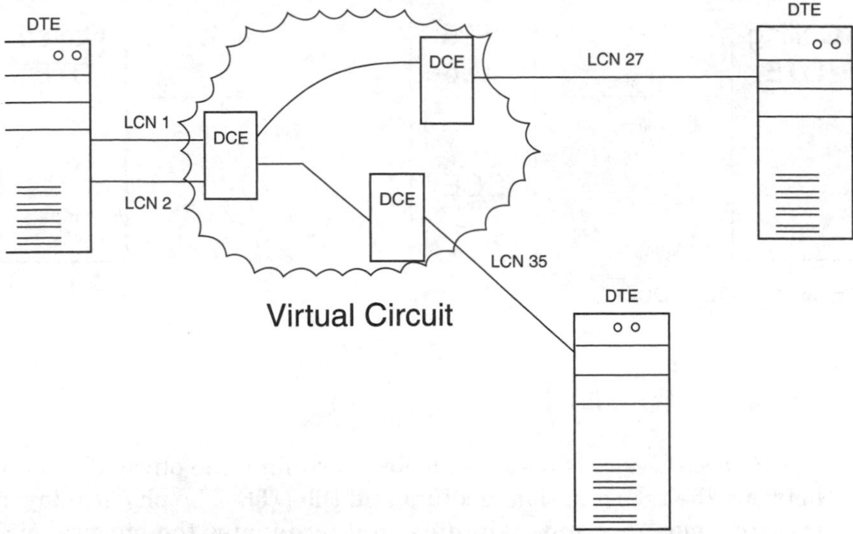

Figure 10 X.25 Virtual Circuits and Logical Channels

duration of a single connection, and then terminated afterwards. During set up, SVCs may take the form of one of three types:

- Incoming: The DTE is in receive mode only.
- Outgoing: The DTE is in transmitting mode only.
- Two-way: The DTE is capable of transmitting and receiving.

After connection set up, the channel always assumes two-way operation. Depending upon restrictions placed on connection set up at either end of the connection, one or more of the above forms may be used.

X.25 is a connection-oriented protocol, establishing a circuit before transmission, monitoring and maintaining the circuit during transmission, and terminating the circuit afterwards. After termination, the virtual circuit is released for use with other devices. Applications, which are running on two machines on the network, require communication over a virtual circuit.

X.25 defines a node as DTE (Figure 11). A DTE is a computer that is attached to and communicates with the network. The DTE connects to the network using data circuit-terminating equipment. Within the network, routing is performed by data switching equipment. X.25 does not concern itself with the particulars of the intervening network; it is not an "end-to-end" protocol in the sense that it defines all steps of transmission. Within the network, many different protocols and media may be present, X.25 defines the relationship between the DTE and the DCE.

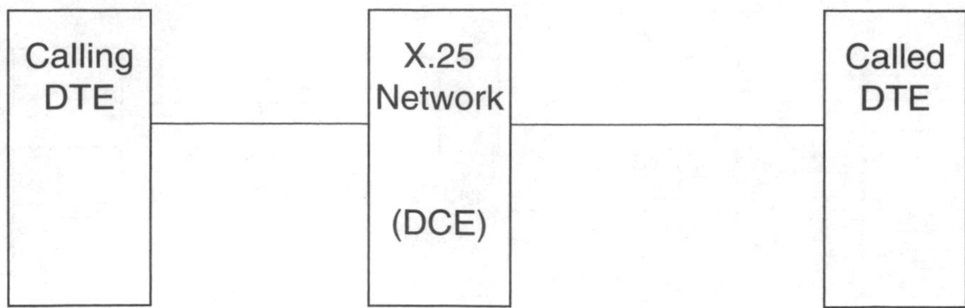

Figure 11 DTEs & DCEs

X.25 defines three levels.

1. *The Physical Layer* is responsible for controlling the physical circuit between the transmission medium and the DTE. The physical layer initiates, monitors and maintains, and terminates the physical circuit between DTE and DCE. The physical layer defines four characteristics:
   - Physical: The type of equipment interface; this could be a D-type connector with 15, 25, or 34 pins.
   - Electrical: The signal level or type.
   - Functional: The actual function of each wire, or the definition of the function of each pin on a connector.
   - Procedural: The setup, maintenance, and termination of a circuit. Also the handling of errors.

The X.25 physical layer is analogous to the physical layer in the OSI Model.

2. *The Frame Layer,* also called the link level, is concerned with the communication between the DTE and DCE. Three levels of procedure for data exchange between the two entities are defined:

TABLE 1 X.25 Layers

Packet layer	Sets up and maintains the connection across the network.
Frame layer	Responsible for the communication between the DTE and the DCE. Also called the Link Layer.
Physical layer	Controls the physical circuit between the transmission medium and the DTE.

- Initialization/Termination
- Error handling and correction
- Flow control

This layer provides services that allow the orderly and efficient exchange of information between the DTE and the DCE. Information is transmitted in the form of Frames. Frames are called by different names, such as Packets and Packet Frames. A frame is a series of bytes, or octets, which are arranged in an agreed upon manner (protocol) in order to be received and decoded. Three types of frames are defined in X.25:

- Information: This is a frame that contains the data to be transmitted. All X.25 formatted data packets are transmitted within I Frames. I Frames are numbered sequentially.
- Supervisory: These frames monitor and maintain the circuit once established. S Frames provide such information as the acknowledgement of the reception of I Frames, requesting retransmission of I Frames, and the suspension of transmission of I Frames. S Frames are numbered sequentially for ordered transmission.
- Unnumbered: These frames send information on the mode of operation required for the link.

The frame layer of the X.25 stack is equivalent to the data-link layer of the OSI Model.

3. *The Packet Layer*'s primary function is to set up and maintain a network connection for the purpose of transmitting frames across the network between DTEs. Connections to the network are made through this layer by higher level protocols such as the Transmission Control Protocol (TCP) operating in the Transport Layer. This layer defines the structure of packets and generates control packets that establish and maintain virtual circuits between DTEs. The packet layer provides services that configure virtual circuits into logical channels for multiplexing data onto a single virtual circuit. Data transmission and reception services, error detection and correction, and data flow control are other primary functions of this layer.

Routing protocols

A routing protocol is any protocol that has been developed for use in the routing of data between two nodes. Internetworks fall into two divisions. Routers and bridges make up what is referred to as Intermediate Systems (IS) and end users, or nodes, are referred to as End

Systems (ES). Within internetworks, smaller networks exist that are referred to as Autonomous Systems (AS), or routing Domains. A routing domain is a part of the larger internetwork that is administered or owned by a single organization. This smaller network is connected to other networks to form the larger internetwork.

Three types of communication are possible between the three entities that require different, specialized protocols:

- IS-IS communication: Occurs within an AS and uses an Intradomain Routing Protocol or Interior Gateway Protocol (IGP). This means that a single protocol is used inside the network.
- ES-IS communication: Occurs between the node and the router. This is facilitated by the network layer protocol being used, such as IP or IPX.
- IS-IS communication: Between AS. This communication occurs between routing domains and utilizes an Exterior Gateway Protocol (EGP) or an Interdomain Routing Protocol.

Modulation protocols

Modulation protocols are used primarily when data is to be transmitted over a telephone line. This line can be a public, or dial-up line; or it can be a dedicated, or leased line. Whatever type is used, a modem is typically used to convert the data from a computer to varying voltages that are then impressed upon the line. Modems are discussed in the next section. The voltage changes result in the familiar tones that are heard during data transmission. The method for converting data to tones is called modulation. A modem, or MOdulation/DEModulation, is the interface between the sending and receiving equipment, and the transmission line.

Modems

Modems, or MOdulation/DEModulation, come in many flavors. Class of service, modulation method, signaling method, error-correction capability, and location classifies modems. Modems convert the digital signals from the sending computer to audio tones that represent voltage variations. At the receiving end, the modem converts the audio tones back into binary form. In the parlance of the local connection standard, RS-232C, the modem is a Data Circuit-terminating Equipment, or DCE. The DCE is connected to and dependent on the computer (PC). The PC is called the Data Terminal Equipment, or DTE. The DTE sends instructions and binary data to the DCE for processing and conversion prior to transmission. At the receiving end, the DCE converts the signal back into binary and sends the data to the DTE (Figure 12).

Class of service. Modems are available for several classes of service. These include Narrowband, Voice-grade, Wideband, or Short-haul. Each class of service is specialized and optimized for a specific type of connection.

In narrowband service, the modem uses a subvoice grade bandwidth, typically 100–300 bps or slower. This class of service is, or was, used for Teletype transmission. Voice-grade modems are for use on circuits with an audio bandwidth of 300–3400 Hz. Wideband modems operate in the bandwidth range above the voice-grade bandwidth. Short-haul modems are capable of transmission speeds in excess of 1 MBs or short distances. These modems are typically used intracampus on dedicated internal communication circuits.

Modems can be further subclassified into leased line and dial-up modems. The major difference is the method of connection. In the leased-line modem, the connection is of a permanent nature; that is, the connection is set up and dedicated to that communication path and is not expected to be broken under normal operation. This type of modem is generally used for communication between computers owned by the same entity. Leased lines can offer several distinct advantages over other types.

Leased lines arc directly wired through a switching apparatus, guaranteeing that the transmission travels the same path, through

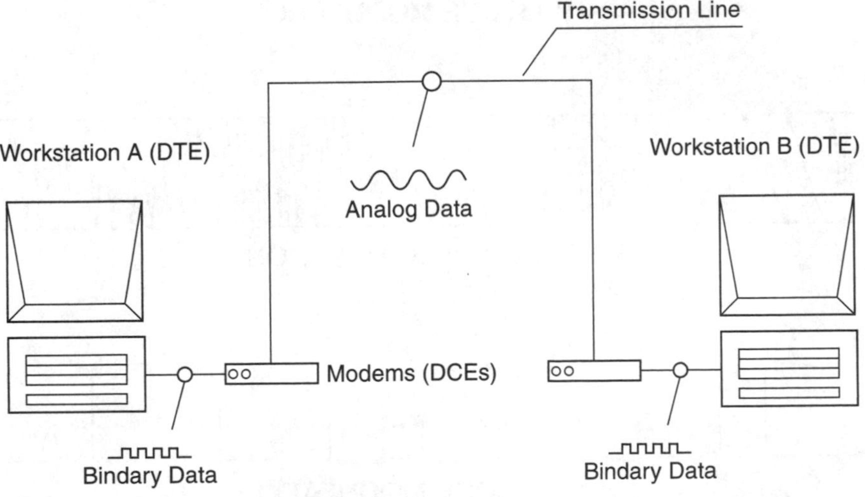

Figure 12 Modems, DTEs & DCEs

the same equipment, at all times. Switching decreases reliability. Leased lines also offer significant cost savings over other types of connections.

Dial-up modems are the type of modem most users are familiar with. These modems communicate with the sending computer, adhering to a set of standards known as the AT command set, developed by the Hayes Corporation. Dial-up lines are slower than leased lines and suffer the additional disadvantage of having to negotiate switching centers and traffic restrictions on the negotiated transmission path. This type of modem is used for infrequent traffic where the need for communication is not immediate. These modems are a staple of the PC industry and have become standard equipment in new PCs. Dial-up modems have allowed the expansive growth of the Internet.

Modulation methods. Modulation may occur in several ways. All methods require a change in the Carrier signal, whose frequency is the mean of the extremes of any modulation. To represent, or convert, a 0 or a 1 into a tone, the modem may change the amplitude, frequency, or phase of the carrier. Often, these techniques are combined (Figure 13).

In Amplitude Modulation (AM), a 1 may result in a higher magnitude of the carrier signal, while a 0 will result in a comparatively lower signal. The carrier frequency is held constant. This method of

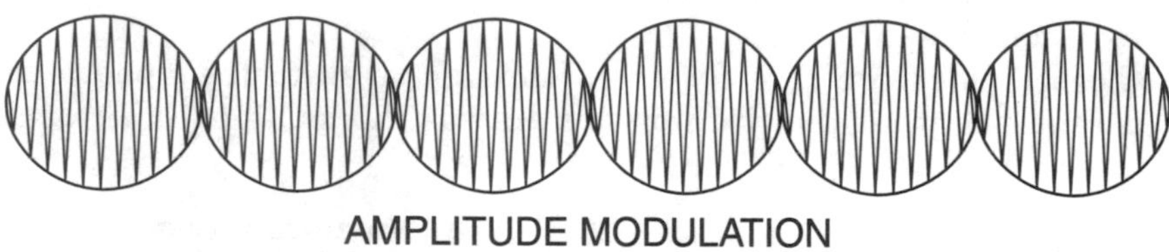

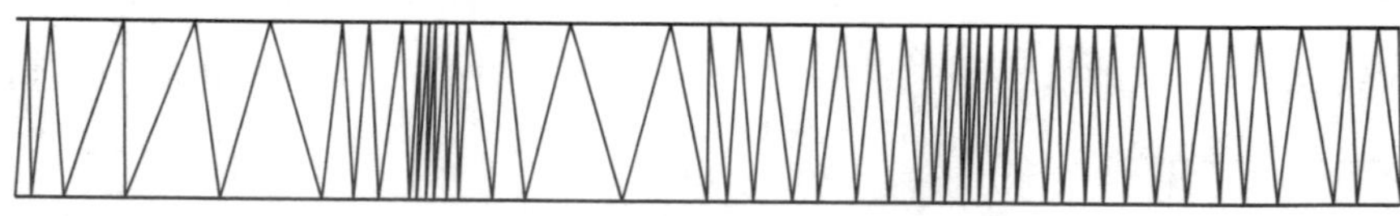

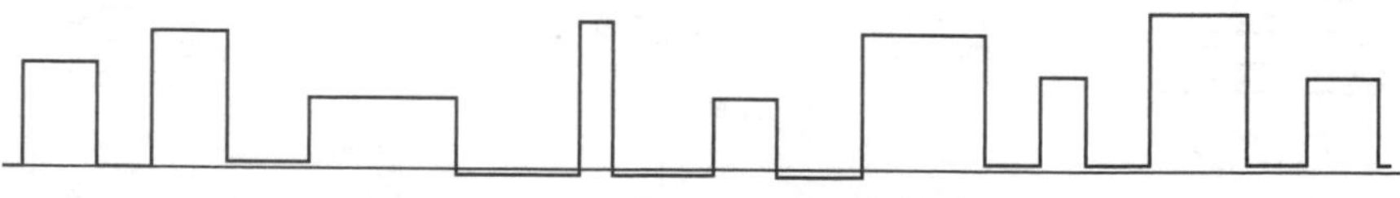

Figure 13 AM, FM & PM

modulation is susceptible to noise and is not widely used by itself. This method is also called Amplitude Shift Keying (ASK).

Frequency Modulation (FM) varies the frequency, or pitch, of the carrier signal. FM changes a 1 to a high frequency and a 0 to a low frequency relative to the carrier frequency. The amplitude of the carrier is held constant. Frequency Modulation is also called Frequency Shift Keying (FSK).

Phase Modulation (PM) is a method by which a bit is encoded at different points of the carrier's sine wave. A 1 may be encoded as starting at 45 degrees out of phase with the carrier, a 0 zero could start at 180 degrees, and so on.

Other methods of modulation are Quadrature Amplitude Modulation (QAM) and Trellis Coding. QAM utilizes both PSK and AM. Trellis Coding Modulation uses a technique similar to QAM, but adds error-correcting functions to improve transmission reliability.

Signaling methods. Two sets of standards govern the operation of a modem with regard to the type of signal used, transmission speeds, and the type of connection or interaction between machines. The Bell Standards define guidelines for the operation of lower speed modems. CCITT Standards cover high-speed modems and error-correction methods. Three types of connections are possible (Figure 14).

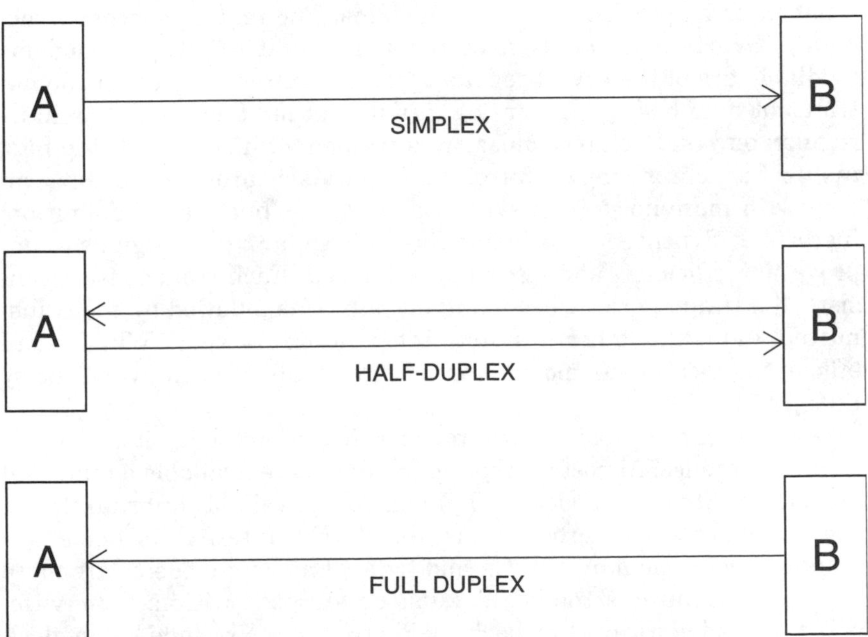

Figure 14 Simplex, Half-duplex & Full-duplex Transmission

The first type, a Simplex connection, allows data transmission in one direction only. A basic cable television connection is a simplex connection in that it delivers a signal to the subscriber with no transmission back to the head end. With newer and more sophisticated cable systems being developed, this type of connection is rapidly becoming extinct except in highly specialized applications. Simplex transmission actually requires major reconfiguration of transmission equipment and is not at all cost-effective in today's data communication market.

The second type of connection possible is called half duplex. Half duplex allows communication in both directions, but not simultaneously. A rough analogy is a telephone conversation using speakerphones. Only one person can talk and be heard at a time. This annoyance is due to inadequacies in the telephone equipment rather than the telephone lines. This type of connection is rarely, if at all, used. It is, in effect, wasting half of the capacity of the transmission line.

The third type of connection is known as Full Duplex. Full duplex connections allow simultaneous, two-way transmission. This has become the standard for all types of modem-based communication. Full duplex connections make use of the entire bandwidth of the transmission lines. Full duplex connections allow the simultaneous exchange of data between two machines, greatly improving the efficiency and throughput of the transmission path.

Before we review the standards that govern modem operation, two other factors are used to classify modems. The first is whether a modem is Synchronous or Asynchronous. The vast majority of modems available are of the asynchronous type. Asynchronous modems do not transmit data based upon timing signals, rather, they send data in irregular bursts. Each transmission is framed with start and stop bits, in effect synchronizing each transmission. Asynchronous modems can deal with more noise than synchronous types, but are, by definition, inefficient. Synchronous modems rely on regular timing signals to improve line efficiency and transmission speed. Each transmission consists of a timing, or synchronizing bit, which is followed by the transmitted data. No other framing information is sent. While more efficient, synchronous modems must slow transmission over a noisy line.

Location of a modem is not a major concern and lends itself to personal preference. Almost all types of modems are available as internal modems; that is, the modem will plug into an available internal slot on the computer's motherboard (as in a PC). Internal modems are standard equipment on all PCs sold today. External modems are more expensive because of the box it must be enclosed within. Otherwise, operation is identical. Drawbacks aside from price include loss of desk space and a serial port.

Standards

While all standards supporting dial-up connections also apply to leased lines, the opposite is not necessarily true. Table 2 is a brief overview of common standards that are still in use:

TABLE 2 Modulation Protocols

Standard	Data rate (bps)	Duplex capability	Modulation	Baud	Sync/Async
Bell 103A/113*	0–300	Full	FSK	0–300	Async
Bell 201	2400	Full**	DPSK-4	1200	Sync
Bell 202	1200	Half	FSK	1200	Async
	5 (reverse)	Half	On/Off Keying	5	n/a
Bell 208	4800	Full**	DPSK-8	1200	Sync
Bell 209A	9600	Full	QAM-16	2400	Sync
Bell 212A	1200	Full	DPSK-4	600	Sync
	300	Full	FSK	300	Async
V.17 fax	14400	Half	Trellis-128	2400	Sync
	12000	Half	Trellis-64	2400	Sync
	9600	Half	Trellis-32	2400	Sync
	7200	Half	Trellis-16	2400	Sync
V.21	0–300	Full	FSK	0–300	Async
V.22	1200	Full	DPSK-4	600	Both
	600	Full	DPSK-2	600	Both
	0–300	Full	FSK	0–300	Async
V.22bis	2400	Full	QAM-16	600	Both
	1200	Full	DPSK-4	600	Both
V.23	1200	Half	FSK	1200	Both
	600	Half	FSK	600	Both
	75 (reverse)	Half	FSK	75	Async
V.26	2400	Full	DPSK-4	1200	Sync
	75 (reverse)	Full	FSK	75	Async
V.26bis	2400	Half	DPSK-4	1200	Sync
	1200	Half	DPSK-2	1200	Sync
	75 (reverse)	Half	FSK	75	Async
V.26ter	2400	Full	DPSK-4	1200	Both
	1200	Full	DPSK-2	1200	Both
V.27	4800	Full	DPSK-8	1600	Sync
	75 (reverse)	Full	FSK	75	Async
V.27bis	4800	Full**	DPSK-8	1600	Sync
	2400	Full**	DPSK-4	1200	Sync
	75 (reverse)	Full**	FSK	75	Async
V.27ter	4800	Half	DPSK-8	1600	Sync

(Continued)

TABLE 2 Modulation Protocols Continued

Standard	Data rate (bps)	Duplex capability	Modulation	Baud	Sync/Async
	2400	Half	DPSK-4	1200	Sync
	75 (reverse)	Half	FSK	75	Async
V.29	9600	Full	QAM-16	2400	Sync
	7200	Full	QAM-8	2400	Sync
	4800	Full	QAM-4	2400	Sync
V.32	9600	Full	Trellis-32	2400	Sync
	9600	Full	QAM-16	2400	Sync
	4800	Full	QAM-4	2400	Sync
V.32bis	14400	Full	Trellis-128	2400	Sync
	12000	Full	Trellis-64	2400	Sync
	9600	Full	Trellis-32	2400	Sync
	7200	Full	Trellis-16	2400	Sync
	4800	Full	QAM-4	2400	Sync
V.33	14400	Full	Trellis-128	2400	Sync
	12000	Full	Trellis-64	2400	Sync
V.FAST	28800	Full	TBD	TBD	Sync

* This standard represents an entire family of modems.
** Full Duplex operation only on four-wire systems/half Duplex on two-wire systems.

File Transfer Protocols

XMODEM

XMODEM is a file transfer protocol originally developed by Ward Christensen. There are two basic types of XMODEM protocols, XMODEM and XMODEM/CRC. XMODEM offers the advantage of error checking on a block-by-block basis to assure that the data sent contains no errors. It does this by adding a checksum byte to the end of each 128-byte block of data; the receiver calculates its own checksum and compares it to the one received. If an error is detected in the transmission, XMODEM will request that the sending PC retransmit the block of data. In addition to the above checksum comparison, XMODEM/CRC adds another level of error detection using a complex Cyclical Redundancy Check (CRC) algorithm.

XMODEM and XMODEM/CRC are slow protocols compared to many others available, but they are quite reliable and available in almost all communications packages. They should only be used when your software supports no other protocol. XMODEM/CRC is preferable to XMODEM because of its greatly improved error checking. 1K-XMODEM performs exactly like regular XMODEM/CRC, but increases the block size to 1024 bytes, hence the name 1K. It is slightly faster (on fairly clean phone lines) than regular XMODEM due to a smaller number of blocks being sent, and therefore, fewer block checks being made. 1K-XMODEM/G, a version of 1K-XMODEM, makes use of Microcom Networking Protocol (MNP) hardware error correction to do away with the block-by-block checking in the normal version. The result is a very fast, single-file transfer protocol for use if YMODEM/G is not readily available.

YMODEM

YMODEM block sizes are variable at 128/1024, but 1K (1024 bytes) is the usual size. Error checking is performed, and is accurate to 99.99%. By definition, all YMODEM transfers are capable of sending multiple files at one request, with the file size and date included in the "header block" sent prior to each file. YMODEM supports multiple file transfer (both down and up) of up to 50 files in a "batch." Use of YMODEM, if supported by a caller's software, is recommended over XMODEM and 1K-XMODEM for speed, reliability, and features. This variation of YMODEM is usually only available when callers are using modems supporting MNP or the U.S. Robotics ARQ hardware error checking. MNP is a hardware-based system in which the modems perform the actual error checking and correction, if needed. The software simply sends the information blindly from one system to the other using the protocol for block-sorting information only. YMODEM/G is among the fastest protocols with the exception of the newer versions of ZMODEM. YMODEM/G also supports multiple file transfer (both down and up) of up to 50 files per "batch."

ZMODEM

ZMODEM is a "streaming protocol," one which sends variable-sized blocks of data with error checking for an accuracy of 99.9999%, but does not wait for an acknowledgement from the receiving computer. The sending system assumes data received is OK unless a repeat request is sent for a specific block. This streaming activity makes ZMODEM one of the fastest protocols available. ZMODEM supports multiple file transfer capability. This is commonly referred to as "batch transfers." ZMODEM also has the unique capability to resume file transfers that have been aborted for some reason and thus only partially completed. This crash recovery facility is usually not needed, but is very handy when it is.

KERMIT

This protocol's main claim is not speed, but rather its ability to interact with many types of computers from mainframes to micros. It can cope with systems limited to seven-bit characters even when the data to be transmitted is in eight-bit form. All characters to be sent are translated into standard printable characters and reconstructed on the receiving end. While not terribly efficient, it is sometimes an absolute necessity for data transfer involving different types of systems and terminal types. It is not recommended for PC-to-PC transfers. This protocol is not widely used.

Communication Reference Models

Why a model is needed

In all industries, standardization of methods and equipment is necessary for the continued, efficient operation and consistent production of the industry. In the automotive industry, parts for a particular model of automobile are standardized first by the model of vehicle it is a part of; it is further standardized based upon the type of part it is or the function it performs.

A headlight may conform to a standard for brightness as decided by the governing agency for automotive standards (SAE, or the Society of Automotive Engineers). The headlight would also conform to the size and shape as dictated by the manufacturer's internal standard for the model or model family. In this way, uniformity is achieved both in form and function. Methods of manufacture are then developed around standard parts, allowing the development of specialized, time-saving tools and techniques of fabrication.

Standards allow for a base upon which further development can be accomplished. Building on methods or equipment that are established and proven, allows greater strides in improving efficiency and productivity in the least amount of time. With a body of work to support the development of a new product, development time itself is inherently faster and more efficient. Mistakes that would bog down a researcher are already incorporated into the standard design, allowing the developer to solve new problems without "reinventing the wheel."

The ISO Model is also a type of standard. Several other organizations have also developed data communication standards. The Institute for Electrical and Electronic Engineers (IEEE) developed a set of widely used standards. These standards have been adopted by the ISO, and deal primarily with the methods of medium access and control for local and metropolitan area networks (LANs, MANs). This family of standards falls under the heading of IEEE-802. The ISO designation is ISO/IEC-8802. The methods of access and control fall largely within the lowest two layers of the ISO model, within the framework that the 802 standards were developed. This family of standards exhaustively defines Ethernet, Token-ring, Token-bus, MAN practice, and Broadband networking.

Other organizations that have developed communication standards are the American National Standards Institute (ANSI) and the Military establishment. Manufacturers develop standards, which are offered as enhancements to their products. Adherence to standards is entirely voluntary and market-driven. Often, a standard has become widely accepted and implemented before any attempt is made to define an open standard for the device or technique. That the standard lacks an official stamp or definition is of no consequence. The industry or consumer has found this *de facto standard* to perform a useful function. Most of the *de jure standards,* that is, official standards defining modern communication, had their beginning as a widely accepted method of accomplishing a needed task.

The OSI model

The International Standards Organization (ISO) has developed what has come to be known as the Open Systems Interconnect (OSI) model (Table 3). The model provides a "map" of the theoretical basis for data communication within and between computers. The model tries to identify the operations required for successful transfer of data from sending node to receiving node. In so doing, seven operational layers have been identified, each describing unique functions replete with specialized protocols, protocols performing different services. These layers are

- Application
- Presentation
- Session

TABLE 3 The OSI Reference Model

Application	Graphical user interface; primary user interface with communication system.
Presentation	Supports the functionality of the application layer by providing sevices such as formatting and translation of data.
Session	Maintains the transmission path by synchronizing packets and controlling access to the medium by the application layer.
Transport	Ensures the quality of transmission and determines the best route for transmission of data using the network layer below.
Network	Finds a route for transmission of data and establishes and maintains the connection between two connected nodes.
Data link	Creates, transmits and receives packets. Controls the physical layer.
Physical	Converts data into bits for transmission and converts received bits into usable data for the layers above it.

- Transport
- Network
- Data Link
- Physical

These layers are described in detail in the following paragraphs. Other communication models have been developed by organizations that have the desire to enhance or specialize methods of data communication, or to complement a product line. These alternate models will be discussed later.

The OSI model describes two types of communication. First, the model describes the path of information from the point of input from the operator through the equipment and to the physical communication medium. Referring to the model, this action occurs either up or down the stack. Theoretically, each layer will interface seamlessly with the layer above and below it. Using this method, protocols can be developed and incorporated into a network protocol suite in order to provide such services as transmission control and internetworking. The user, at the application layer, enters data at the terminal interface. The data is subsequently passed down the stack. It should be noted that entire layers may be ignored and the data passed through if the functions provided by those layers are not needed. In traveling down the stack, each layer will examine the data from instructions that have been incorporated by the preceding layer. If the instructions call for a service unique to that layer, the process or application is implemented and the data, along with any instructions, is passed down to the next layer.

At each layer, a new header and trailer is "wrapped" around the transmitted data. The header is unique to a certain protocol and is recognized only by the protocol by which it is defined. A good analogy would be to consider the myriad electrical plugs and sockets. A plug for one voltage or signal is not compatible with the socket for a different voltage or signal. This is a loose analogy at best, but essentially describes what a protocol does. The header is "read" by the next receiving layer. If the data, or packet, contains instructions, the process is repeated. If the header contains nothing for that layer, it is passed on. The data is ultimately passed through to the lowest or physical layer after each layer has appended or passed through the data. The packet is encoded and placed onto the transmitting medium with whatever access control protocol is being used for the medium. The data-link and physical layers of the model control this process.

The data-link and physical layers of the model provide the services needed to establish and control communication between sending and receiving nodes. The data-link layer provides facilities for the control of the transmission of data blocks and recovery from error. This layer also controls the rate at which packets are received from the physical layer. Control of the physical layer is continued or shared in this layer in order to achieve an error-free interface with the network layer.

The physical layer supplies and controls the physical communication medium used for the transmission of data. This layer is responsible for the actual signal transmission over the network. Upon receiving data, the physical layer is charged with maintaining proper interfacing with the data-link layer and effecting the transfer of data from the physical medium to the data-link layer. The data-link layer uses the physical layer to form its connection.

At the receiving node, the packet is acquired from the medium if the acquiring node is addressed in the packet's header. Packets can be addressed to one node or be "broadcast" over the network. The receiving node receives the packet into the physical layer and immediately discards the headers and trailers dealing with the medium access control method or protocol used. The packet is passed up the stack to be decoded and routed. Higher layers of the model perform the essential functions of routing, translation, and formatting for ultimate presentation to the operator.

Layer groupings. The seven layers describe, in an ordered fashion, the interaction between the various services available for delivery of information from one user at a specific node to another at a different specific node. The top layer, the Application layer, is where information is input and provides service that allows the user to communicate with the network. The bottom layer, the Physical layer, is where the actual transmission of bits takes place. It is important to notice how the data changes from the time it is input at the keyboard for transmission to the point at which it is actually transmitted.

The purpose of the model, as previously stated, is to order these tasks and provide a framework to work within. There are distinct differences between the layers and groups of layers. The top three layers, Application, Presentation, and Session, are primarily involved with information processing, formatting, and control functions that are accessible to the user or the user-application program. The bottom three layers, Network, Data Link, and Physical, are concerned with the control and successful transmission and reception of data from the physical network. These layers perform the control of the physical interface and provide services that directly communicate with other nodes over the cable or fiber. These layers are also called the "subnet" layers. Services such as timing and proper data formatting are performed at this layer.

In between the upper and lower layers of the model is the Transport layer. This layer represents the "border" between the upper data processing layers, and the lower data communication layers. The Transport layer performs the control and arbitration between the two types of services, which is crucial to the proper operation of the system. Data and commands are passed through this layer during every session; other layers may be bypassed or ignored if the services available at that layer are not needed. Every data transmission uses the services available at the Transport layer. It is at this layer that the Transmission Control Protocol (TCP) resides and functions.

In the following sections, the OSI layers are explained from the top down; that is, from the operator to the network. In so doing, the reader is urged to apply a theoretical data transmission to the explanation. It is felt this will aid the understanding of the concepts described. Note that the same services are performed on a received transmission.

Application layer. The application layer is the uppermost layer of the OSI model. The user controls the application layer directly, providing the Human–Machine Interface (HMI). Through this layer, the user is able to access the lower layer services available. Programs such as e-mail and file transfer services reside within this layer. Network management applications run from inside this layer. The application layer is the primary interface between the user and the communication system.

Application Service Elements, or ASEs, provide the application and communication services at this layer. ASEs are grouped into two classes:

- Common Application Service Elements (CASE)
- Specific Application Service Elements (SASE)

CASEs provide services to many applications while SASEs provide services to one type of application only. ASEs are available using an Application Programming Interface (API). APIs provide programmers with "handles" or function calls for interfacing with available services and link them to applications. Applications running on a desktop will use the services found in this layer to properly format and transfer data for action at the next layer. Examples of services provided at this layer are Electronic Mail (E-mail), File Transfer Protocol (FTP), and Hypertext Transfer Protocol (HTTP).

The application layer passes data and instructions to the next layer, the presentation layer, for further processing.

Presentation layer. The presentation layer provides services such as special formatting and character conversion; in short, translation and formatting of received and transmitted data. The services are used primarily to support the operation of the application layer. This layer rarely, if ever, exists as a pure layer. It is often working in tandem with the session layer directly below it. Software at this layer controls printers and plotters and various other peripherals. A common example of a presentation layer service is Microsoft Windows. Windows provides a system of shared files and services that allows a larger packet payload without the burden of additional control characters, which would cause the packet to be inordinately large.

These layers, and the services provided within, are actually encompassed by the application layer above and the session layer below. It must provide support for compatible application programs running above as well as provide properly formatted data to layers below it for subsequent transmission.

The presentation layer concerns itself with the syntax of the packet. Three types of syntax may be described. Originating syntax is formatting that is passed from the application layer to the presentation layer. The presentation layer must understand this syntax. The transfer syntax is the result of the reformatting of data from the application layer. This syntax is largely dependent upon the applicable transfer protocol being used by the lower layers of the model. The receiver syntax is formatting that is sent from the presentation layer to the application layer on the receiving side. This may or may not be the same. The receiving node may be running the same programs but may have a dissimilar operating system or local architecture.

Centered on this layer are services that make this layer the best suited for protocol translation and cryptography. As this layer concerns itself with the semantics of the bits in a packet, it is well suited for the work of stripping away the unnecessary information surrounding the data in a packet. The data is formatted for proper presentation and sent to the application layer for use. Cryptography, or the coding of data for security and reliability, is a service for which this data translation and formatting layer is used.

Session layer. This layer is responsible for the setup and maintenance of the data transmission path. It synchronizes the path and sequences the data packets for proper transmission. Application process access to the communication medium is controlled at this layer. Monitoring of the connection status is also performed. Incoming packets are isolated and subsequently grouped to allow for seamless release to the presentation and application layers. Error-checking and correcting services are performed at this layer. Essentially, this layer maintains continuity of transmission to the upper layers, while providing access to the lower levels by the application layer.

Three types of interaction is defined at this layer:

- Two-way Simultaneous: communicating entities may send and receive simultaneously.
- Two-way Alternate: entities alternately send and receive.
- Monologue: one-way transmission of data.

The session layer controls the transport layer in the establishment of a connection. It will identify and attempt to recover lost or corrupted data. It also provides and monitors the necessary synchronization between transmitting and receiving entities.

Transport layer. It is the transport layer that straddles the line between data processing that is done in the upper layers, and data communication that is done in the lower layers. The transport layer receives data and instructions from the upper layers through the session layer. At this point in the movement of data down the stack, the data being transmitted has been completely formatted and is in the form that will be delivered to the receiving node. No further change to the core data will occur. The transport layer provides services that the upper layers use to establish and maintain connections. It also serves to control the unreliable service that is provided at the subnet layers. By unreliable, it is meant that the subnet layers do not provide services that ensure quality of service. The transport layer assures that a given connection is established at an agreed transmission speed and level of error correction. In this way, a level of service and transmission quality is established and guaranteed for that connection.

The session layer passes packetized data to the transport layer, which then assigns a sequence number to each packet. This information is included in the packet header and is included in the addressing used at the lower levels of the stack. The sequence numbers are used at the receiving node to determine if all packets have been received and in the correct order. If the packets are not received correctly, the transport layer services request a retransmission of the data. If the packets are received out of order, this layer reassembles the packets in the correct sequence. In all cases, these services are performed in tandem with the session layer above it, where the packets are reassembled into a complete packet.

While the transport layer uses, and is dependent upon, the services of the session layer above it, it uses and controls the services of the network layer below it in order to establish the best connection. In so doing, this layer finds and establishes the best route for the transmission. To reiterate, the transport layer is crucial in the whole process of data transmission by occupying the position between the upper data-crunching layers and the lower communication-oriented layers.

In operation, the transport layer performs three steps, or phases, to transmit data between two computers. In the first phase, the establishment phase, a connection is established between the transport layer protocols running on both machines. Parameters such as transmission speed, service class, and transmission size are agreed upon. The network layer is manipulated to form a network connection and a connection address is set. The connection has been established and the level of quality for the connection is known and consistent.

The second phase is the data transfer phase. During this phase the data is not only transferred, but also monitored and checked for errors. The data is transferred between the two transport layer entities. Smaller packets from different processes may be combined at this time to take advantage of the established connection, but may not go to the same machine, but rather to another with which a connection has been established.

After the transfer of data, the connection is terminated during the termination phase of transmission. This action will free up transport layer resources for the negotiation of another connection, either incoming or outgoing.

Services classes. The transport layer is dependent upon the quality of the three lowest layers of the stack. While extensive error-checking is performed over preestablished connections, service on the physical medium ranges from very reliable to unreliable and will affect the speed and quality of the transmission and transmitted data. The OSI model defines three classes of service for connections established through the three subnet layers:

- Type A: very reliable, connection oriented.
- Type B: unreliable, connection oriented.
- Type C: unreliable, may not be connection oriented.

Because of the range of quality available, the transport layer must adjust the transmission service level to compensate for the type of subnet available for transmission. The ISO model defines five classes of transport layer service protocols, which provide varying degrees of quality assurance for the transmission. These levels are described in the following list:

TP0, Transport Protocol Class 0: This level provides a minimum of support for the lower layers, assuming a Type A subnet. This level of service assumes that the subnet layers will do most of the work in transmitting data and does not perform any error checking or correction. The connection is established, so the packets are not numbered prior to transmission.

TP1, Transport Protocol Class 1: This level assumes a Type B subnet, which may be unreliable. TP1 provides error detection and the capability to request retransmission from the sending transport layer.

TP2, Transport Protocol Class 2: Assumes a Type A subnet. TP2 can multiplex transmissions, supporting multiple transport connections over a single network connection.

TP3, Transport Protocol Class 3: Assumes a Type B subnet, but can support multiplex transmissions. This class has the capabilities of TP1 and TP2.

TP4, Transport Protocol Class 4: This class of services assumes nothing about the subnet and provides support for connectionless services. This is the most powerful class in the sense that it provides all necessary services to compensate for a completely unreliable connection, of which the quality is completely unknown.

The most widely used protocol at the transport layer is the Transmission Control Protocol, or TCP, which performs all of the functions just described.

The subnet layers. In the OSI Model, the three lowest layers of the stack are referred to as the subnet layers. These layers are primarily responsible for the interface to, and control of, the physical medium. Controlling the medium means controlling the method of access and arbitration for the use of the medium. The three layers described in the following sections are the mechanisms by which the physical medium is manipulated in order to get data from the sending node to the receiving node.

Routers and Bridges operate at these levels, as do specialized protocols such as X.25. The service provided at these levels may or may not be reliable; that is, the quality of the system and services present at these levels may not be consistent. This places an additional burden on the layers above as previously discussed. Services also fall into connection-oriented or connectionless categories.

Service at these layers falls into three classes: A, B, and C. The reader is invited to review the section on the transport layer for a discussion of these distinctions.

Network layer. At the network layer, the uppermost subnet layer, packets are properly routed and transmitted to the receiving node. At the network layer, also called the packet layer, the actual physical routing is set up, and the quality of the service over this routing is evaluated. Parameters such as error rate, availability of connections, any delays, and the effective throughput over the connection are passed to the transport layer. If the routing of the transmission is over several different networks, quality will vary. The transport layer determines how much, if any, interaction and control is needed for the transmission to occur successfully.

The network layer controls the actual physical transmission of the packets, which is done at the data-link level (discussed next) and governed by the commands received from higher layers as passed through the transport layer. Services provided to the transport layer include address resolution between the two transport layers, packet sequencing, and flow control. At the end of transmission, the network layer performs the physical termination of the link.

As part of setting up the link between the two transport layer entities, this layer also provides routing and switching services. These services are essential for transmitting packets to a specific address. In a connectionless environment, these services are not needed as packets are broadcast to all receiving nodes.

Data-link layer. The data-link layer receives data and commands from the network layer, and wraps the data into a new type of packet, now called a frame. The frame created at the data-link layer is unique to the architecture and protocol being used for that particular network. Control of the next layer, the physical layer, is continued from this layer. Higher layers use the data-link layer to control the rate of transmission. The terms Medium Access and Medium Control become the driving concepts behind the services provided at this layer. The data-link layer detects and corrects errors at the physical level and acknowledges transmission and receipt of data.

Data-link layer sublayers. The data-link layer consists of two sublayers as defined by the IEEE-802 Standards. The first is the Logical Link Layer. The layer provides services that mediate between the higher level network layer protocols and the second sublayer, the Medium Access Control (MAC) sublayer. The MAC provides services that interface higher level protocols with the physical layer.

Logical link sublayer. This sublayer communicates with the higher layers through a concept known as Service Access Points, or SAPs. SAPs are mailboxes for the upper level protocols, allowing a higher layer to request services by, in effect, leaving a message. If more than one type of network layer protocol is being used, the use of this system keeps all requests for service separate and queued. A SAP uniquely identifies a particular protocol, allowing it to be handled properly. This layer provides three types of delivery services:

- Type 1: Connectionless service without acknowledgement of transmission or receipt of data. This is the fastest method of transmission, but the least reliable. Connectionless service does not set up a route for transmission and does not track frames after transmission. This level of service does not provide acknowledgement of successful transmission of data. Type 1 service provides no mechanism for determining if a frame has reached its intended destination. While this level of service is inherently unreliable, it is widely used. IP, or Internet Protocol, is a connectionless protocol operating at the network layer and is part of the TCP/IP suite.

- Type 2: Connection-oriented service. In connection-oriented service, a route is set up and established prior to the transmission of data. Other services such as flow and error control are also possible. Frames are sequenced and reassembled into the proper sequence if required. Transmission Control Protocol, which operates at the transport layer, is a connection-oriented protocol and part of the TCP/IP suite. Two flow control methods are used as part of this service type:
  - Stop and Wait: Each frame must be acknowledged by the receiving node prior to the next one being sent.
  - "Sliding Aperture": Several frames may be sent and acknowledged by the receiving node prior to transmission of several more. The number of frames sent is dependent upon the size of the aperture.
- Type 3: Connectionless service, but with the capability of acknowledging or confirming frame delivery.

Frames developed at the LLC are called Protocol Data Units, or PDUs. The structure of these frames is described in IEEE-802.2. The document describes four components of a PDU:

- Destination Service Access Point (DSAP): This eight-bit value identifies the higher level protocol that is being used and is the first eight bits of the PDU.
- Source Service Access Point (SSAP): An eight-bit value that follows the DSAP, which identifies the local machine or user. Frequently, the SSAP and the DSAP are the same.
- Control: This is a one- or two-byte value that identifies the type of frame. Three types of frames are possible:
  - Information (I) frames: These are used for transmitting data.
  - Supervisory (S) frames: Used for monitoring the transmission of I frames. S frames are found only in Type 2 or connection-oriented service. S frames do not contain any data.
  - Unnumbered (U) frames: The vehicle by which network links are set up and broken in Type 1 or Type 2 service. U frames are also used to transmit data in Type 1 or Type 3, connectionless service.
- Data: The last component of a PDU is the data, or information, to be transmitted. The length of this field is variable depending upon the amount of data passed down from the network layer. The size of the data field is determined by the type of medium access method used.

Medium access control. Medium Access Control (MAC) the lower of the two sublayers provides services that allow various medium access methods to be used. The basic function of the MAC sublayer is to transmit and receive LLC frames. The MAC sublayer also provides the LLC with status information about the transmission. MAC interfaces with the physical layer directly. Services such as error detection, framing, and timing are provided to the physical layer. Data is passed to this sublayer from the LLC sublayer and framed to conform with the architecture and protocol being used on the local network.

Physical layer. The physical layer is concerned with the actual transmitting medium being used. Standards at this level describe methods of signaling, transmission, and physical interface. This layer receives data from the data-link layer above it, and converts the data, in the form of 0's and 1's, into a series of electrical signals. These electrical signals are transmitted to the receiving node's physical layer for conversion back to 0's and 1's. The data is gathered into frames and passed up to the data-link layer for further processing.

The type of medium being used is defined at this level. Several types of coaxial cable are described in the standards. Other media are radio, twisted pairs, and fiber-optic cable. The IEEE-802 family of standards describes the physical characteristics for different methods of medium access and control at the physical layer.

Connector types are also defined, but are dependent upon the type of cable being used, and to a lesser extent, the type of signal. The method or format of the transmitted signal is another parameter defined by physical layer standards. The voltage levels or method of data encoding is also defined. Briefly, below are listed several examples of physical layer standards:

- EIA-232D: This standard defines the physical interface and electrical characteristics of the signal. This standard describes a serial connection between two computers and has been used for many years as a basic means of connecting two computers. The standard was known as RS-232 prior to formal standardization (RS stands for Recommended Standard).
- EIA-422 and 422A: Defines the electrical characteristics of balanced and unbalanced communication circuits for a digital interface.
- IEEE-802.3: Describes the physical and electrical characteristics of Ethernet, including physical interface and signaling.
- IEEE-802.5: Defines Token-ring interface and signaling methods.

TCP/IP

The TCP/IP reference model is another network communications reference model. It is the model upon which the current Internet architecture is based. The theory behind the reasons and operation of the model are the same as in the OSI model. The difference lies in the number and functions of the layers. In the TCP/IP model, only four layers are defined (Table 4):

- Application: This layer contains all of the higher level protocols as in the OSI model. Presentation and session layer protocols and services are not needed and are not present in this model.
- Transport: Performs the same services and functions as the OSI model.
- Network: Also called the Internet layer, it is the layer that controls the transmission of data.
- Host (to network): Analogous to the data-link and physical layers of the OSI model. While this model does not define a physical interface or medium, it does specify that the applicable protocol is to communicate with this layer.

This model is an illustration of how specialized protocols reduce overhead by bypassing unneeded services and functions in some layers. The four layers defined in this model correspond roughly to a combination of all seven OSI layers, the transport and network layers corresponding directly.

TCP, or Transmission Control Protocol, operates at the transport level of the model. TCP provides full duplex, connection-oriented transmission control, with acknowledgement of receipt of transmission from the remote node. It also provides flow control.

An alternate protocol used at this layer is UDP, or User Datagram Protocol. This is a much simpler, connectionless protocol without acknowledgement of receipt of transmission. It is faster than TCP, but less reliable. Upper layer processes such as SNMP (Simple Network Management Protocol) use UDP. A UDP packet format is shown in Figure 15.

At the Network, or Internet layer, the predominant protocol is called the Internet protocol (IP). IP is responsible for moving data from node to node. It was developed by the Department of Defense (DOD) to al-

TABLE 4 The TCP/IP Stack

Application	This layer contains all of the higher level protocols in the OSI model.
Transport	Performs the same services and functions as the OSI model.
Network	Also called the Internet layer, it is the layer that controls the transmission of data.
Host	Analogous to the Data-link and Physical layers of the OSI model.

----------------32 bits----------------	
Source Port	Destination Port
Length	Checksum
Data	

Figure 15 UDP Packet Format

low communication between disparate computer systems. IP is a connectionless protocol, handling transmission without guarantee of delivery. IP receives data and commands from the higher levels, such as the transport layer, where delivery acknowledgement is performed.

SNA

Systems Network Architecture is a model developed by IBM to allow any IBM machine to communicate with another. SNA was originally developed for use with mainframe computers. This arrangement consisted primarily of a large central computer with many dumb terminals. SNA was developed to allow any peripheral to communicate with the central computer. SNA also has seven layers (Table 5), which also perform distinct services and functions at each level. Originally SNA had but five layers; however, two additional layers were added, one at the top and one at the bottom of the stack, in order to make comparisons with OSI simpler. SNA's seven layers are

- The Transaction Services Layer is the highest level and provides remote database access, exchange of documents between nodes, and e-mail. Note that machines using this model and the architecture based upon it, are also using proprietary protocols. To simplify the discussion, the listing of these protocols will be avoided.
- The Presentation Layer, where formatting, compression, and decompression occur.
- The Data-Flow Control Layer defines the characteristics of the connection. Such characteristics include full or half duplex transmission, error correction and recovery, and acknowledgement of packet delivery. Other services provided are the proper grouping and sequencing of packets.
- The Transmission Control Layer is responsible for establishing, maintaining, and terminating a connection between nodes. In IBM parlance, a connection is called a session. Routing and the successful delivery of packets is done at this layer. Also, data encryption and decryption is performed here.

TABLE 5 The SNA Stack

Transaction services	Sessions are requested and initiated at this layer.
Presentation services	Responsible for getting the data to the receiving node in the proper form including: format conversions, data compression and decompression of packets.
Data-flow control	Defines the general characteristics of the connection such as: error recovery, packet grouping and acknowledgement.
Transmission control	Establishes, maintains and terminates the sessions between nodes.
Path control	Creates the logical connection between nodes.
Data-link control	Lowest layer of SNA. Responsible for reliable transmission of data.
Physical control	Defines the physical interface and the medium used (not defined as part of SNA).

- The Path Control Layer provides services that create the logical connections between nodes. This layer contains three sublayers as defined by IBM:
  - Transmission Group Control, which identifies and monitors the links.
  - Explicit Route Control, which actually performs the routing of the packet.
  - Virtual Route Control, which manages the logical link once established.
- The Data-Link Control Layer is the lowest specified layer in SNA. This layer is responsible for providing services to reliably transmit data across a physical medium. Various open protocols are supported at this level, including X.25.
- The Physical Layer is actually not defined by SNA, but exists to provide congruency with the OSI model. At the physical layer, all that exists is the medium. SNA will operate over a variety of media and interfaces. Serial and parallel interfaces, coaxial cable, fiber-optic cable, or twisted-pair cable may be used.

The layers can be grouped into two categories using the middle five layers of the model. The Path Control Network is made up of data-link control and path control layers and is responsible for moving data through the network. The Network Addressable Unit (NAU) Network is made up of three of the upper four layers, the Transaction, Presentation, and Data-Flow Control layers. This network is responsible for the control and management of the network.

Fiber distributed data interface

Fiber Distributed Data Interface, or FDDI, is a fiber-optic based network operating at 100Mbps. FDDI is implemented as two counter-rotating token-passing rings (Figure 16). The first ring is referred to as the Primary Ring, the second being called the Secondary Ring. FDDI was developed by the American National Standards Institute (ANSI). FDDI Networks can span a geographical area of 100 Km, with up to 100 nodes attached. Depending upon the type of fiber-optic cable used, station spacing of 2 to 40 Km apart can be supported. FDDI networks are extremely reliable, as the dual, or redundant, ring arrangement allows extreme fault tolerance.

The FDDI specification defines several devices for use on this type of network. A typical FDDI network consists of attached nodes, called Stations; a Network Interface Card (NIC); fiber-optic cable and connectors, which are described in the following chapter; concentrators, which are wiring centers for an FDDI network; and couplers, which actually split the optical signal.

FDDI stations are one of two types:

- Double Attachment Stations (DAS): A DAS has two transceivers that communicate with both rings. The attachment is a dual-fiber cable. On the first cable, one fiber carries the incoming signal from the primary ring, while the other carries the outgoing signal for the secondary ring. The second dual-fiber cable carries the incoming signal for the secondary ring and the outgoing signal for the primary ring. DAS may be attached to a backbone network or multiple backbone networks. This type of station is also called a Class A station.

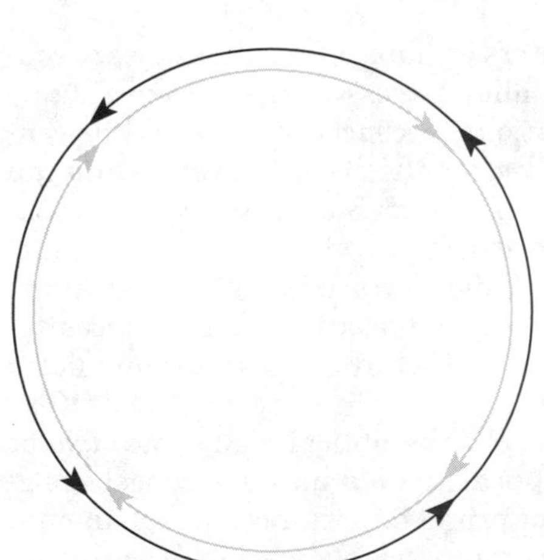

Figure 16 FDDI Rings

- Single Attachment Stations (SAS): This type of station has only one transceiver and is connected to the primary ring only. These stations cannot be used to communicate with backbone networks. SAS must be connected by use of a concentrator, described later. While the advantage of the dual ring is lost through the use of these stations, speed is retained and the failure of a node is well isolated from the main network. These stations are also called Class B stations.

Stations on an FDDI network communicate through Ports. Four types of ports are defined by the ANSI standard:

- Port A: Defined for DAC and DAS only. It connects to the primary ring input and the secondary ring output.
- Port B: Defined for DAC and DAS only. It connects to the primary ring output and the secondary ring input.
- Port M: Defined as the master port, it is used to connect two concentrators. The concentrators can be single or double attachment units. The M port can also be used to communicate with both single and dual attachment stations.
- Port S: The slave port, it is defined for use with single attachment devices. It is used to connect two SAS, or can connect an SAS to a SAC.

Network Interface Cards (NICs) are the physical interface to the fiber and are located within the computer or router. NICs may have either one or two optical transceivers which conform to the FDDI standard.

Optical Cables and connectors are discussed in detail in the next chapter. Briefly, an optical fiber may be single- or multimode cable. These designations refer to the amount of traffic a fiber can transport and also to the maximum distance it may be transmitted. Connectors are of the standard fiber-optic ST type, but are keyed to prevent incorrect connection of nodes.

Concentrators are similar to routers or hubs. Concentrators are connected to both rings and serve to allow access to the second ring by Class B or Single Attachment Stations. A concentrator could be considered a common second transceiver for the SAS. Concentrators can be Dual Attachment Concentrators (DAC) or Single Attachment Concentrators (SAC). A DAC is used to attach any of the four entities just discussed (SAS, SAC, DAS, DAC). Single Attachment Concentrators, in contrast, are used to connect only single attachment devices together. SACs must be connected to a DAC in order to communicate with the larger network.

Fiber-optic couplers are used to split the optical signal into two or more signals or to gather many optical signals into one cohesive signal. The efficiency of a coupler is of prime importance. When an optical signal is split, the signal becomes appreciably weaker. Each signal is half as strong as the original signal. Couplers are classified in terms of the number of inputs and outputs. A coupler with three inputs and five outputs is referred to as a 3 × 5 coupler. Couplers are subject to cross talk, particularly if signals are multiplexed.

Couplers can be defined based upon the manner in which an optical signal is processed. Some types of couplers are listed here:

- Active Couplers: These devices are capable of receiving an optical signal, converting it to an electrical signal, boosting it, and then converting back to an optical signal for retransmission. Passive Couplers, by contrast, do not change the signal in any way besides splitting the signal into separate paths. Passive couplers always introduce a loss into the signal path; the signal leaving a passive coupler is always weaker than the signal that was received.
- Passive Star Couplers (Figure 17): Several fibers are fused together at the point at which they meet. A signal transmitted from any node will be transmitted to every other node. As the number of fused fibers grows, so does the loss, or attenuation, of the optical signal.
- Tee Couplers: One input and two output ports.
- Directional Couplers: A directional coupler can split signals in one direction only. A Bidirectional Coupler can handle two-way traffic.
- Combiners: Called Combiner Couplers, these devices handle traffic that is modulated at different wavelengths. Multiple input signals are combined into a single output signal. This is accomplished by using Wavelength Division Multiplexing (WDM).
- CSR Couplers: Centro-Symmetrical Reflective Couplers use a concave mirror to distribute incoming optical signals to outgoing fibers. The mirror is adjusted to control the direction or fiber to which the light signal is sent.

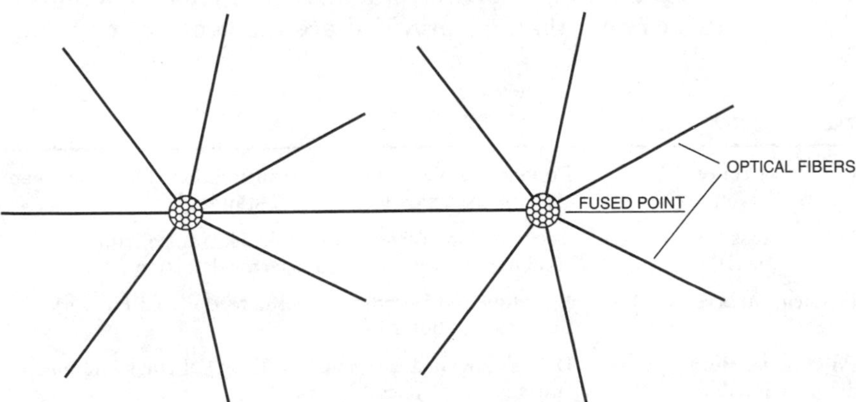

Figure 17 Passive Star Couplers

FDDI defines four layers (Table 6) that correspond to the physical and data-link layers of the OSI model. Beginning at the physical interface, these layers are listed here:

- PMD, Physical Medium Dependent: This is the lowest layer in the FDDI standard. PMD specifies connector characteristics, optical power sources, and optical transceivers. Connectors are called Medium Interface Connectors (MIC). The MIC serve as the interface between the electrical and optical portions of the network. ANSI has specified a unique connector for use with FDDI, called, appropriately, the FDDI Connector. This is the only optical layer in the standard. Cabling is also specified at this layer. Two counter-rotating optical rings are defined, the first being the primary ring and the main transmission path; the secondary being redundant and providing a backup transmission path that is capable of carrying all network traffic should the primary ring fail. The secondary ring is generally idle during normal operation.
- PHY, Physical Layer: This layer provides interfacing between the MAC layer above and the PMD below. This layer is the lowest electronic layer in the specification. The PHY layer also provides signal encoding and decoding as defined by the ANSI specification. The PMD and PHY layers correspond to the OSI physical layer.
- MAC, Medium Access Control: This layer is analogous to the MAC sublayer of the data-link layer of the OSI model. In FDDI, this layer defines and formats data frames. It also manages the method of medium access, the token. Data from higher layers are passed to the MAC sublayer by the LLC sublayer above it.
- SMT, Station Management: SMT is not technically a separate layer. SMT provides monitoring and management of a node's activities. The SMT will also dynamically adjust the network's bandwidth. The SMT straddles the three FDDI layers in order to effect this control. Some of the services that are provided are the generation of diag-

TABLE 6 FDDI Layers

Link layer control (LLC)	Provides services to enable transmission of data between stations. Not part of the FDDI standard.
Media access control (MAC)	Construction of frames and tokens, transmitting, receiving frames and tokens, error-detection.
Physical sublayer	Encoding and Decoding, error detection and recovery, filtering and buffering.
Physical medium dependent sublayer	Defines how nodes connect to the FDDI rings and how stations are physically connected.

nostic frames; control of access to the network; and ring management, which actively troubleshoots the network. SMT is responsible for routing traffic around a bad node or faulty ring section. SMT has no distinct counterpart in the OSI model.

FDDI networks are susceptible to three modes of failure. All three are recoverable because of the high degree of fault tolerance inherent in this type of network. The first type of failure is the loss of a node. FDDI defines a type of coupler that will actively bypass the signal around the bad node. The second mode of failure is a failure of part of the primary ring. In this case all traffic is routed to the secondary ring. SAS will be effectively isolated from the ring because concentrators provide connectivity to the primary ring only. For this reason, SAS should be used for noncritical traffic.

The third type of failure is when a node is lost, the rings between nodes fail, or a node is isolated because of multiple-ring segment a failures. All DAS are capable of "loopback," which allows a node to route traffic back out of the node using a single dual cable. This routing is done dynamically and is handled by the Station Management Facility.

SONET

Synchronous Optical Network (SONET) is another high-speed optical network defined as a standard by ANSI. As in FDDI, different layers are defined that control the functions necessary to facilitate the transmission of optical signals. SONET operates at speeds ranging from approximately 52 Mbps to 2.5 Gbps. SONET was conceived as the carrier for such networks and services as ATM (Asynchronous Transfer Mode) and ISDN (Integrated Services Digital Network).

SONET networks consist of three parts: (1) Sections, which are the cables between the sending and receiving nodes. (2) Endpoints at either end of a section, which may be transceivers, repeaters, or multiplexers and are the source and destination nodes; and (3) Lines, which connect two Multiplexers. Repeaters and Multiplexers are integral parts of SONET. As in FDDI, optical signals must be converted from electrical to optical for transmission at the sending node, and converted back to electrical signals at the receiving end. In SONET parlance, Synchronous Transport Signal (STS) and Optical Carrier (OC) are the terms used to describe electrical and optical signals, respectively.

The SONET standard defines four layers Table 7:

- Photonic: Defines the physical interface characteristics of the network such as cable, signal parameters, and physical interface. It is at this level that electrical to optical and optical to electrical conversion takes place; that is, where the STS to OC and OC to STS

TABLE 7 SONET Layers

Photonic	The physical characteristics of the network are defined here. Signals are converted between electrical and optical form at this layer.
Section	Performs error correction and creates frames for transmission over the physical media.
Line	Responsible for getting frames from one node to another. Controls timing of transmission.
Path	Responsible for establishing and maintaining the transmission path.

conversion takes place. The Photonic layer provides services that transport information on and across the physical medium.

- Section: SONET frames are created at this layer. Any scrambling or encryption is also performed at this level. Error correction is performed at this layer, and the section is monitored for proper operation.
- Line: This layer provides services that ensure the delivery of frames over the physical medium. This layer performs synchronization of the transmitted data and performs multiplexing for the path layer.
- Path: This layer is responsible for getting the data from source to destination. From source to destination is the path. The path layer formats data for proper framing by the line layer below it.

SONET frames are simple in structure, but are complicated functionally. SONET frames are 810 bytes in length, broken up into nine 90-byte portions. Each portion is transmitted sequentially. Three bytes in each portion are overhead, containing information for control and parity checking; the remaining 87 bytes are data (payload). The payload is called the Synchronous Payload Environment (SPE).

SONET defines eight levels of channel capacity. These channels are described in Table 8.

Asynchronous transfer mode

ATM is a packet switched, broadband network that is capable of transmitting voice, data, and video signals on the same physical media. ATM is a connection-oriented protocol. ATM is fast; the first implementations operated at 155 Mbps, increasing to over 600 Mbps. Gigabit speeds are planned as the technology becomes more widespread.

ATM packets are called Cells. Cells are a fixed length of 53 bytes consisting of a header 5-bytes long followed by a 48-byte payload (Figure 18). The header contains information on routing and flags that make the cell expendable if network traffic levels make it necessary to

TABLE 8 SONET Channel Bandwidths

Transmission rate	SONET level
51.84 Mbps	OC-1/STS-1
155.52 Mbps	OC-3/STS-3
46.56 Mbps	OC-9/STS-9
622.08 Mbps	OC-12/STS-12
933.12 Mbps	OC-18/STS-18
1.244 Gbps	OC-24/STS-24
1.866 Gbps	OC-36/STS-36
2.488 Gbps	OC-48/STS-48

shed loads. In this case, expendable cells would be deleted, creating room for essential traffic.

ATM uses fiber-optic based networks to transmit data. Such services as ISDN rely on ATM for the operation of the service. Transmission can be local or long-distance, over public or private lines. ATM is a

CELL CONTENTS

Bit 8	Bit 7	Bit 6	Bit 5	Bit 4	Bit 3	Bit 2	Bit 1	OCTETS
GENERIC FLOW CONTROL				VIRTUAL PATH IDENTIFIER (VPI)				0
VIRTUAL PATH IDENTIFIER (VPI)				VIRTUAL CHANNEL IDENTIFIER (VCI)				1
VIRTUAL CHANNEL IDENTIFIER (VCI)								2
VIRTUAL CHANNEL IDENTIFIER (VCI)				PAYLOAD TYPE		CLP		3
HEADER ERROR CONTROL (HEC)								4
PAYLOAD								5
PAYLOAD								6
PAYLOAD								7
PAYLOAD								~
PAYLOAD								51
PAYLOAD								52

Figure 18 An ATM Cell

TABLE 9 ATM Layers

ATM adaptation layer	Provides protocol translation between ATM and other communication, such as video, data, or voice.
ATM layer	ATM Cells along with headers and trailers are created here.
Physical layer	Defines the physical medium and interfaces.

three-layer protocol that is further divided into planes, which define domains of activity (Table 9). The layers are

- Physical: Defines the physical characteristics of the transmitting medium and interfaces. This layer has two sublayers that further define services at this level:
  - Physical Medium, PM: provides the definition of the physical medium and timing.
  - Transmission Convergence, TC: This is the higher of the two sublayers and is responsible for the integrity of the cells being created at higher levels. This sublayer performs routine checking on cells in the data stream from the ATM layer above.

The User Network Interface defined within the ATM specification allows for several different types of physical interfaces. SONET is one type of network used by ATM. Though ATM is not tied to any particular interface, optical fiber is the only medium defined for use with ATM.

- ATM Layer: This layer forms the cells to be transmitted by the physical layer. Virtual circuits are defined and executed. Using data passed from the next higher level, the ATM Adaptation Layer, the ATM layer adds the 5-byte header information and sends the completed cell to the physical layer for transmission.
- ATM Adaptation Layer, AAL: This is the highest layer defined by the ATM specification. The AAL performs translation services from various received protocols such as higher level data protocols, voice or data, to standard 48-byte ATM cells. These cells are passed to the ATM layer for further processing. The AAL supports two sublayers:
  - Convergence sublayer: This is the upper sublayer of the two and provides interface services to higher level protocols. Access to these services is made through Service Access Points (SAPs).
  - Segmentation and Reassembly, SAR: This sublayer accepts the data passed to it from the CS sublayer as variable length packets and reformats the data into fixed length 48-byte cells. The SAR is also responsible for putting data back to the original format at the receiving node. The SAR sublayer is also responsible for maintaining the proper sequence of the transmitted cells and correcting transmissions that are received out of sequence.

The AAL defines four types of service, a Protocol Data Unit (PDU) being defined for use at each service level. Each PDU carries 48 bytes, which consist of the header, trailer, and payload. The four service levels are

- Class A: Constant bit rate (CBR) service. Class A service is most appropriate for use on voice data. The protocol used at this level is AAL 1, which defines a PDU payload of 47 bytes. This PDU carries the largest amount of data.
- Class B: Variable bit rate (VBR) service. Class B service is used for video data. The protocol used at this level is AAL 2, which defines a PDU with a 45-byte payload.
- Class C: Connection-oriented service. The protocols used at this level are AAL 3, which defines a PDU payload of 44 bytes.
- Class D: Connectionless service. The protocol used is AAL 4 with a payload of 44 bytes.

ATM also defines three domains of activity, called planes. Three planes are defined by the ATM specification:

- User: On this plane, users exchange information. Flow control and error-correction and recovery are also performed on this plane.
- Control: Connection set up and management are performed on this plane. Connection set up, maintenance, and termination are handled here.
- Management: This plane includes two types of management functions:
  - Layer management: Operation and maintenance of each layer function and information flows are managed here.
  - Plane management: Coordinates functionality between planes.

Hubs

A hub (Figure 19) is a device that serves to connect many nodes to a central point in the network. Signals are transmitted from a node to the hub, which then retransmits the information to every node connected to it. An example of a hub is the MAU (Multistation Attachment Unit), which was described in a previous section. Hubs are media specific. Hubs will not operate as protocol converters. These devices operate with one protocol and one type of media. All hubs provide connectivity between nodes and groups of nodes. Hubs operate wholly within the confines of a single network, providing connectivity between the nodes of that network. A hub may be connected to another hub, creat-

Figure 19 A Hub

ing an extended network. Certain types of hubs are capable of operating using multiple architectures.

At the most basic level of operation, hubs simply pass on information they receive. Higher end devices will also act as filters or routers, further blurring the distinction between the various network connection devices. MAUs provide more features in terms of packet routing and network management. With this in mind, hubs can operate on any of three levels:

- Physical: Providing connectivity only.
- Data-Link: Providing simple routing and filtering capability.
- Network: Providing intelligent routing of packets.

Types of hubs

Peer hubs and stand-alone hubs are two classifications of this device. A peer hub is located within a server, and uses the server's power supply and data buses. A stand-alone hub exists outside of any machine, providing its own power and providing physical interfaces for the connected nodes. The interfaces are media specific; that is, the connection is specialized for the media being used. Coaxial cables require a certain type of physical interface, or connector. Twisted pairs utilize quite a different physical interface. Some external hubs provide separate high-speed connections for interconnecting two hubs.

Intelligent hubs. Intelligent hubs provide more than simple switching of packets between nodes. This type of hub will also incorporate management capability into the operation of the device. Some of these features may be statistical record keeping, error correction, automatic reconfiguration of the network to isolate bad nodes, or the ability to be remotely controlled. These devices require more interaction with the network manager and a more complex configuration.

Multiarchitecture hubs. This distinction is wholly within the gray area between the definition of a hub and the definition of a concentrator. A concentrator is a device that is capable of communicating with several different network architectures both by protocol and by physical interface. The difference between the two devices is that a concentrator outputs less signals than it receives. A multiarchitecture hub is more properly within the definition of a concentrator, and the terms are used interchangeably.

Active vs. passive hubs. The last classification of the hub has to do with whether a hub acts upon the transmitted data in any way. An active hub will perform various tasks upon the data as it passes through the hub. Timing of the packets, and the boosting of the signal strength are two of the functions an active hub may perform. A passive hub, in contrast, does not change the signal in any way. A passive hub acts as a wiring and retransmission device only.

Routers

Routers (Figure 20) actively determine a transmission path for packets being transmitted between nodes. The communicating nodes, on separate networks, may be separated by large distances and may be accessible only through several interconnecting and switching devices. A router first determines the routing and establishes a path prior to commencing transmission of the data packet.

Routers operate at the network layer of the OSI model and are dependent upon the applicable network layer protocol for the correct addressing format. Most routers are multiprotocol routers, capable of communicating using several different addressing formats. These routers also have the capability of filtering traffic utilizing the address format to determine which network is the recipient of the data.

Types of routers

Routers fall into several broad categories. Most routers fall into multiple categories or straddle multiple capabilities. Listed below are some of the capabilities and features that are available.

Single and multiprotocol routers. As mentioned, routers operate at the network layer and are dependent upon the protocol being used at that level. Single protocol routers are capable of handling only one addressing format as specified by a single protocol. An IP (Internet Protocol) router is designed for use with the IP address format and does not recognize any other format. A multiprotocol router is capable of processing several types of network layer protocol addresses. It may also handle IPX (Internetwork Packet Exchange) or IGP (Interior

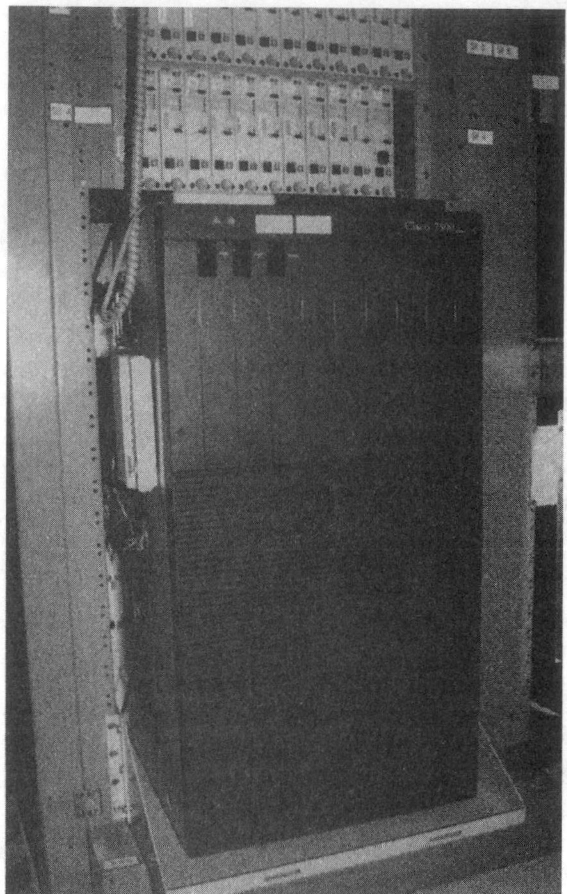

Figure 20 Router. (From J. Clayton, *McGraw-Hill Illustrated Telecom Dictionary,* 1998)

Gateway Protocol) in addition to IP and X.25. As more protocols are used and processed, a slowdown in traffic throughput can be expected.

Central and peripheral routers. A router may be the main, or central, point of interconnection for many separate networks. These central routers provide support for many different architectures and protocols. A peripheral router is a device that is limited to a single network and/or protocol. A peripheral router is typically used to connect a small network to a larger interconnected network, or internetwork (Figure 21).

LAN and WAN. Routers can also be grouped by the range over which a router must determine a route. LAN routers are routers that operate over a small range geographically. Routing is a relatively simple matter compared to the extensive route discovery that must be performed

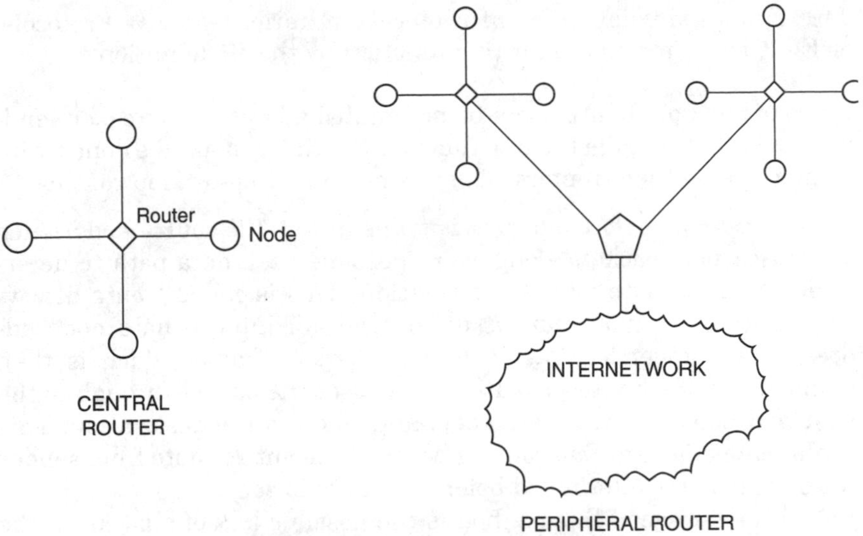

Figure 21 Central and Peripheral Routers

by a WAN router, which must determine a transmission path over widely dispersed internetworks existing many miles apart. Also, WAN routers must support protocols used for long-distance transmission, such as X.25.

Routing concepts

Routing is simply the determination of a path between sending and receiving nodes. It is a major function of the network layer of the OSI model. In any network there will be several smaller networks that are interconnected to form a larger network. Routers allow information to be passed between these and any external network that is interconnected.

A router may use a connectionless or connection-oriented protocol to accomplish delivery. In connectionless service, protocols such as IP make a "best effort" to deliver information, but does not guarantee delivery. In connection-oriented service, the transmission path is set up prior to data transmission. Data is transmitted and the connection is maintained and monitored. Acknowledgement of delivery is received, and the link shuts down.

Routing domains are smaller networks that are part of the whole, but administered, or owned, by a specific organization. Within the bounds of the organization's network, the routing protocol used is called an Intradomain Routing Protocol. This concept is called Interior Gateway Protocol in IP parlance. Connecting two or more domains requires an Interdomain Routing Protocol, which is capable of routing data formatted using different protocols. Exterior Gateway Protocols, or EGP, is the term used for this function by the IP definition.

Types of routing. Routing can be accomplished by the source, or sending, node. This is called source routing. Routing can also be done by intermediate nodes or routers. This is known as Hop-by-Hop routing.

Source routing. Before a packet is transmitted, the source node sends out "discovery" packets along every possible open data path to determine the best route for the information. This is called "route discovery." Once a suitable route is discovered, all intermediate node addresses are appended to the data packet, or frame, which is then transmitted. Each intermediate node passes the packet through to the next addressed intermediate node, and so on to the destination. This is also called End-to-End routing because the entire route from sender to receiver is predetermined before the data is sent.

A disadvantage of this method is the possible loss of a packet if the path is closed between the arrival of the discovery packet at the source node, and the transmission of the data packet.

Hop-by-hop routing. In source routing, intermediate nodes do not perform any work, data is merely passed through to the next intermediate node. Hop-by-hop routing tasks intermediate nodes with responsibility for the delivery of the packet. The packet contains source and destination addresses only. Intermediate nodes must contain enough information in their routing tables to find the best route for the data. An advantage of this type of routing is that routers can move data around closed links and networks that are experiencing high traffic volume.

Bridges

Bridges (Figure 22) are devices providing interconnection between separate networks. The interconnected networks may be on either side of a building or may be in different cities or even countries. Bridges are capable of transmitting packets between networks, but perform no change to the information transmitted. A bridge sees every packet that is transmitted over each of the connected networks. One of the primary functions of the bridge is to reduce unnecessary traffic on a local network. A bridge filters out unnecessary traffic by discarding packets addressed to nodes located on other networks. A bridge examines the physical address contained in the packet frame to determine the location of the receiving node, as opposed to a router, which uses the network address. Traffic of this sort is internal traffic, which is handled within the confines of a remote network.

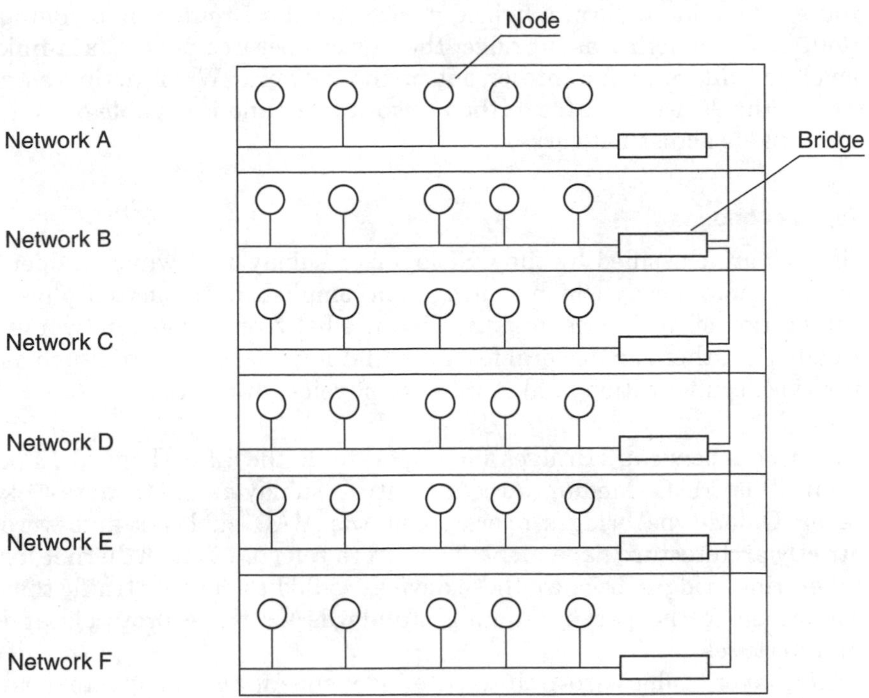

Figure 22 Bridges

Bridges are protocol independent in that different higher level protocols can use the same bridge to transmit information over the network. To higher level protocols, the bridge is transparent.

Layer of operation

Bridges operate at the Data-link layer of the OSI model. A bridge merely examines a packet and determines if the traffic is for a particular network. Higher level devices, such as routers, are concerned with the exact routing of the packet to a particular node on a particular network. A bridge is both a medium, in that it is a connecting device for the network media, and a filter, because it rejects any traffic which is not meant for a node on the network to which it is attached. A bridge examines the hardware address and makes this determination.

Bridges vs. routers

In the early days of networking, the terms bridge and router were used interchangeably. As explained above, bridges and routers perform distinctly different functions. In some applications, however, bridges and routers combine to form a hybrid device called a Brouter, or Bridging Router. When acting as a bridge, the device operates at the data-link level and filters traffic into or out of the network. When acting as a router, the device operates at the network layer and is capable of routing packets across networks.

Types of bridges

Bridges are classified by the OSI layer or sublayer at which it operates, the manner by which routing is accomplished, the actual physical location of the device, and the distance between bridged networks. In addition, there are several features that must be considered such as the type, configuration, and number of physical interfaces.

OSI layer dependent. Bridges may operate at the LLC (Logical Link Control) or MAC (Medium Access Control) sublayers of the data-link layer. Older type bridges operated at the MAC sublayer and were strictly architecture dependent. They were referred to as Ethernet, or token-ring bridges, because these devices could handle only traffic that conformed to the packet framing standards for those protocols and architectures.

LLC layer bridges are architecture independent and constitute most newer types of bridges. This type of bridge is capable of handling packets framed by different standards, the unique framing information having been stripped from the packet by the MAC sublayer below it.

Routing. The manner by which packets are routed is dependent upon the type of network being bridged. In an Ethernet based network, transparent routing is used to deliver a packet to a particular node. This routing is done without any interaction from the user and is somewhat "hit and miss." Transparent bridges locate the receiving node and store the location and routing for future use.

Another type of routing, as described earlier, is called Source Routing. Source routing is used in token-passing networks. In source routing, the sending node must provide routing information as well as addressing. The source node performs a technique called "route discovery" by sending out a "discovery packet." This type of routing uses an algorithm, or instruction set, to determine the routing between nodes. The algorithm for this task is called a Spanning Tree Algorithm, and is used to compute the available routes from source to destination. IEEE-802.1 defines the Spanning Tree Algorithm.

Source routing is much more complex than transparent routing, but provides an explicit path between nodes. Most bridges offer source routing as an option; all bridges are capable of transparent routing alone.

Internal vs. stand-alone. A bridge may be located internally to the computer or as a stand-alone device. An internal bridge plugs into an expansion slot in the server, utilizing the machine's internal power and data paths. An external bridge is a separate device that connects to the server using cables and supplies its own power.

Several items must be considered for both types of bridge:

- Type of media supported. Twisted pair, coaxial cable or fiber.
- Number of ports.
- Transmission speeds.
- Level of network management provided. Does the bridge collect statistics such as amount of traffic handled or error rate.
- Compatibility with any common carrier long-distance services being considered.

Local and remote. A local bridge connects networks that exist within the same, typically small, geographical area (Figure 23). These devices are connected using whatever media the network is using. Local bridges are used to segment networks to allow for reduction of traffic on any one network. Bridged networks may exist in the same room, or within the same building or campus.

A remote bridge connects networks which exist a considerable distance apart from one another. These devices are specialized; they provide connection to packet-switched networks such as those using the

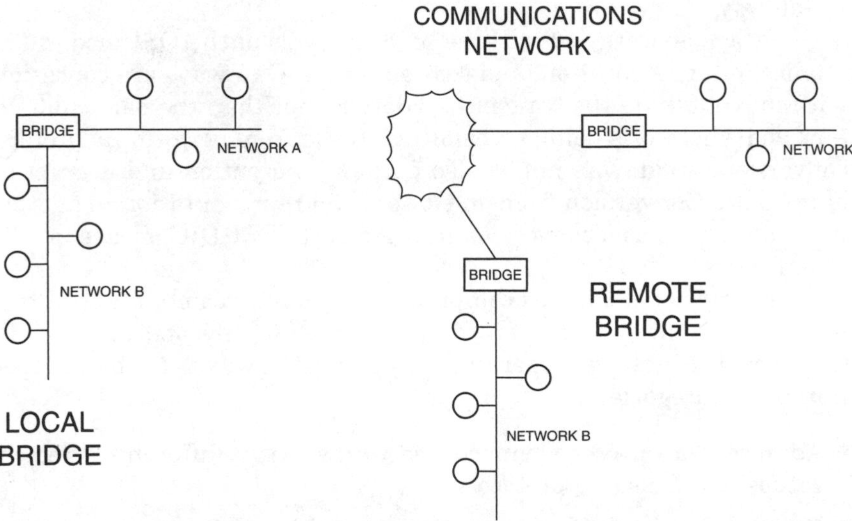

Figure 23 Local and Remote Bridges

X.25 protocol. Remote bridges also provide access to ISDN (Integrated Services Digital Network) or high-speed long-distance optical networks. Remote bridges are used in pairs, one at each end of the connection. Physical interfaces differ, and it is important to determine the method of connection to the common carrier. One side of a remote bridge may have connectors that are compatible with an Ethernet or token-ring network, the other end may need a RS-232 interface for connection to the long-distance service equipment.

Learning bridges. Learning Bridges build a table of node locations and store this information for future use. Whenever a node logs onto a network, certain information is transmitted over the network. This information is read, and stored for use in routing of information. As would be expected, the performance of this type of bridge improves over time. The bridge is constantly updating added and dropped addresses. Static bridges do not perform this function and do not build a table of addresses. For this type of bridge, addresses must be entered by hand.

Gateways

A Gateway can be considered to be a multiprotocol, multiarchitecture router. The purpose of a gateway is to connect two or more different network types. For example, it might be desired to connect an Ethernet-based, bus-type network to a token-ring network, and then connect this combination to a factory LAN using token-bus. The device that specializes in this type of protocol and architecture conversion is called a Gateway.

Gateways operate at the three highest levels of the OSI model: The application, presentation, and session layers. Gateways are concerned with the content of the transmitted data rather than the particulars of how and where it is being transmitted. Aside from performing protocol conversion, a gateway might also perform encryption and decryption of the data. Conversion from one lower level representation to another may occur, such as a conversion from ASCII to EBCDIC for use in IBM mainframes.

A gateway is a dedicated computer and software combination. Other interconnection devices operate, in contrst, on only one layer of the OSI model. A gateway operates on three. Gateways can be grouped into three categories:

- Address Gateways: Connects networks with different directory setups but the same protocols.
- Protocol Gateways: Connects networks with different protocols.

- Format Gateways: Connects networks that format data differently, such as the aforementioned ASCII to EBCDIC conversion.

Gateways have been developed for special applications. E-mail gateways provide e-mail services to a LAN, Internet gateways provide internet access and mainframe gateways provide an interface between LANs and mainframes.

Repeaters

Repeaters are physical layer devices that extend the physical range of a network by extending the medium. A repeater receives the signal, boosts it and then retransmits the signal. A repeater links segments of a network together. All segments connected by repeaters are considered to be on the same network. Repeaters perform no action on the data; the data remains unchanged except for the signal boost it receives. Repeaters perform no routing or bridging functions. Repeaters may be connected to repeaters for the purpose of extending a network. The inter-repeater link is just a cable with no attached nodes. This is useful for connecting networks in different parts of a building, particularly on different floors.

Chapter

3

The Medium

Electrical Basics

Most users tend to take the medium, or wiring, for granted when communicating over the network. In fact, the medium is no less important than any other component of a data communication network. Increasingly, radio, particularly cellular communication, is taking on the task of communicating data. We will touch briefly upon this subject at the end of this chapter. The idea behind this chapter is to give the reader an idea of what is involved in making the physical connection that makes networks possible. The concepts presented here carry into wireless networking in the sense that interfacing to the carrier is basically the same as interfacing to a wire strung between two computers. It is simply the method of transport that changes.

Many different types of wiring exist for the purposes of communication. The most familiar type is the wiring that connects your telephone to the wall. This is the most extensive data communication network in active use. It is also the oldest. As described later on, it is relatively simple. The wiring used for this system is made of copper and is in almost universal service. Copper wire exists in many different thicknesses, known as "gauges." The thickest wire made has a gauge of 0000 or 4/0, the thinnest 40. The gauge numbers are AWG, or American Wire Gauge. This is the standard for wire in the United States and describes a range of wire thicknesses from approximately .530 inch for # 4/0 wire to .0031 inch for # 40. Each size of wire has different characteristics that allow engineers to simplify the design of a system. These characteristics include weight per unit length and electrical current carrying capacity. These factors weigh heavily in designing power handling systems (Table 1).

TABLE 1 American Wire Gauge

Gauge no.	Diameter in mils at 20°C	Cross section at 20°C		Ohms per 1000 feet			
		Circular mils	Sq. inches	0°C (32°F)	20°C (68°F)	50°C (122°F)	75°C (167°F)
0000	460.0	211600	0.1662	0.04516	0.04901	0.05479	0.05961
000	409.6	167800	.1318	.05695	.06180	.06909	.07516
00	364.8	133100	.1045	.07181	.07793	.08712	.09478
0	324.9	105500	.08289	.09055	.09827	.1099	.1195
1	289.3	83690	.06573	.1142	.1239	.1385	.1507
2	257.6	66370	.05213	.1440	.1563	.1747	.1900
3	229.4	52640	.04134	.1816	.1970	.2203	.2396
4	204.3	41740	.03278	.2289	.2485	.2778	.3022
5	181.9	33100	.02600	.2887	.3133	.3502	.3810
6	162.0	26250	.02062	.3640	.3951	.4416	.4805
7	144.3	20820	.01635	.4590	.4982	.5569	.6059
8	128.5	16510	.01297	.5788	.6282	.7023	.7640
9	114.4	13090	.01028	.7299	.7921	.8855	.9633
10	101.9	10380	.008155	.9203	.9989	1.117	1.215
11	90.74	8234	.006467	1.161	1.260	1.408	1.532
12	80.81	6530	.005129	1.463	1.588	1.775	1.931
13	71.96	5178	.004067	1.845	2.003	2.239	2.436
14	64.08	4107	.003225	2.327	2.525	2.823	3.071
15	57.07	3257	.002558	2.934	3.184	3.560	3.873
16	50.82	2583	.002028	3.700	4.016	4.489	4.884
17	45.26	2048	.001609	4.666	5.064	5.660	6.158
18	40.30	1624	.001276	5.883	6.385	7.138	7.765
19	35.89	1288	.001012	7.418	8.051	9.001	9.792
20	31.96	1022	.0008023	9.355	10.15	11.35	12.35
21	28.45	810.1	.0006363	11.80	12.80	14.31	15.57
22	25.35	642.4	.0005046	14.87	16.14	18.05	19.63
23	22.57	509.5	.0004002	18.76	20.36	22.76	24.76
24	20.10	404.0	.0003173	23.65	25.67	28.70	31.22
25	17.90	320.4	.0002517	29.82	32.37	36.18	30.36
26	15.94	254.1	.0001996	37.61	40.81	45.63	49.64
27	14.20	201.5	.0001583	47.42	51.47	57.53	62.59

Table 1 American Wire Gauge (*Continued*)

Gauge no.	Diameter in mils at 20°C	Cross section at 20°C Circular mils	Sq. inches	Ohms per 1000 feet 0°C (32°F)	20°C (68°F)	50°C (122°F)	75°C (167°F)
28	12.64	159.8	.0001255	59.80	64.90	72.55	78.93
29	11.26	126.7	.00009953	75.40	81.83	91.48	99.52
30	10.03	100.5	.00007894	95.08	103.2	115.4	125.5
31	8.928	79.70	.00006260	119.9	130.1	145.5	158.2
32	7.950	63.21	.00004964	151.2	164.1	183.4	199.5
33	7.080	50.13	.00003937	190.6	206.9	231.3	251.6
34	6.305	39.75	.00003122	240.4	260.9	291.7	317.3
35	5.615	31.52	.00002476	303.1	329.0	367.8	400.1
36	5.000	25.00	.00001964	382.2	414.8	463.7	504.5
37	4.453	19.83	.00001557	482.0	523.1	584.8	636.2
38	3.965	15.72	.00001235	607.8	659.6	737.4	802.2
39	3.531	12.47	.000009793	766.4	831.8	929.8	1012
40	3.145	9.888	.000007766	966.5	1049	1173	1276

When designing data-handling systems, several other factors come into play. In every electrical system, the magnitudes and directions of electricity may change in frequency from none to several million times per second. As the signal characteristics change, the electrical characteristics of the wire also change. These changes can impair or even destroy the signal. Knowing the characteristics of the wiring that is being used is one way of enhancing the data- or signal-handling capabilities of the system. Even better, knowing the characteristics of the wiring available allows for the use of the most correct and well-suited type of wiring for the system.

What follows is a discussion of electricity and the electrical characteristics of wire. These concepts require that the reader exercise some imagination in order to visualize how and why these characteristics interact. The material is presented without resorting to obscure mathematics and terse technical discussions.

Electricity

Electricity is simply the movement of tiny particles called electrons. Electrons make up a part of every atom. In the atom, which is usually described as resembling a small solar system such as the one we exist in, orbiting electrons surround a nucleus (Figure 1). Some atoms

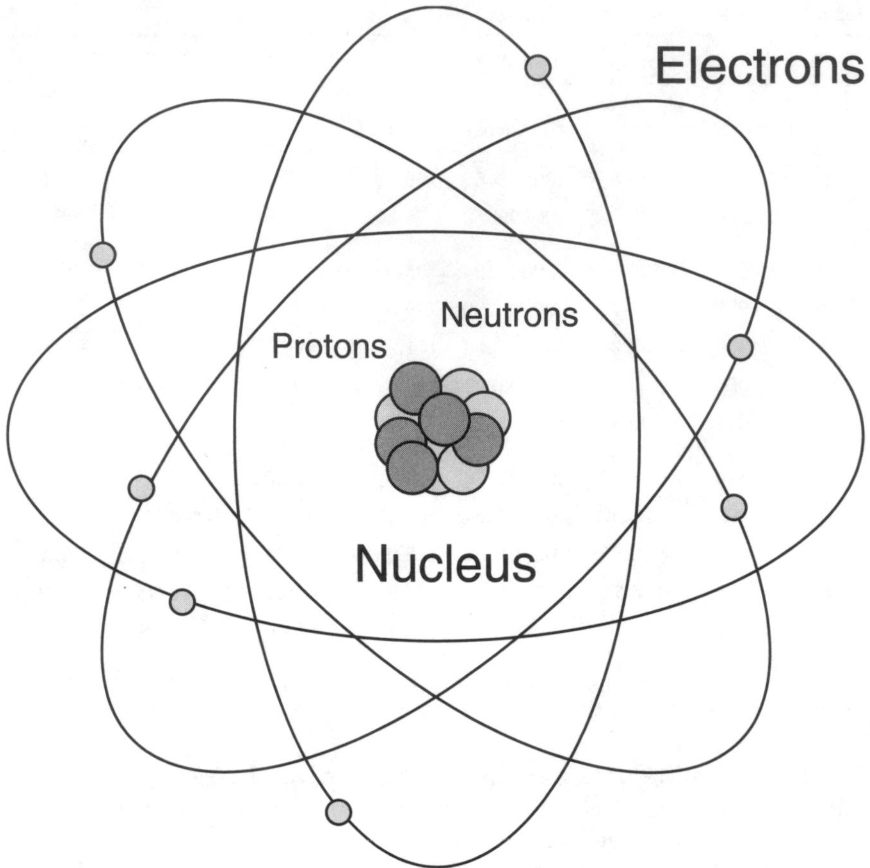

Figure 1 The Atom

exhibit a strong hold on their electrons, resisting the pull of outside forces that attract their electrons. The atomic structure remains intact and the atom retains its characteristics. Other atoms exhibit a weak holding force on their electrons and readily give up their electrons to an outside force.

Materials that exhibit a strong hold on their electrons are called insulators. Without the presence of easily acquired electrons from the material's constituent atoms, no movement of electrons is possible. Therefore, the material has an infinite resistance to the flow of electricity (electrons). Materials such as rubber and glass have no free or weakly held electrons and are therefore insulators.

Conversely, materials that exhibit a weak hold on their electrons are called conductors. As a force is applied to a conductor, electrons are attracted to the force. As the number of electrons build up they form current. Copper and iron are examples of conductors.

Units

Voltage

The force that acts to free electrons in a conductor is known as Voltage. The volt is named in honor of Alessandro Volta, an Italian inventor who dabbled in early data communication. Voltage can be thought of as being a pressure, as previously stated. A voltage is called a difference of potential. Imagine a system of pipe connected to a pump. The pump can be compared to a battery, which is a common source of voltage. If a valve is opened water flows through the pipe. The valve can be compared to a switch in an electrical system. If the valve is closed, no water flows. The pump still runs and attempts to force water through the system, but has no path, or circuit, to flow through. The pump, or battery, provides the force by which the water flows through the system. Each molecule of water can be thought of as an electron. If no force is applied to the molecule, it will remain in place. It requires a force to move it.

In the electrical circuit, this force is exhibited as a difference in potential. In the water analogy, the pump creates a difference in pressure with high pressure being created at the pump output and low pressure at the pump input. A battery creates an abundance of electrons at one terminal, or pole, and a deficiency of electrons at the other. These are called the negative and positive poles, respectively. Electricity will flow in the direction of the absence of electrons, or from negative to positive. Electrons also have the tendency to repel other, adding to the force moving the electrons. One volt is measured across one ohm of resistance whenever one ampere of electricity flows.

Current

When enough electrons are moving, a current flows. Returning to our water example, the valve is opened and water flows around the loop. We have established that there is pressure in the system and that the water is being pushed through the pipe as a result. Now we want to determine how much water is moving through it. Perhaps we are concerned that the pressure is too great for this system and is forcing too much water through too small a pipe. Eventually, the system will fail if this is the case. The amount of water being forced through the system is the flow. In the electrical system, the amount of electrons flowing through a system is called the current. Current is measured in Amperes. As it is with voltage, the ampere is named in honor of an early experimenter in electrical technology named André-Marie Ampère.

For those who simply have to put a number on things, one ampere is equal to 6.023×10^{23} electrons passing any given point under the influence of a difference of potential of one volt.

Resistance

Resistance is defined as the property of the medium to hinder, or hold back, the flow of current. If we return once more to our water example, we see that the water is flowing freely around the pipe as long as the valve is opened. If the valve is completely closed, the flow stops. This is an infinite resistance. An infinite resistance does not usually exist in an electrical circuit. Resistance exists in degrees from none to infinity. If we now insert a section of pipe into the circuit that is half the diameter of the rest of the pipe, the flow not only slows down through that section, but also causes the water to back up behind it, raising the pressure. Referring to our electrical example, a resistor is placed in the circuit. The resistor now slows the flow of electrons and causes the voltage to build up around it. The resistor is now offering a restriction to the voltage, causing it to expend the electrical energy as heat. In our hydraulic circuit, the pump is working harder and also expending the energy as heat.

In any wire, there exists a characteristic resistance. This is due to the molecular makeup of the material. In fact, every wire has a typical resistance per unit length. This is very important to the designer of a communication system. If a signal is impressed upon a wire and is required to be intelligible two miles away, it may not be possible without some intermediary remediation. This is because the signal will degrade, or be attenuated, by the resistance of the wire. Other factors will act upon the signal to degrade it, particularly if the signal is alternating at a high frequency. This will be discussed in a later section.

The third basic electrical parameter is called the Ohm after Georg Simon Ohm, a German mathematician who figured out the relationship between voltage, current, and resistance. This is known as Ohm's Law, and is the basis for every electrical relationship in existence. At some stage of electrical design, whether it be for the lighting of a building or for a high-speed data communication network, Ohm's law is used.

Ohm's law

We have just reviewed the three basic parameters that make an electrical circuit possible. Every electrical circuit contains these three elements. It could not exist otherwise. Voltage creates the force by which electricity moves. Electricity itself is quantified as current, and work is performed as some form of resistance to the movement of current. Put another way, current is transformed into a means of performing work by the circuit resistance. A motor converts electrons into movement by a selective resistance to the flow of electrons. An electric heater offers a carefully selected resistance to electricity in order to heat the coils and expend heat. A television uses various combinations of resistance with other components to receive electrical impulses from the airwaves and present them to the viewer.

In order to properly define the system in terms of capacity and requirements for voltage supply and work required, Ohm's law is used to determine the relationship between voltage, current and resistance. Ohm's law, in its most basic form, defines this relationship as E = IR, or Voltage (E) equals the product of Current (I) and Resistance (R). Voltage is dependent on the flow of current and the amount of resistance in the circuit (Figure 2). Various permutations of this formula describe the relationship in terms of the other two elements. The mathematically adept reader is urged to work through some other forms to help illustrate this relationship. A full discussion of Ohm's law is entirely beyond the scope of this book.

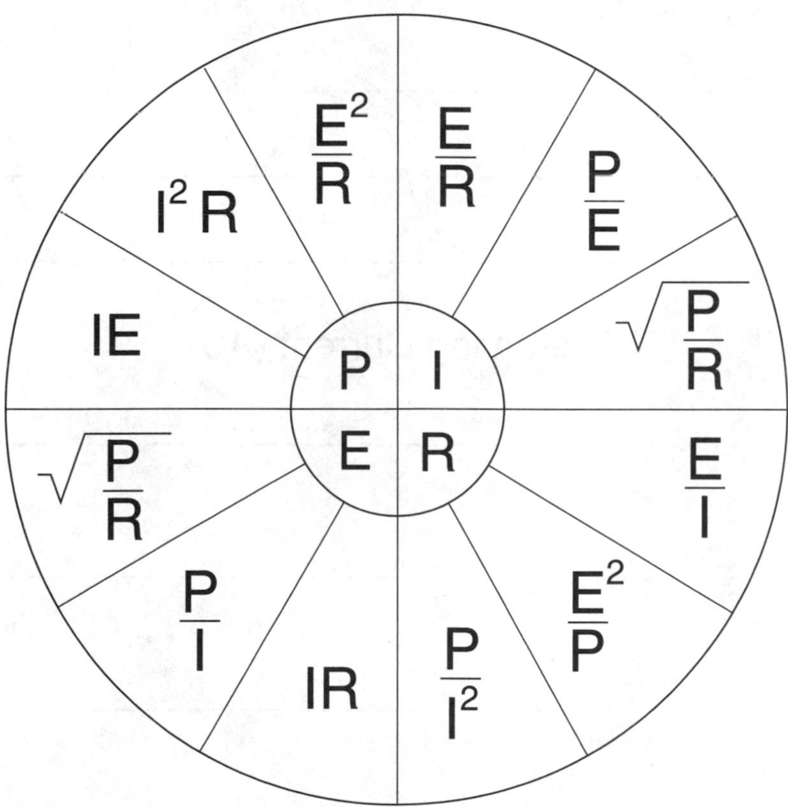

Figure 2 Ohm's Law Wheel

Alternating and direct current

In the previous discussion, we reviewed three electrical parameters that affect the behavior of an electrical circuit. In our discussions, we used the example of a pump and associated piping as an analogy for an electrical circuit. This circuit represented a direct current circuit in that it used a battery to provide the voltage source for the circuit. Direct current describes a system in which the voltage and the current remain at the same polarity at all times as part of the circuit (Figure 3).

As illustrated, the battery's poles are connected to the circuit. The positive pole never changes to the negative pole and vice versa. The voltage remains at the same potential at the poles at all times (Figure 4). Further, the current always flows in the negative to positive direction. It is not possible for the current to reverse direction. This type of current makes up an entire class of electrical devices known as DC devices. These devices are designed for use exclusively with Direct Cur-

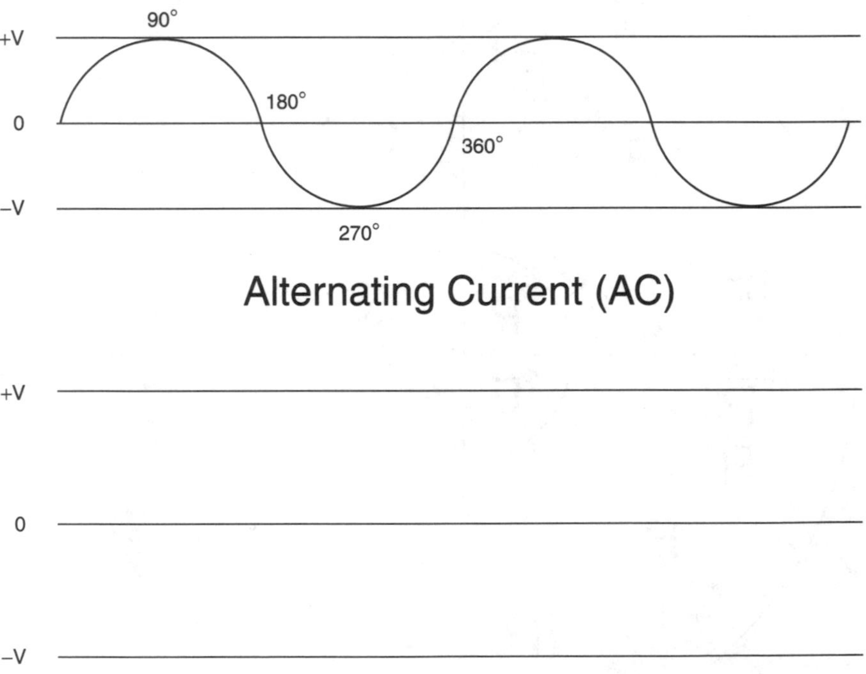

Figure 3 AC & DC Voltages

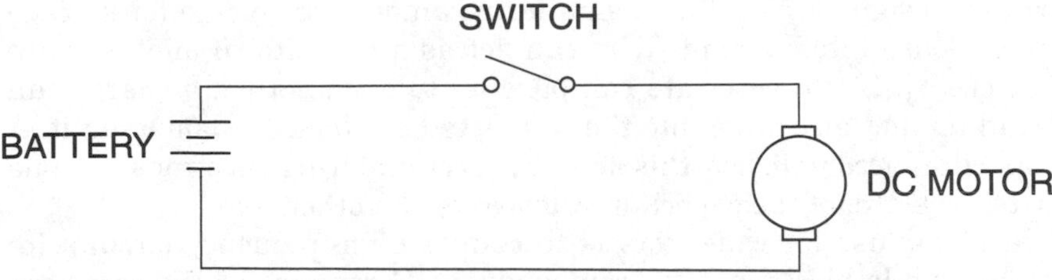

Figure 4 A Direct Current Circuit

rent. While DC is very efficient over short distances and within small systems, it is not effective as a long-distance power source.

Alternating current, or AC, is the prevalent system for power transmission and supply. We are not interested in the power supply aspects of AC, however. All data transmission systems, with the exception of telephony, use a form of AC for the transmission of data. In the use of AC, a broad spectrum of frequencies is available for the transmission of information. It is the characteristics of the transmission medium in response to changes in frequency that we will discuss next.

Capacitance

Capacitance is the capability of a material to store a charge of electrons within itself. Capacitors are used in many different applications because of this. A capacitor is an electronic component (Figure 5) that

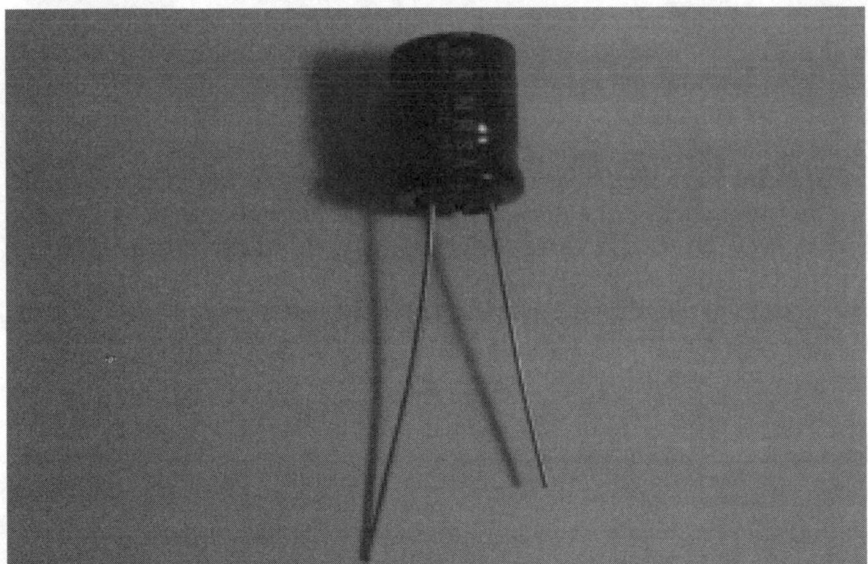

Figure 5 A Capacitor

can be designed to contain a significant amount of voltage for storage and release into a circuit. Wire can act as a capacitor if another wire is close by. If two wires are run parallel to one another, a charge can build up and discharge into the connected equipment, damaging it. A capacitor accomplishes this feat by accumulating electrons on one wire, or plate, of the capacitor, relative to the other.

Another use for capacitors is to couple high-frequency circuits together while blocking a DC component. DC remains at the same polarity and does not change. A capacitor offers an open circuit to DC. AC, on the other hand, changes polarity at the frequency at which it is operating. This has the affect of passing AC around the capacitor. Following this explanation, it can be seen that as the frequency gets lower, as it gets closer to zero, the ability to pass AC around it becomes less and less efficient. At some higher frequency, the capacitor begins to look like an open circuit to the AC. This change in the relationship between the capacitor and the impressed current serves to slow down the flow of electrons. This change in flow is proportional to the frequency. This is known as capacitive reactance and is inversely proportional to the frequency.

Inductance

Another characteristic of AC circuits is inductance. Inductance is associated with coils of wire and their magnetic fields. As an AC signal is passed through a wire, a magnetic field expands around it. If another wire is in the vicinity, the magnetic field from the first wire "cuts" the second wire and causes current to flow in it. If a number of wires are wound in a coil and an AC signal is passed through it, all of the wires will exhibit expanding and contracting fields around them due to the rising and falling of the AC signals. Because of the induced current flows from the other wiring, the coil will alternately aid and resist the flow of current.

An inductor, which is what this coil of wire is called, behaves opposite to the capacitor just discussed. An inductor will offer no resistance but that of the wire itself to the flow of DC. It will, however, offer resistance to the flow of AC. This resistance is directly proportional to the frequency (Figure 6). As the frequency gets higher, so does the resistance. As the frequency gets lower, the resistance follows. This resistance is called inductive reactance.

Impedance

When Resistance, Capacitive Reactance and Inductive Reactance are combined after lengthy calculation, a characteristic circuit Impedance is obtained. Impedance is defined as the total resistance to the flow of

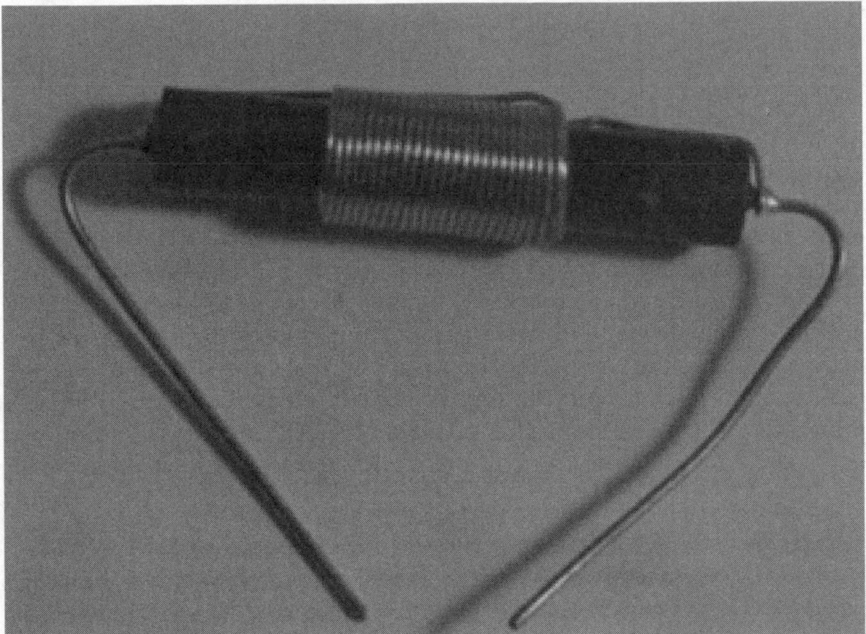

Figure 6 An Inductor

alternating current. By manipulating the values of capacitance and inductance in a circuit, high frequencies can easily be handled. In most high-frequency circuits, maximum energy transfer in the form of the strongest possible signal, cannot be accomplished without the careful matching of impedance of the transmitting and receiving equipment and the transmitting medium between them. Signals can be reflected or deflected depending upon the degree of mismatch. Techniques that serve to maximize the efficiency of the transfer of signal have been developed. One such technique is described below.

Tuned circuits

A special class of electronic circuit takes advantage of the properties of inductive and capacitive reactance. These are called tuned circuits. What is meant by tuned is the ability to control the impedance of the circuit by changing the values of the constituent capacitance and inductance. The circuit is tuned to a characteristic frequency and thereby can be made to pass or block sections or entire bands of frequencies. By allowing the ability to manipulate one of the constituents, the frequencies passed or blocked can be continuously changed. This is exactly how an older type of radio tuner worked. In that case, a variable capacitance that was attached to the tuning knob was used. As the knob turned, the "tuner" selectively maximized transfer of the correct frequency into the receiver section of the radio.

A tuned circuit can exist in one of two configurations: Parallel or Serial (Figure 7). This means that the capacitor and inductor are either wired side by side or end to end, respectively. Each has an advantage over the other. In the radio tuner example, a parallel tuned circuit was used. The key to understanding the operation of a tuned circuit is to understand the concept of resonance. Resonance is the property of the circuit to resonate. When a circuit resonates, it achieves a state in which, depending upon configuration, impedance is either maximum or minimum. This is determined upon the design of the circuit. In a parallel circuit, the impedance is zero at resonance; in a serial circuit, it is maximum. This principle is used in the design and use of coaxial cables, one of the types of media used for data transmission.

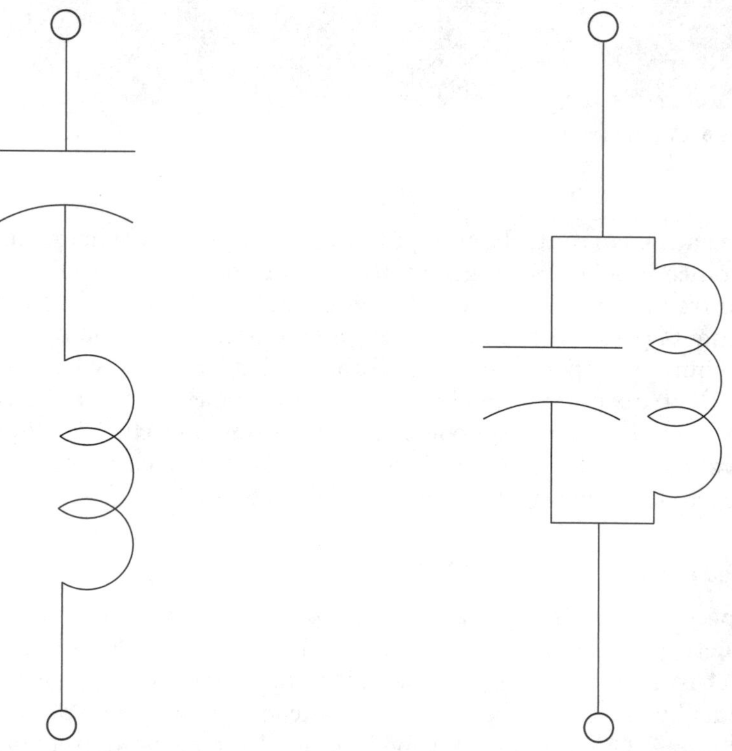

Figure 7 Parallel and Series Resonant Circuits

Wire and Cable

Before we get into the specifics of cabling for data communication, a brief discussion of the telephone communication system is in order. The telephone system is widely used for the communication of information. The means by which this system is used varies depending upon the required speed of transmission. The cost varies also. Two types of telephone line are in common use: the public lines, which constitute the dial-up network, and leased lines, which are dedicated wires that are leased by organizations for fast and reliable communication between multiple sites. Leased lines move data over multiple channels at speeds exceeding one million bits per sec (mbps). Dial-up lines transfer data in the thousands (kbps).

In practically every home there exists at least one telephone line. This line is being used to transmit data via modem at speeds up to 128,000 bits per second. This is done by converting digital data to analog data and then transmitting the information to another converter at the receiving end to be translated back to digital information. The translation devices are called modems and are described in greater detail in Chapter 2. The important thing to remember is that data transferred by this method relies upon the data being in analog form. This can be very inefficient and limits the speed by which data can be transferred.

At this writing, digital dial-up communication is not widely available. Some cable systems offer the service, and use of the Integrated Services Digital Network (ISDN) is slowly being introduced into the home market by local telcos. As the demand for faster data communication by the consumer grows, digital communication will become more widely available as well as affordable. Presently the service and connection equipment are priced beyond the means of most people. As the speed of analog modems begins to reach its limit, the use of digital communication will become widespread. One other problem is the availability of bandwidth, or the capacity of the lines for traffic.

In the early days of communication, telegraph companies actively sought ways to increase the capacity of their lines. If more than one message could be sent at the same time, the cost of operation was cut in half. In time, systems were developed that could transfer a message in both directions on the same pair of wires. This was known as the duplex. A young inventor by the name of Thomas Edison developed the Quadraplex, which was capable of transmitting four messages in either direction at the same time. This greatly increased the capacity of the lines.

As the telephone system became more popular, the Bell System began to experiment with transmission systems that relied upon "carriers" much like the method used to transmit radio waves. The wire pair carried a number of separate frequencies that had information impressed upon them at the transmitting end. At the receiving end, a device listened for that frequency and decoded the information. In this way multiple conversations or transmissions could be carried over the same pair of wires, vastly improving their capacity. This system is known as multiplexing.

The Bell System gradually improved the traffic capacity of its lines, particularly on the long-haul intercity lines. The first high capacity leased line went into operation between Pittsburgh and Baltimore in 1918 and was designated a "Class A Carrier." Over the next five decades, refinements in technology, most notably the invention of the transistor, improved the speed and capacity of the system.

In 1962, Bell introduced the first of the "T" type carriers. These carriers were and still are all digital trunk lines. The T1 carrier was introduced and had a capacity of 1.544 million bits per second (mbps). As designed, this allowed the simultaneous transmission of 24 different voice circuits, or channels. Each of these voice circuits requires 64,000 bits per second (kbps) per channel to operate. These circuits are called a channel bank. When using copper wire, the signal needs to be regenerated every 6000 feet.

A T1 can be extended to about 200 miles but typically is much shorter. The T1 line terminates at either end in a device called a Channel Service Unit, or CSU. The CSU interfaces with the data transmission equipment at either end of the T1 line. Other T-type carriers are used that have greater channel capacity. T1C, and T2 carriers have channel banks of 48 and 96 circuits (3.152 and 6.312 mbps), respectively. T3 and T4M state their capacity in terms of T1 channels of 28 and 177 (44.736 and 274.176 mbps), respectively.

T1 channels are further broken down into "fractional T1," which allows higher bandwidth communication from 56 kbps to 766 kbps (Table 2).

Types of wire for data

Coaxial cable. Coaxial cable is widely used in the transmission of information. Coax, as it is commonly called, eliminates the problems in-

TABLE 2 T Carrier Rates

Carrier	#T1	Signal level	Data rate
T1	1	DS1	1.544 Mps
T1C	2	DS1C	3.152 Mps
T2	4 (2 × T1C)	DS2	6.312 Mbps
T3	28 (7 × T2)	DS3	44.736 Mbps
T4	168 (6 × T3)	DS4	274.176 Mbps

herent in regular wire (Figure 8). As described earlier, a wire can induce a current into another, adjacent wire. But another, spurious signal can also be impressed upon a wire. This has the effect of introducing "interference" into the signal that the wire is carrying. In the worst cases, the interference will render the signal useless. In most cases, time and capacity are wasted by the system during retransmission of the message. This is the situation when a bad telephone connection is made and the speaking parties have to repeatedly ask the people to repeat themselves.

Another problem is a property of the wiring known as "self inductance." This means that as the current travels down the wire, the magnetic field that develops around the wire tends to resist the transmission of energy. This resistance, or reactance, becomes greater as the frequency of the current becomes greater. If an additional wire is added as part of the circuit, the problem becomes twice as great, with a simple tuned circuit being created. Without any consideration of the capacitance or inductance of the wiring, the impedance of the wiring over the range of frequencies of the information will vary unpredictably. The information transfer will not be effective if it occurs at all. Attenuation, or the degradation of the signal, will be excessive at some parts of the transmission and nonexistent at others. This will have the effect of rendering the information unintelligible.

In the late-nineteenth century, Werner Von Siemens developed the first cable which controlled the capacitance and inductance of a cable reliably and allowed engineers to design fixed characteristics of attenuation into a transmission line. He called it the coaxial cable. The cable contains a solid or stranded wire surrounded by an insulating material, which is in turn surrounded by a "shield." The shield is a foil or braid that is covered again with a plastic insulating material. The inner insulator is known as the dielectric. The outer insulator is known as the jacket.

What makes the cable so efficient is the optimization of the electrical characteristics for what is called the "attenuation profile." This profile resembles a bell curve. It describes graphically the signal degradation characteristic of the cable. Typically, most coax peaks at either 50 or 75 ohms for a given frequency spectrum. This allows the design of a system that can compensate for the known attenuation.

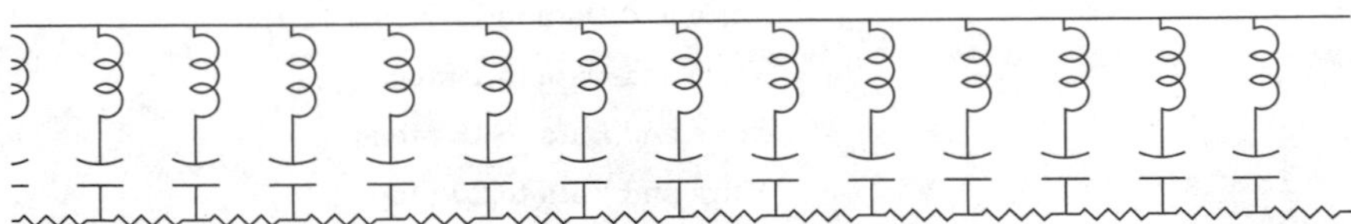

Figure 8 Electrical Equivalent of a Coaxial Cable

A coaxial cable can be thought of as an array, or string, of series resonant circuits (Table 3). This is so because of the arrangement of the conductors and insulators. All wires have inductance as previously stated. The coaxial nature of the conductors creates one large capacitor. The inductance coupled with the capacitance serves to block the frequencies most often associated with interference. These frequencies most often occur at radio frequencies and are referred to as RFI, or Radio Frequency Interference. The shield is usually connected to a common terminal known as a ground. The shield is also referred to as a drain, implying the other function of the shield—to shunt, or drain the spurious signal to a point at which it can be disposed.

A shorthand has been developed that allows the user and the engineer to describe the type of cabling used. Please refer to Figure 11. 10Base5 is a coaxial cable used in IEEE-802.3 networks. The cable is referred to as trunk or "thickwire" because of its relatively large diameter of .405" (Figure 9). The designation describes the cable as having a bandwidth of 10 Mbps utilizing baseband transmission on cable segments of 500 m.

10Base2 describes a cable with a diameter of .188" and is referred to as "thinwire" or Thinnet (Figure 10). Following the naming convention, this cable supports a 10 Mbps throughput using baseband transmission. This cable is limited to a segment length of 200 meters.

100BaseT and 100BaseVG are emerging technologies being developed in order to effect faster data transfer over existing cabling.

100BaseF describes an optical-fiber cable supporting 100 MbPS data

TABLE 3 Cable Types and Characteristics

Type	Description
10Base5	"Thickwire" – Segments are limited to 500 m and 100 nodes
10Base2	"Thinwire" – Segments are limited to 200 m and 30 nodes
10BaseT	Unshielded Twisted Pairs (UTP) Segments limited to 100 m
100BaseT	UTP – "Fast Ethernet"
100BaseVG	UTP – Half-duplex Ethernet at 100 Mbps
100BaseF	Fiber Optic Cable – Segments up to 2 Km
Category	**Bandwidth**
1	Voice and Data to 20 Kbps, Telephone
2	Voice and Data to 1 Mbps
3	Voice and Data to 16 Mbps
4	Voice and Data to 20 Mbps
5	Voice and Data to 100 Mbps

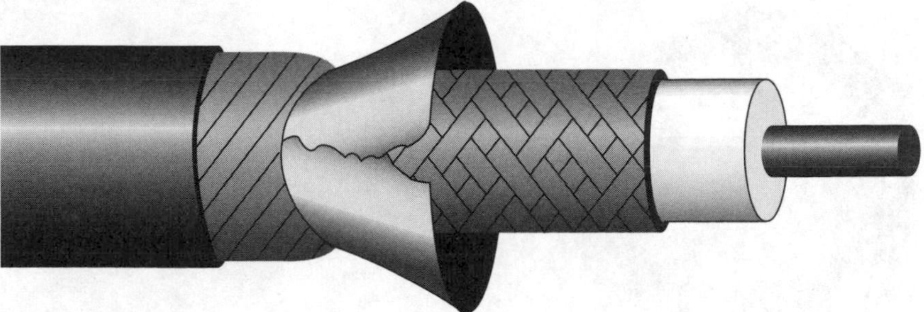

Figure 9 10Base5 Cable (Courtesy Belden Electronics Division, Inc.)

throughput. An addendum to IEEE-802.3 that describes optical transmission is being prepared.

Twisted pair cable. 10BaseT cable designates unshielded twisted pair cable (Figure 11). While this cable is used with the Ethernet, or IEEE-802.3 standard for data-frame format, the physical medium specification widely used is EIA/TIA 568, which specifies cable characteristics and physical interfaces such as RJ45 (Figure 12) modular plug terminations. Segment lengths are limited to 100 meters and connections are point to point. This type of cable is used in star-type networks. 10BaseT does not support broadband transmission. The cable specification is further broken down into categories based upon throughput or bandwidth.

Category classification. Level 1 or category 1, cable is analogous to telephone cable commonly known as "quad," or unshielded 4-conductor twisted pair, and is commonly used in residential telephone installations. Bandwidth is <5000 Hz. Later implementations of this definition do not consider quad as a suitable medium. Quad can be used, however, if the cable is already installed. Category 3 cable is an unshielded eight-conductor twisted pair commonly used in small office networks, which supports a bandwidth to 10 Mbps. Category 5 cable is

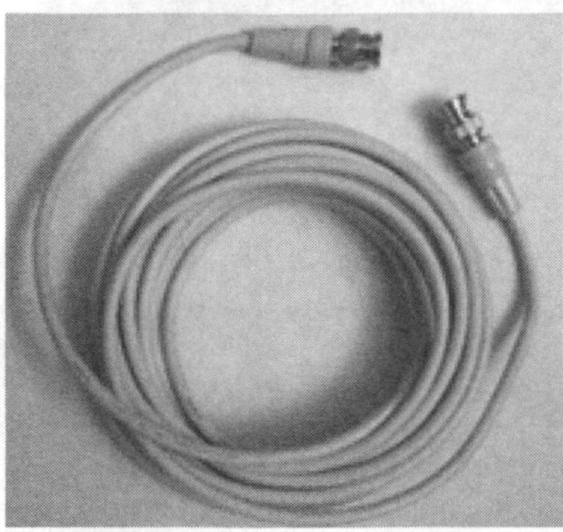

Figure 10 10Base2 Cable.

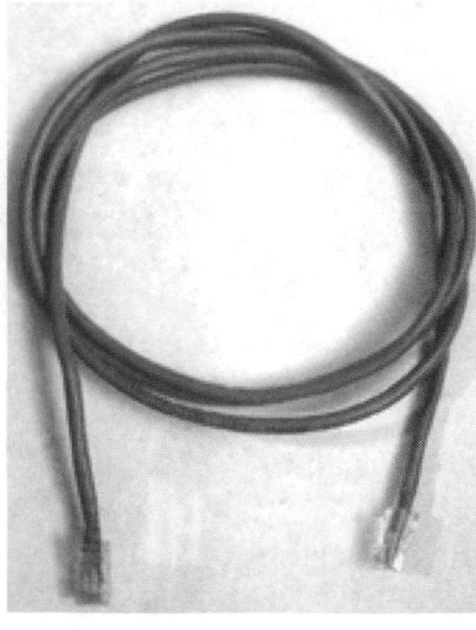

Figure 11 10BaseT Cable

physically similar to category 3 cable but allows throughput of up to 100 Mbps, allowing real-time video services such as conferencing, surveillance, and commercial video feed.

Fiber-optic cable. In 1854, John Tyndall established that light could be transmitted by a jet of water streaming from a container. Later in the nineteenth century, other experimenters used parabolic mirrors and internally metallized tubes to guide light into areas removed from the light source. In the 1920s, several patents were filed for devices using bundles of glass tubes that transmitted crude, but recognizable, im-

Figure 12 RJ45 Connector (Courtesy Belden Electronics Division, Inc.)

ages. The next forty years brought drastic improvements in the manufacture of high-quality optical fibers. The technology has been continually improved, and optical fibers of the highest quality are routinely produced, making possible communication networks capable of transmitting over twenty-five billion bits of information per second.

Fiber-optics based communication systems are rapidly replacing wire-based systems. Wide bandwidth, noise immunity, and low attenuation are making fiber optics an attractive option for not only telephony and LAN applications, but also for process control. Factory or plant environments present many problems to the process engineer when it comes to data reliability and security. Noise is the most common problem encountered. High voltage and motor feeders, motors and motor drives, even fluorescent lighting, can generate EMI that is sufficient to corrupt wire or radio-based systems. In a DCS or SCADA installation, this problem will at least render the system unreliable. An unstable factory net can result in the uncontrolled operation of equipment and possible personnel injury or facility damage. If such an environment is unavoidable, fiber optics will effectively eliminate this aspect. The following material is presented for anyone who is becoming involved with the process or communication industries and who would like to gain a basic understanding of the technology. References have been included, which will allow the reader to get a more in-depth treatment of the subject.

Index of refraction. Basic to the understanding of fiber-optic technology is the concept of the "index of refraction." In order to effect the most complete transfer of energy through a medium, a means must be found by which the light energy is completely contained within the medium. This "containment" of light energy is accomplished by making use of the light transmitting medium's index of refraction. Every material capable of transmitting light has a characteristic refractive index. The definition of the refractive index is "the ratio of the velocity of light in the material to the velocity of light in a vacuum." When a ray of light encounters the boundary between materials of differing refractive indices, for example, air and water, part of the light is bent while a portion is reflected. If the angle at which the light ray (also referred to as a "mode") striking the boundary exceeds some critical angle, the "critical angle of total reflection," the entire light ray is reflected. None of the light from the ray enters the other material. This critical angle is also called the maximum angle of refraction.

A fiber-optic cable consists of three parts (Figure 13). The core is the center portion of the fiber and transmits the light energy. All of the light energy is contained within the core. The core has a high refractive index. The cladding surrounds the core and serves to contain the light energy within the core. The cladding has a lower refractive index

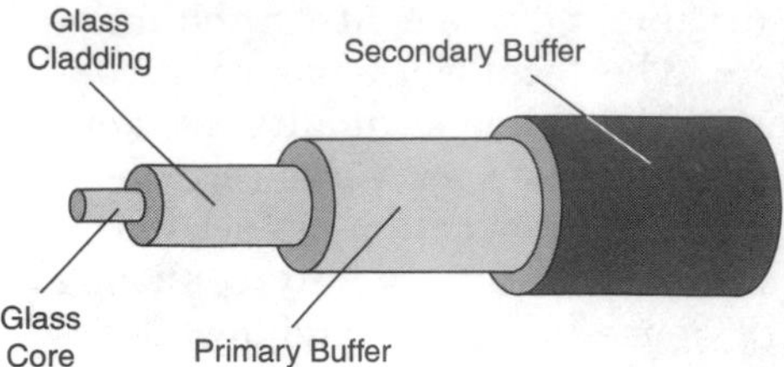

Figure 13 Parts of a FiberOptic Cable

and causes all of the light rays, or modes, to reflect back into the core. This phenomenon is known as "total internal reflection" and is the basic underlying principle of the operation of optical fibers. It has also been termed the "Fiber-Optic Effect."

Optical fiber is typically made from high-purity silica glass, a discussion of the manufacturing of which is beyond the scope of this book. Plastic fiber of varying configurations is also available, but the attenuation of light energy can approach one thousand times that of glass fiber. Optical fiber is identified by its core/cladding size. A single-mode, step-index fiber with a core diameter of 8 μ and a cladding diameter of 125 μ would have a designation of 8/125. Similarly, a multimode, step-index fiber could be designated 200/230. These numbers will identify each type of fiber described in the following section (Figure 14).

The third component of a fiber-optic cable is the coating, an opaque fluoropolymer that protects the core/cladding assembly and provides resistance to water.

Bandwidth, that is, the range of frequencies which the optical fiber can transmit, is affected by the length and integrity of the transmission path and the core/cladding arrangement. Bandwidth is expressed in Hz-Km.

Attenuation in optical fiber is due mainly to scattering and absorption within the core material. Dirty or poorly made-up connections will have an effect on the intensity of the light energy. Attenuation, or loss, is expressed in dB/Km. Loss is, therefore, introduced over distance and at connections. Note that any connection, regardless of quality, will introduce a loss, known as the "insertion loss."

Types of fiber. Three arrangements are in common use. Each arrangement exhibits characteristics that affect the bandwidth and the application for which the fiber can be used. Basically, the refractive indices and diameters of the core and cladding are manipulated relative to each other, producing fibers that are optimized for implementation in specific applications. The single-mode, step-index fiber-optic cable has the highest bandwidth and the lowest loss. A typical fiber of this type

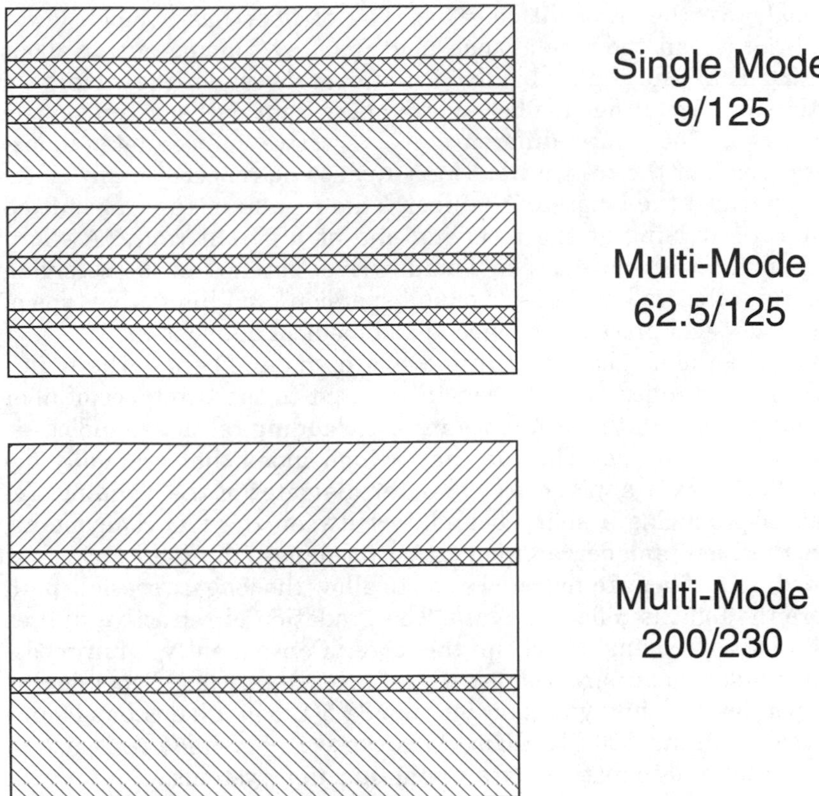

Figure 14 Types of Fiber Optic Cable

is designated 8/125. The single-mode fiber core allows one mode of transmission. This means that one ray of light is transmitted. The term "step-index" refers to the sharp change in refractive index from the core to the cladding. This type of fiber has enormous bandwidth, exceeding 25 GHz-Km. Bandwidth is limited by the change in the velocity of the transmitted light over the wavelength of the transmitted light. This is called chromatic dispersion. Chromatic dispersion is caused by the absorption and scattering of the light within the core and increases with distance. Single-mode cables are difficult to connect or splice, and the small core diameter makes the alignment of the fibers critical. Attaching connectors or splicing requires specialized equipment and trained personnel. Single-mode fibers find the most use in long-distance communications.

At the other end of the bandwidth scale is the multimode, step-index fiber. A typical designation is 200/230. The core diameter is large relative to the light rays transmitted, therefore, more than one ray, or mode, can be transmitted. This type of fiber has a much lower bandwidth than the single-mode fiber, usually topping out at about 20 MHz-Km. The usable bandwidth is even lower, in the vicinity of 5 MHz-Km. A multimode fiber allows many light paths, as previously mentioned. The sharp difference in refractive indices (step-index) causes some of the transmitted light rays to be reflected many times as they travel the length of the fiber. This results in the transmitted light rays arriving at the receiving end at different times. A sharp pulse at the transmitted side would arrive "spread" at the receiving end. This effect is known as "modal dispersion" and limits this type of fiber to low-end process and specialized medical applications.

Between the single and multimode, step-index fibers is the multimode, graded-index optical fiber. In contrast to the two types of fiber already discussed, the difference in core/cladding refractive indices is gradual, not stepped. The core has a high index that gradually becomes lower as it approaches the outer diameter of the cladding. Instead of producing a sharp boundary between the two components, with the resultant degradation in performance, the light rays in a graded-index fiber are reflected so as to allow the longest possible path before the light is reflected again. The gradation of refractive indices produces a focusing effect in the core. Consequently, multimode, graded-index fibers are not greatly affected by modal dispersion. A common fiber of this type is designated 62.5/125 and has a bandwidth between 100 and 800 MHz-Km. This type of fiber finds wide use in commercial and industrial data communication systems.

Light sources for multimode fibers can be either lasers or high-performance LEDs. Because of the large core size of this type of fiber, alignment when connecting is not as critical as in single-mode fibers.

Each type of fiber described previously will only transmit light that enters the fiber core within the core's "cone of acceptance." If a light ray intercepts the core at an angle that is outside of this cone, the light will not be internally reflected. This parameter is called the fiber's Numerical Aperture, or NA. The larger the NA, the larger is the cone of acceptance. The NA is also an indication of the relative light gathering power of the fiber. A 200/230 fiber, then, would have a larger numerical aperture than an 8/125 fiber.

Optical modems. An optical modem converts electrical energy into light energy. The MOdulator/DEModulator receives voltage levels from the data terminal and translates these voltages into light pulses of varying duration arranged to conform to one of several data communications protocols. These protocols were discussed in detail in Chapter 2.

For single-mode fiber, used mostly in long-distance communications, a semiconductor laser is used to transmit light pulses in the

1300–1550 nm range. These devices produce high quality, coherent light sources and can support transmission rates over 2 Gbps at distances in excess of 30 Km.

For multimode fibers, high-performance LEDs are used. Light Emitting Diodes in the 850–1300 nm range are generally employed in the transmission of low to medium transmission rates over low to medium distances. This type of modem is used in most LAN applications. Typical bandwidths for 1300 nm LEDs are about 100 Mbps/10 Km. IEEE-802.3J (Ethernet) will support 10 Mbps/4 Km and FDDI (Fiber Distributed Data Interface) will run at 100 Mbps/2.2 Km.

Cost differences between laser and LED modems are significant. For the majority of LAN or factory applications, LEDs are used; lasers are simply unnecessary. Adequate bandwidth, reliability, and lower cost make any of the high-quality, LED-based modems suitable for most communications applications.

Detectors come in two flavors: the PIN Photodiode and the Avalanche Photodiode (APD). The PIN Photodiode (named for the semiconductor layers that comprise it) generates charge carriers when exposed to light. These charges, or photocarriers, form a photocurrent, which is then translated into usable data. PIN devices require low bias voltages and are abundantly available, matched to the common LED source ranges.

The APD also generates photocarriers when struck by light. These charges cause an "avalanche" of charge carriers and are thus capable of forming a much larger output current. APDs are much more sensitive than PINs. APDs are used primarily for laser sources over long-distance links. APDs are more costly, require bias voltages in excess of 100 volts, and are very sensitive to temperature variations.

Chapter

4

Practical Consideration

This chapter will put all you have learned in the previous three chapters into practical use. You may have a small office with two computers or you may have many hundreds of computers. The principles are basically the same except for scale. The following information will guide you through the process of determining exactly what your need for data communication is and how to go about implementing it. In any event, this chapter will try to point out some of the common approaches and misapproaches to the design of a data communication network. Throughout this chapter, sidebars are offered that present a discussion of relevent National Electrical Code articles. It is important to note that communication circuitry has been the subject of several recent code revisions. These sidebars should be reviewed thoroughly, as the impact of the code, and failure to follow it, will have a significant impact on your installation. A broad discussion follows, it is applicable to the office as well as to the factory floor. The differences in the media have been exhaustively detailed and the different wiring standards have been presented and explained in preceding chapters. You are no doubt asking how to apply this information. This book makes a wonderful paperweight after all is *read* and done. But aside from collecting dust it has to have a practical purpose. Leaving theory behind, we turn now to the practice.

Defining Your Needs

The first step is determining exactly what you do and do not need in terms of communicating. If you have time yell loudly and throw a disk full of data at the person on the other machine. Sound silly? This was the preferred method of data transfer before the advent of affordable and convenient networking and was referred to as the "five-finger network." It was more refined in that the disk was simply put into someone's inbox. This system also had its drawbacks. The disk could be lost or stolen and the data could not be shared conveniently among several users. And that was only within an office existing within several connected rooms. If the file was needed in another city, express mail companies filled the gap and everyone waited two or three days for the information to disseminate. Recall the situation Drs. Metcalf and Boggs found themselves in that led them to develop Ethernet (see Chapter 2).

The needs of an office fall into the categories described in the following three sections.

- Size of the network: The number of machines that need to communicate and share files is a major determining factor in deciding what architecture to use. Any one of several available solutions can easily accomplish simple communication between two or three personal computers.
- Method of connection: The simplest is to connect two machines using a serial cable. Using one of several available programs, one user can have the capability of accessing files on the other user's computer or completely take over the operation of the other machine and control it remotely. This solution works well only when two machines are connected together within a small area, the same room, ideally (Figure 1).
- Network operating systems: Modern operating systems afford the user the capability of easily networking one computer with many others. A popular user interface offers basic networking capability for small office environments, while the enhanced, network-oriented version is suited for networks of unlimited size situated over large geographical areas. These newer operating systems are competing head-to-head with the established network operating systems which

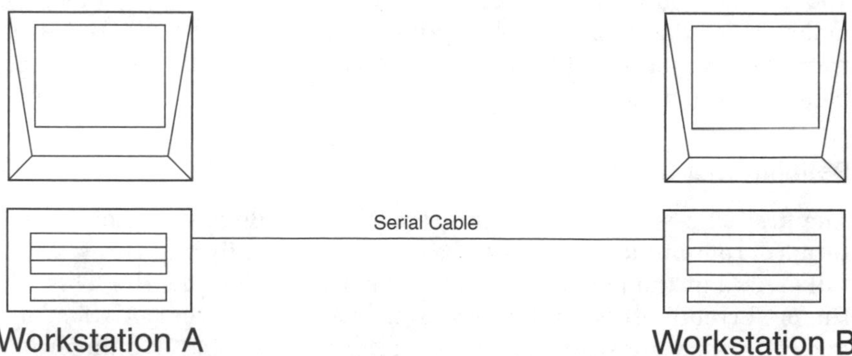

Figure 1 Method of Connection

have been, and still are, the workhorses of modern networking. A discussion of the various operating systems is definitely beyond the scope of this book, but is the subject of many other fine books. A listing of some of these books is included at the back of this book.

The type of machine is also a big factor. This book is primarily concerned with the subject of data communication on a PC-type platform, by far the most common because of cost and availability. Machines based on other than PC architecture typically have some inherent networking capability. There are several different manufacturers of computer equipment, each with their own specialized type of architectures and protocols. The principles explained here can easily be applied to other machines. I will resist the temptation to indulge in comparative architectures. It, too, is beyond the scope of this book. Suffice it to say that most other architectures and protocols are variants of the standards we have already explored, with various nuances designed to enhance the capabilities of the unique machines they serve. Each one must be evaluated upon its own merit. The author chooses to make no recommendation, but instead leaves it up to the reader to take the information presented here and see what works best for his or her needs.

The distance between the machines must be taken into consideration also. This was touched on previously. Of course, communication needs will vary greatly if one wishes to communicate with the machine in the next room one day and a machine one thousand miles away the next. Large geographical distances require specialized software and hardware in order to effectively communicate. They also require the use of sophisticated long-distance and telephone circuitry such as the T3 and the X.25 networks explained in Chapter 2. This will add significant cost to the installation but will serve to effectively bring the distant offices closer together. The data shared over this type of network is available with the click of a mouse as though it were just another drive on your machine. Shorter distances require somewhat less sophisticated technology. These installations use more common Local Area Network technology such as the Ethernet and Token-Ring systems explained in Chapter 2.

The type of data or service to be shared or transferred is an important parameter to be defined. Actually, this need is probably the *most* important and the one that can get you into the most trouble. There are a great many services available for the modern office. The trick is to strictly define what your office needs. File and resource sharing are a given. Without these two needs, there would be no need to connect computers together. Another item to consider is printer sharing. If several users have to print documents, it can create a bottleneck when more than one wants to use the printer. While this need may seem obvious, it is often overlooked. Also, the misinformed or misguided will leave this connection out of the specification based on cost. The author's opinion is that this is a vital need in the modern office. Several different types of printers can be networked as well as multiple printers of the same type. Plotters are also an essential part of some offices and can be networked. While not technically printers, fax machines can be networked and documents can be faxed from the desktop instead of having to be printed and then carried to and fed through the machine. The same holds true for modems. A modem or modems can be placed on a network and connections can be established directly from the desktop.

Of course, you may not want to give everyone these capabilities. Most networking software gives one the capability of assigning rights to each user. The upshot of this discussion is that any type of information or service is capable of being shared among many users, it is up to the designer to determine which are appropriate. As stated previously, a network operates as a virtual computer, behaving as though all of the resources are right there on your desktop.

Once you have defined exactly who and what will be communicating, the next step is to decide upon what medium you will be using. The various choices were discussed extensively in Chapter 3. As important as the medium is, it is just as important to decide how to install it. Network architectures, or topologies, were discussed in Chapter 1, but will be revisited here. In a small office, a simple bus architecture using Ethernet will do nicely. In a large facility, existing on several floors or spread out over several acres, a bus architecture runs into limitations. Combinations of two or more architectures and possibly protocols may be required. In order to extend the reach of the network and to handle traffic more effectively, devices such as routers or bridges are required. These devices move data more efficiently over a given section of the network. Other devices, called Gateways, perform protocol translation and serve to connect machines and networks of differing configurations.

The Practical Network

In Chapter 1, the concept of the Local Area Network was introduced. There is not one "typical" implementation, however. Every installation requires a thorough evaluation based upon the parameters outlined previously. When the scope of the network is defined, the physical layout must be designed and implemented. Some basic concepts do come into play and will allow the designer to build upon what is already proven.

In an installation larger than a few machines in a small office, the resources of one central machine, called a server, are dedicated to the storage and archiving of files, spreadsheets, graphics, presentations, etc. This machine is typically the heart of a network. It is on this machine that the network operating software (the operating system) resides and runs. The server's job is to *serve* the machines that are attached to the network. The server, or file server, is usually a dedicated computer that monitors and controls the operation of the network. The server contains two or more large hard disk drives that store the organization's files and the network operating system (NOS).

The NOS is known as a "multitasking" operating system in that the OS is capable of performing many different tasks seemingly at the same time. This feat is accomplished by the careful management of computer resources on an active basis. The server manages the resources of the network not only from within the server, but also on the networked machines. Computer time is based upon priority and usage history. If one machine on the network is performing a high-priority task, it is given priority access to the server's and network's resources. This machine also drops to the bottom of the priority list when it is done using the server's resources. This all happens at incredible speeds, so the delays in accessing the server will never be noticed. In this way, bottlenecks are avoided and every machine on the network gets a crack at the system resources. If a machine is idle for several computing cycles, it too will drop to a lower position on the priority list. The server is primarily responsible for the proper operation of the network.

Another duty of the server is the management of files. All of the organization's files are stored inside the server and in specialized computer storage units called RAID arrays. Several storage schemes are in use that guarantee data integrity and quick retrieval of data when needed. The drives on the server appear as just another drive on the user's machine. The server also keeps track of files on other user's machines. This allows a user in one department to access files on a machine in another department.

If, for instance, Machine A in the accounting department wants to check on last month's receivables, the requesting machine simply has to access the directory in which the file resides. The beauty of this arrangement is that the system makes the file server appear as though it is just another drive on Machine A's computer. The file is copied onto the requesting computer's hard drive and then opened by the appropriate application program. When the person in the accounting department is done with the file, it can be sent back to the directory it came from just as easily as it was retrieved: If Machine A also wants to print the file for distribution, the file is sent to the department printer via the network. Or maybe Machine A will just e-mail the document to everyone on the distribution list via the network.

Not everyone requires this degree of flexibility. A simple printer-sharing switch may be all that is required to keep a small office running smoothly. As in all things, there are many degrees of sophistication. And with sophistication comes an equal level of technical sophistication. Described previously are just the extremes in installations. The majority of installations are small- to medium-sized offices with less than thirty computers. The principles remain the same for larger installations. Smaller installations require more thought than the larger ones because most network operating systems deal with ten users or more.

Wiring and connectors

Basic to all networks are the wire and connectors that make it all possible. Of course, wireless LANs are changing data communication as we know it, but here we will concern ourselves with the traditional wired LAN. The most common types of wire are coaxial cables and shielded and unshielded twisted pairs. For a more complete discussion of these components, please refer to Chapter 3. It must be remembered that LANs communicate using radio frequencies, which have a tendency to weaken, or attenuate, easily. One must think of the cable as a pipe that contains the energy and keeps it from leaking out.

Coaxial cable. Coaxial cable or coax, acts as just such a pipe, keeping the needed signals in and the corrupting signals out. Coaxial cable is very well suited to environments that are subject to a high level of electrical noise. With proper installation, that is, observing the recommended practices as governed by code, coaxial cable can be run in close proximity to high-voltage cabling with no detriment to the signal it carries. The outer shield of the cable drains away the offending interference and keeps it from corrupting the signal being carried in the core component. Coaxial cable is used for bus-type networks. Aside from fiber-optic cabling, this type of cable is most suited for medium- to large-sized networks. Two common lengths for these cables are 185 and 500 meters per network segment (607 and 1650 feet approximately). Ethernet allows up to five segments. These cabling schemes are called 10Base2 and 10Base5, respectively (the 185 meter designation is rounded off to two hundred for clarity on the 10Base2 designation). These cable types are capable of a high data throughput due to the superior shielding of the signal.

Coaxial cables are extremely durable. They can put up with a substantial amount of abuse and have been observed to work well even under water. Due to the manner in which the cable is constructed, it

SIDEBAR G Coaxial Cable Types

Type	NOM (mm)	O.D. (inch)	Impedance (ohms)
RG-5/U	8.40	0.331	52.5
RG-6/U	6.70	0.264	75
RG-7/U	9.40	0.370	97.5
RG-8/U	10.30	0.406	52
RG-8A/U	10.30	0.406	52
RG-11/U	10.30	0.406	75
RG-11A/U	10.30	0.406	75
RG-13/U	10.70	0.421	74
RG-13A/U	10.70	0.421	74
RG-14	13.80	0.543	52
RG-14A/U	13.80	0.543	52
RG-15/U	13.80	0.543	76
RG-17/U	22.10	0.870	52
RG-17A/U	22.10	0.870	52
RG-19/U	28.50	1.122	52
RG-19A/U	28.50	1.122	52
RG-22/U	10.30	0.406	95
RG-22A/U	10.30	0.406	95
RG-22B/U	10.30	0.406	95
RG-29/U	4.70	0.185	53.5
RG-34/U	15.90	0.626	71
RG-34B/U	16.00	0.630	75
RG-54A/U	6.10	0.240	58
RG-55/U	5.00	0.197	53.5
RG-58/U	4.95	0.195	53.5
RG-58A/U	4.95	0.195	50
RG-58C/U	4.95	0.195	50
RG-59/U	6.15	0.242	73
RG-59A/U	6.15	0.242	73
RG-59B/U	6.15	0.242	75
RG-62/U	6.15	0.242	93
RG-62A/U	6.15	0.242	93
RG-63/U	10.30	0.406	125
RG-63B/U	10.30	0.406	125

SIDEBAR G Coaxial Cable Types Continued

Type	NOM (mm)	O.D (inch)	Impedance (ohms)
RG-71/U	6.40	0.252	93
RG-89/U	16.00	0.630	125
RG-108/U	6.20	0.244	78
RG-108A/U	5.97	0.235	78
RG-122/U	4.06	0.160	50
RG-130/U	45.90	1.807	95
RG-133A/U	10.30	0.406	95
RG-142B/U	4.95	0.195	50
RG-164/U	22.10	0.870	75
RG-174A/U	2.54	0.100	50
RG-178B/U	1.80	0.071	50
RG-179B/U	2.54	0.100	75
RG-180B/U	3.68	0.145	95
RG-187A/U	2.79	0.110	75
RG-188A/U	2.79	0.110	50
RG-195A/U	3.94	0.155	95
RG-196A/U	2.03	0.080	50
RG-213/U	10.30	0.406	50
RG-214/U	10.80	0.425	50
RG-216/U	10.80	0.425	75
RG-217/U	13.80	0.543	50
RG-223/U	5.40	0.213	50
RG-316/U	2.59	0.102	50
RG-385/U	16.80	0.661	50
RG-393/U	9.90	0.390	50
RG-400/U	4.95	0.195	50
RG-401/U	6.35	0.250	50
RG-402/U	3.58	0.141	50
RG-405/U	2.18	0.086	50

has a high pull strength. The recommended installation is in conduit, even though it will work just as well strapped to a pipe (not recommended). Office installations find it thrown over dropped ceilings replete with fluorescent fixtures. One consideration, particularly with the 10Base5 cable, is the maximum bend radius. This cable has a diameter of slightly less than ½" and is very stiff. It has a bright yellow insulating jacket and is referred to a "yellow hose" or "frozen yellow hose." As with all cable, the proper bending radius must not be exceeded. The rule of thumb is that the bend radius must not exceed ten times the diameter of the cable. 10Base5 is approximately ½" in diameter (actually .457"). The maximum bend radius should be about five inches. This cable can be bent further, but only when it is absolutely unavoidable, as when only a ninety-degree angle is possible. It must be understood that the cable cannot be reused if it is ever pulled out. The 10Base2 cable is less of a problem, being around ¼" in diameter. This cable is black, resembling that used for cable television.

Just as important are the connectors and fixtures that terminate the cables. If any chain is only as strong as its weakest link, then any network is only as reliable as the most poorly installed connector. Many networks have been put out of commission because one strand of shield has found its way to the core and sapped away the signal. Connectors for this type of coax are available with silver or gold plating. For the few extra pennies it costs, it is worth it to install a better quality connector. Connectors are available using a threaded nut as is commonly seen in the cable television industry. More commonly a nut and ferrule system is used (Figure 2). The installation procedure for each type is described next. Each type of connector can be wall mounted for ease of connection to a desktop PC. Multiple configurations are available from many manufacturers. Custom configurations are also available.

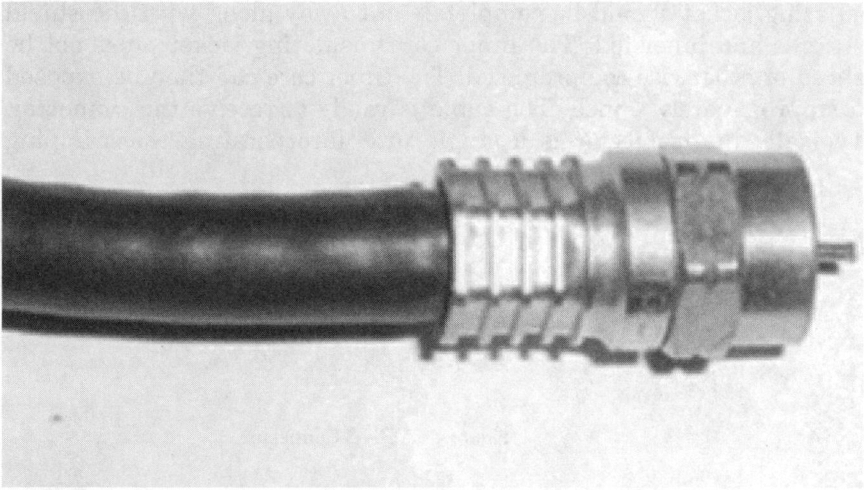

Figure 2 Coaxial Connectors

Figure 3 10Base2 Connectors.

As just described, two types of coax is common for networks. For the smaller, more flexible, 10Base2 cable, the following rules apply to the layout of the network.

There are only five segments of 200 meters each allowed. Both ends of the network must be terminated. On this type of cable, it is common to use a "T" connector (Figure 3) to make the network connection to the Network Interface Card (NIC). The terminator(s) may be installed directly on the T connector. There may only be thirty nodes per segment with a maximum of 1024 nodes on the entire network. There must be at .5 meter (1.6') between connectors. If a T connector is not used as would be the case with a wall fixture, the attachment cable can be up to 50 meters (165') in length. Note that a "T" connector is used behind the wall plate to connect to the wall fixture. One end of the network cable must be grounded. If both ends are grounded, differences of potential can develop that will wreak havoc on the network. BNC or Bayonet connectors (Figure 4) are used for terminating 10Base2 cable.

The procedure for installing a connector is as follows. The outer insulating jacket should be completely cut away along with the shield strands and inner foil. The inner core insulating jacket must not be sliced or otherwise compromised. The inner core can then be exposed to approximately ⅛ inch. The cable is ready to receive the connector. Typically the connector is a single unit incorporating the crimping

Figure 4 A BNC Connector

ring. It will appear as two concentric barrels. The smaller of the two barrels is inserted over the inner core insulation and forced between this insulation and the shield strands. The center core should be centered in the connector and extend no further than the connector's twist-lok. The outer barrel will fit neatly over the outer insulation. This outer barrel is the crimping ring. Using a crimping tool, the installation is completed by crimping, or compressing, the outer barrel around the outer insulation. Some connectors use a nut and ferrule system. This is similar to the above except that you must remember to slip the barrel over the cable before stripping and inserting the connector. The barrel is then brought up and threaded into the back of the connector. Both systems work well, offering good physical strength and signal integrity. It is a good idea to check the installation by measuring for continuity between the shield and core. An infinite resistance should be seen.

For the 10Base5 type of coax, the procedure for termination is very different. Splicing of the cable is achieved by using a connector of the nut and ferrule type and a barrel connector is used to make the connection between the two cables. For connection to a node, the cable must be drilled using a special jig. While this cable is losing ground to glass fiber, installations and maintenance of 10Base5 coax are still common.

The procedure for making a node connection using this cable is as follows. 10Base5 cable can only be "tapped" at specific points on the cable two and one half meters apart. These points are marked on the cable by a black line. Making a connection at any other point will cause reflections that will disrupt the signal. There may be up to 100 taps per 500 meter segment. Overall there may be up to 1024 taps over the five segments allowed for this type of network. Note that each segment must be grounded. The connection is made using what is called a transceiver tap, which is commonly referred to as a "vampire" tap (Figure 5). The tap is also called a Medium Attachment Unit (MAU). It operates by injecting the signal into the cable via a small needlelike antenna. The antenna does not make physical contact with any part of the cable but the core conductor. As previously stated, a special drilling jig is used. The cable is clamped into the jig at the insertion point marked on the cable. The jig actually controls the depth to which a special drill penetrates the outer insulation and shield. Without this,

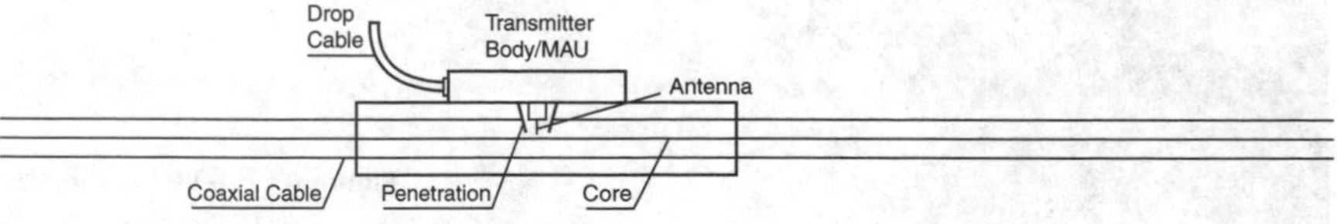

Figure 5 10Base5 Vampire Tap

the drill would not drill through cleanly, leaving strands of shield in contact with the core, making the connection useless. The jig allows a small area of the outer insulation and shield to be cleanly removed. It also prevents damage to the core conductor by limiting the depth to which the drill penetrates the core insulation. The result is usually a clean penetration of the outer and inner insulation, exposing a small area of the core, with no damage. The tap is then clamped to the cable, the antenna being carefully inserted into the hole. The clamping mechanism punctures the outer insulating jacket on either side of the hole in order to make a good electrical connection with the shield.

We are not done yet, though. The final connection to the node is made using a twenty-five conductor cable terminated on either end to a DB-25 connector. The cable and connectors are called the AUI, or Attachment Unit Interface. The other end of the AUI cable is connected to the network interface card (NIC).

Either type of cabling must be terminated at both ends. This is accomplished by installing a connector on the free end of the cable (at an insertion point on 10Base5 cable). This terminator is typically 50 ohms and eliminates reflections and maintains the electrical characteristics of the cable. On 10Base2 cable, the terminators may be installed directly on the T connectors at the back of the network interface card (NIC).

Unshielded twisted pairs, or UTP, do not require a shield because of the phenomenon of common mode rejection. Because of the twist in the wire, the interfering signals cancel each other out. UTP also offers the advantage of using standard RJ connectors in terminating the cables. There is a trade-off, however, in that the maximum distance in a UTP network is limited to 100 meters (330 feet). Shielded twisted pairs (STP) are also used, but are bulky and difficult to use compared with UTP.

Connectors for UTP resemble the familiar telephone connector RJ-11 (Figure 6) but with an important distinction: the RJ-11 handles

Figure 6 RJ-11 Connector

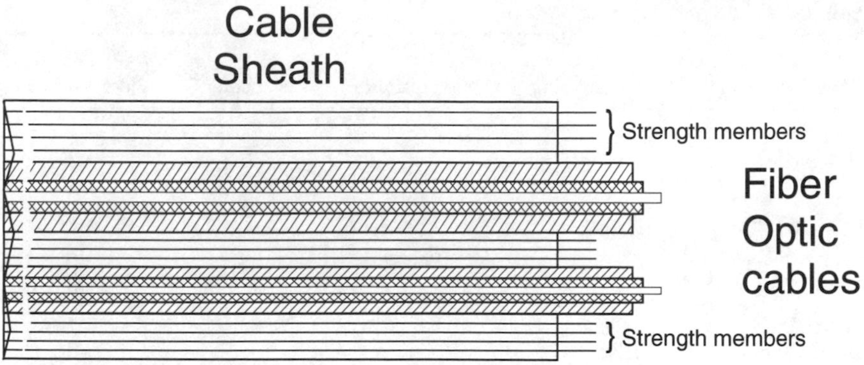

Cable Sheath
Strength members
Fiber Optic cables
Strength members

Figure 7 Tight Tube/ Buffer Fiber Optic Cable

four wires while the UTP connector, the RJ-61, handles eight wires. Due to the lack of shielding, the length of the runs is limited and the prevalent topology for UTP is the star configuration. At either end of the run, an RJ-61 connector is attached to the cable. This method of cabling is much simpler and cheaper than the coaxial system described previously. The most difficult part of the installation is the termination of the cables. With a minimum of practice, the connectors can be properly attached using a specialized crimping tool. Wall sockets using RJ-61 connectors are even simpler, using self-terminating inserts to make the connection. This wiring system has borrowed heavily from the telecommunications industry.

The importance of making a proper UTP connection cannot be overstressed. If a poorly trained or sloppy technician makes up the connections in a hurry or with the wrong tools, the run will fail and have to be done twice. Worse, if the connection is marginal and passes test-

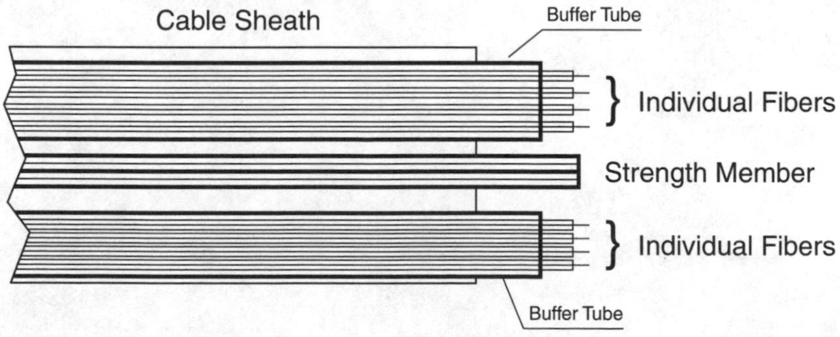

Figure 8 Loose Tube Fiber Optic Cable

SIDEBAR B 66 and 110 Block Wiring

Wire/Color code	Tip and ring	Pair number	50 Pin positions	66 or 110 Block positions
White/Blue	Tip 1	Pair 1	26	1
Blue/White	Ring 1		1	2
White/Orange	Tip 2	Pair 2	27	3
Orange/White	Ring 2		2	4
White/Green	Tip 3	Pair 3	28	5
Green/White	Ring 3		3	6
White/Brown	Tip 4	Pair 4	29	7
Brown/White	Ring 4		4	8
White/Slate	Tip 5	Pair 5	30	9
Slate/White	Ring 5		5	10
Red/Blue	Tip 6	Pair 6	31	11
Blue/Red	Ring 6		6	12
Red/Orange	Tip 7	Pair 7	32	13
Orange/Red	Ring 7		7	14
Red/Green	Tip 8	Pair 8	33	15
Green/Red	Ring 8		8	16
Red/Brown	Tip 9	Pair 9	34	17
Brown/Red	Ring 9		9	18
Red/Slate	Tip 10	Pair 10	35	19
Slate/Red	Ring 10		10	20
Black/Blue	Tip 11	Pair 11	36	21
Blue/Black	Ring 11		11	22
Black/Orange	Tip 12	Pair 12	37	23
Orange/Black	Ring 12		12	24
Black/Green	Tip 13	Pair 13	38	25
Green/Black	Ring 13		13	26
Black/Brown	Tip 14	Pair 14	39	27
Brown/Black	Ring 14		14	28
Black/Slate	Tip 15	Pair 15	40	29
Slate/Black	Ring 15		15	30
Yellow/Blue	Tip 16	Pair 16	41	31
Blue/Yellow	Ring 16		16	32
Yellow/Orange	Tip 17	Pair 17	42	33
Orange/Yellow	Ring 17		17	34
Yellow/Green	Tip 18	Pair 18	43	35
Green/Yellow	Ring 18		18	36
Yellow/Brown	Tip 19	Pair 19	44	37
Brown/Yellow	Ring 19		19	38
Yellow/Slate	Tip 20	Pair 20	45	39
Slate/Yellow	Ring 20		20	40
Violet/Blue	Tip 21	Pair 21	46	41
Blue/Violet	Ring 21		21	42

SIDEBAR B 66 and 110 Block Wiring Continued

Wire/Color code	Tip and ring	Pair number	50 Pin positions	66 or 110 Block positions
Violet/Orange	Tip 22	Pair 22	47	43
Orange/Violet	Ring 22		22	44
Violet/Green	Tip 23	Pair 23	48	45
Green/Violet	Ring 23		23	46
Violet/Brown	Tip 24	Pair 24	49	47
Brown/Violet	Ring 24		24	48
Violet/Slate	Tip 25	Pair 25	50	49
Slate/Violet	Ring 25		25	50

ing, it could become unstable later on and wreak havoc with the network. Near End Crosstalk, or NEXT, is the phenomenon that creates the most problems, second only to not having made a proper physical connection. NEXT will manifest itself in garbled and corrupted signals and the dropping of nodes as unresponsive. While this wiring system is cheap and easy to use, proper technique is still very important. A brief description of the procedure follows.

The UTP cable is cut square. That is, the end is cut at a precise right angle. The outer insulating layer is trimmed back approximately ¼ inch. Do not strip or untwist the wiring. UTP operates only because of the twist. Insert the untwisted wires into the RJ-61 connector. Observe the proper orientation of the wiring with regard to the orientation of the connector. The color code for the plug position varies with the standard used. For the T568A&B standards, the wiring is shown in the sidebar. Carefully insert the connector into the crimper and firmly

SIDEBAR C TIA Wiring Color Codes

TIA 568A [8-position modular plug (RJ-45)]

Wire color code	Pair	Position
White/Green	3	1
Green/White	3	2
White/Orange	2	3
Blue/White	1	4
White/Blue	1	5
Orange/White	2	6
White/Brown	4	7
Brown/White	4	8

SIDEBAR C TIA Wiring Color Codes Continued

TIA 568B [8-position modular plug (RJ-45)]

Wire color code	Pair	Position
White/Green	3	3
Green/White	3	6
White/Orange	2	1
Blue/White	1	4
White/Blue	1	5
Orange/White	2	2
White/Brown	4	7
Brown/White	4	8

TIA 570 (6-pin modular jack RJ-25)

Wire color code	Pair	Position
White/Green	3	1
Green/White	3	6
White/Orange	2	2
Blue/White	1	3
White/Blue	1	4
Orange/White	2	5

crimp the wire into the connector. The connector will puncture the wires without any need for stripping back the insulation.

Fiber-optic cables. Depending upon application, distance involved, and location, several types of cable configurations and connector types are available. Optical fiber is fragile and must be protected, mostly from mechanical stresses such as bending, crushing, thermal effects and pulling during installation. While tensile strength typically exceeds 600 Kpsi, glass fiber can lose a significant amount of strength when exposed to water. Excessive bending will also cause degradation in performance. With a reasonable amount of care, optical fiber is easily installed and has a life in excess of thirty years. The most common mode of failure in optical cable is the backhoe syndrome. As with any underground installation, accurate drawings, close supervision, and the installation of warning tapes will serve to avoid costly outages.

Tight Tube, or *Tight Buffer* design (Figure 7) has a PVC coating that

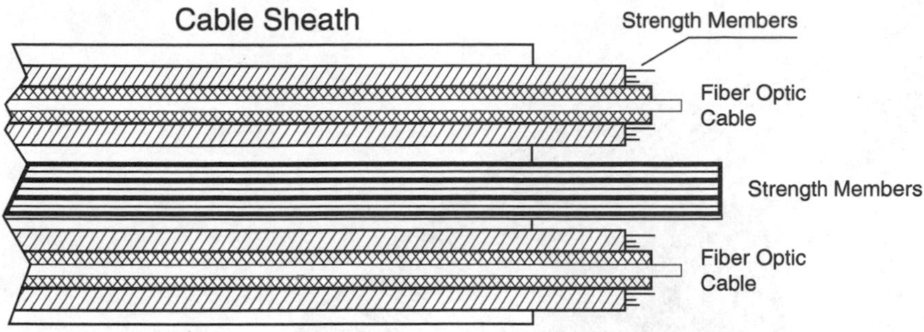

Figure 9 Breakout Cables

is tightly bonded to the fiber and does not allow the free movement of the fiber. This type of cable allows for the inclusion of strength members and can be pulled through conduit and cable trays. However, this design has low crush resistance and is susceptible to deformation due to thermal expansion; hence, it is recommended for indoor use only.

Loose Tube design (Figure 8) allows for the free movement of the fiber, as is implied by the name. Each component of the cable, i.e., the sheath or outer coating, the strength member, and the buffer tubes that carry the fibers, has differing thermal characteristics. By allowing the free movement of the fibers and the components surrounding

SIDEBAR A Fiber-Optic Cable Color Code (EIA/TIA-598)

Fiber no.	Color	Fiber no.	Color
1	Blue	14	Orange/Black Stripe
2	Orange	15	Green/Black Stripe
3	Green	16	Brown/Black Stripe
4	Brown	17	Slate/Black Stripe
5	Slate	18	White/Black Stripe
6	White	19	Red/Black Stripe
7	Red	20	Natural/Black Stripe
8	Black	21	Yellow/Black Stripe
9	Yellow	22	Violet/Black Stripe
10	Violet	23	Aqua/Black Stripe
11	Aqua	24	Rose/Black Stripe
12	Rose		
13	Blue/Black Stripe		

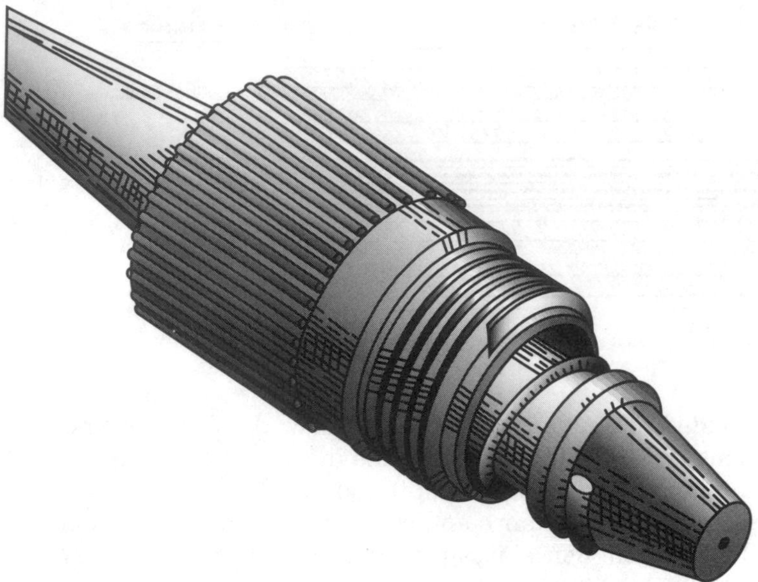

Figure 10 Biconic Connector

them, any deformation is avoided. Loose-tube construction offers much improved crush resistance over tight tube owing to buffer tube protection of the fibers. Loose-tube cables utilize a strength member that is used as the pulling member for conduit installation. Certain installations use the strength member as a replacement for the messenger used in an aerial installation. The strength member is made either of a high tensile strength plastic, such as nylon, or of steel. Note that the addition of a steel strength member renders the cable conductive. Loose-tube cables are usually filled with a gel that surrounds the fibers and increases protection from water. This also improves crush resistance because of the cushioning effect of the gel. This type of cable is used for outdoor applications, though it does find use in harsh industrial environments. A drawback of this type of cable is the difficulty in handling the individual fibers. The fiber coating does not have to be as thick as in tight-tube construction and attaching connectors is difficult.

Breakout kits are available for loose-tube cables and provide each fiber with a sheath that is color-coded and equipped with a strength member. The thicker sheath facilitates the attachment of connectors.

Breakout cables (Figure 9) are a hybrid of the above two types. In a breakout cable, each fiber is treated as a separate unit, complete with sheath and strength member. This design eliminates the need for a breakout kit, the sheath allowing much easier attachment of connectors. Recall that tight-tube construction allowed no free movement and

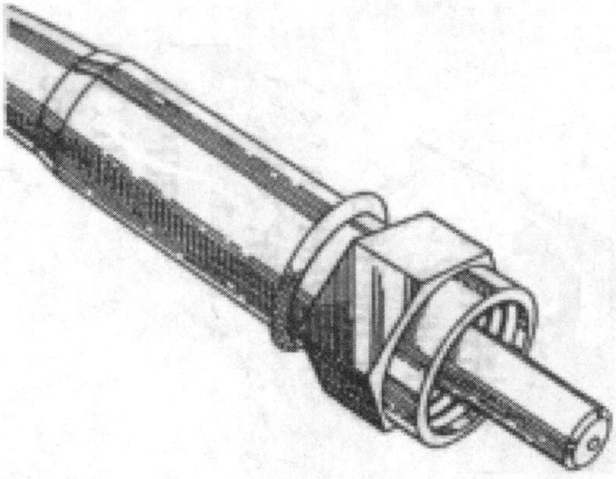

Figure 11 SMA Connector

low protection against mechanical stress, but had a thick coating which allowed ease of handling. Loose tube, on the other hand, allowed free movement and provided a good degree of protection. Breakout cables allow free movement of the fiber subunits and individually protect the fibers by virtue of the thicker coating/strength member arrangement. Each fiber subunit is configured as a tight tube. Breakout cables are also available equipped with a separate strength member as described in use by the loose-tube design.

Several connector types are available. As the technology matures, the cost and difficulty of making a connection will become less of a consideration. At present, attaching connectors requires specialized equipment and trained personnel. The task of attaching connectors is time-consuming and even the best made connectors introduce an optical, or insertion, loss. Connectors are classified by the shape and size of the alignment ferrule and by the means of mating the housings. The ferrule is a sleeve that holds the fiber in position inside the connector assembly. The fiber to fiber alignment inside of the connector coupling assembly is critical for the most complete and efficient transfer of light energy. In a single-mode fiber, with a core of only 8 μ, any misalignment results in a significant or complete loss of signal. In a 200/230 fiber, as much as 15% misalignment can be tolerated. If the connector's mechanical tolerances can be tightly controlled, then the proper installation of the fiber into the ferrule is the only variable. Each type of connector described next requires a different technique for proper installation. Please refer to manufacturer's literature for the latest information on the connector.

Biconic connectors (Figure 10) were the first type of connector developed and were the workhorse of the long distance telcos. Many are

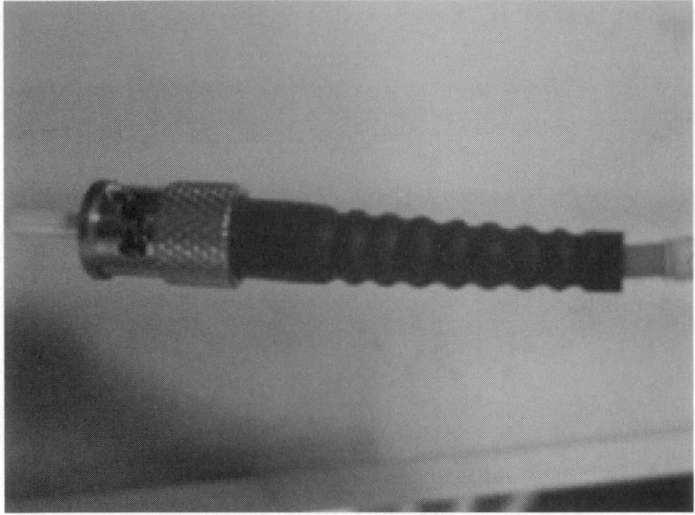

Figure 12 ST Connector

still in use, but are being replaced with newer types. This connector has conical ferrules that seat into a double conical sleeve enclosed in housings that screw together.

The *SMA* connector (Figure 11) is another early connector still in use in older installations. It is very rugged and was the first truly sucessful multimode connector. There are two types. The SMA-905 uses a straight ferrule, while the SMA-906 uses a stepped ferrule. A threaded housing attaches to a coupling and another SMA connector. There is no female connector. This connector is not suitable for single-mode fiber.

Figure 13 SC Connector

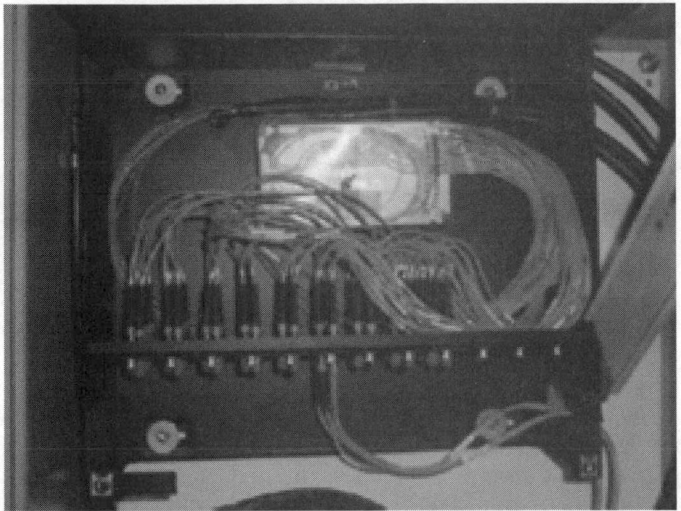

Figure 14 Fiber Optic Patch Panels

ST connectors (Figure 12) use a straight ferrule with a twist-lok or quick bayonet latch connection. This connector is the predominant LAN connector. A variation, the *ST-PC*, provides superior performance but adds cost to the installation. A Japanese variant, the *FC*, is more difficult to install but stands up to more abuse.

The newest type of connector is the *SC* type (Figure 13) and has been selected as the replacement for the ST and ST-PC types. The SC is available in simplex or duplex configurations. The connector is designed to mate with ST connectors at wall or wiring closet interfaces.

When a permanent connection between fibers is desired, optical fibers can be physically spliced. Splicing, while limiting flexibility, is cheaper and results in a stronger, permanent connection. Three methods are used.

A butt splice is analogous to the device used on conventional wiring. The fibers are inserted into the device and the device is crimped. These devices introduce a substantial loss and are only used for temporary repairs. Another type of mechanical splice requires the treatment and precise alignment of the fiber ends. This type of splice is a low-cost, long-term alternative to a connector, introducing less loss in most cases. The above types do not require a significant investment in equipment, but some training is required.

The most reliable and low-loss splice is the fusion splice. The fiber ends are fused together using specialized equipment and trained personnel. This type of splice will stand up to a high degree of physical stress and introduces the lowest optical loss of the three presented.

Patch panels (Figure 14) are used as distribution points, storage locations for spare fiber, and as splice cases. Strategically located patch panels will allow the easy configuration and reconfiguration of the network. As devices are added and removed, the patch panel serves as a convenient point at which fibers have been terminated, tagged, and made available for use. Unterminated, spare fibers can be stored for future use. The panel provides protection from environmental conditions and many include splicing bays.

The right tools

Having explained the basic techniques involved in terminating the cables, it should be apparent that certain specialized tools are required to accomplish these installations. Aside from the normal compliment of hand tools, the technician should invest in several items that will make life easier. It is not enough to say "we suggest you acquire these things." If one is considering doing this work themselves, it is a virtual necessity to have the proper tools. I will not provide a detailed list of the items needed for each particular technology. Nor will I make any recommendations.

I will, however, spend some time on why the correct job requires the correct tool. This short dissertation may be lost on some readers. I would feel that all the hard lessons I have learned from watching or doing the job incorrectly would be lost if I did not pass them on. One cannot describe the frustration experienced from watching someone butcher an installation. I stood by helplessly while a subcontractor with little or no experience used a penknife on a punchdown block, and then denied it. The installation, needless to say, had to be redone. Several positions on the block were rendered unusable.

The proper tools are a must in order to do a proper job. The use of the proper tools in the hands of a technician with the expertise are a formidable combination. Together they produce a superior product that can be depended on to provide quality service from the day it is first used and for many years after. Confidence is instilled in those who use contractors who acquire and learn to use the correct equipment. Nothing frightens an owner more than watching a poorly trained technician struggle with a piece of delicate and expensive equipment. It creates an impression of ineptitude and does not invite repeat business.

Proper training goes hand-in-hand with proper tools. A responsible contractor is one who keeps up with the industry and stays current with the latest methods for installation and commissioning. Staying informed about the latest developments in the industry you have chosen is the surest way to be able to serve your customer and to assure yourself of repeat business. It has been said that once you get a reputation, it is hard to shed. You have shown your desire to do this by purchasing this book. Suffice it to say that if you are doing this work, or are considering doing this work, it is essential that you and your personnel get the right training and equipment needed to perform the work. Your customer will thank you.

Potential problems

Whenever any kind of wiring is done, there is a potential for injury, loss of life, and damage to the equipment or the facility. With good work habits and adherence to the code, these can be minimized if not eliminated. The code will be addressed here. Good work habits have to be learned and practiced. What will be discussed in this section are several common mistakes technicians make when data communication wiring and fixtures are installed.

Mixing power and signal wiring. Any time power and signal wiring is run it is imperative to make sure there are separate runs for each. Signal and power, or low-voltage and high-voltage wiring must be segregated from one another. This applies to both tray and conduit installations. It may go without saying, but it is unacceptable to create mixed bundles of low- and high-voltage wiring. The code strictly defines what constitutes each type. This means, also, that signal wiring cannot be brought through a pull box unless there is a proper barrier. It is definitely unacceptable to mix signal and power in the same wall plate. One of the most noticeable effects of this type of installation is the inducement of sizable and harmful voltages in the signal wiring. The equipment that the signal wiring connects together is designed to see +/− 15V at best. Many network cards have been destroyed by the presence of these voltages. Almost an equal number of technicians, myself included, have been unpleasantly surprised by the presence of an unexpectedly high voltage.

It is up to the contractor to carefully review the specifications and drawings before the bid to determine if the engineer has unfortunately missed this essential detail. Signal and power wiring must never be pulled in the same conduit. Likewise, power and signal wiring must never be laid in the same duct or tray. Much of this problem has gone away with the increased use of fiber-optic cabling, which uses light instead of electrical energy. It is simply not possible to induce a signal into these cables. Each type of wiring must be separate and apart from each other. A careful review of the electrical drawings will reveal whether this important requirement has been overlooked and will

SIDEBAR D NEMA Enclosure Ratings

Type	Description
1	Type 1 Enclosures serve mainly to provide protection against the enclosed equipment or locations in areas where unusual service conditions do not exist.
2	Type 2 Enclosures are used indoors to provide protection against limited amounts of falling water or dirt.
3	Type 3 Enclosures are outdoor enclosures, which provide protection against windblown dust, rain, and sleet.
3R	Type 3R Enclosures provide protection against falling rain and sleet; also designed to be undamaged by the formation of ice on the enclosure.
3S	Type 3S Enclosures are outdoor enclosures, which provide protection against windblown dust, rain, and sleet. Type 3S enclosures are designed to be undamaged by the formation of ice on the enclosure.
4	Type 4 Enclosures provide protection against windblown dust, rain, and sleet; splashing water; and hose-directed water. Designed to be undamaged by the formation of ice on the enclosure.
4X	Type 4X Enclosures provide protection against corrosion, windblown dust, rain, and sleet; splashing water; and hose-directed water. Designed to be undamaged by the formation of ice on the enclosure.
5	Type 5 Enclosures provide protection against falling dirt, settling airborne dust, flyings, lint, and fibers. Provides protection against limited splashing or dripping of light liquids. Superseded by NEMA Type 12 in many installations.
6	Type 6 Enclosures are designed for use indoors or outdoors in locations where occasional submersion is expected. Type 6 enclosures must protect the enclosed equipment from occasional submergence in 6 feet of water (static) for 30 minutes; from falling or hose-directed water; lint, seepage, splashing or condensation of noncorrosive liquids. Enclosure is also designed to be undamaged by the formation of ice on the enclosure.
6P	Type 6P Enclosures are designed to protect enclosed equipment from prolonged submergence in six feet of static head; from falling or hose-directed water; lint, seepage, splashing or condensation of noncorrosive liquids. Enclosure is also designed to be undamaged by the formation of ice on the enclosure.
7	Type 7 Enclosures are designed for use in indoor areas classified as Hazardous Class 1, Division 1, Groups A, B, C, or D. Type 7 enclosures are designed to contain pressures resulting from an internal explosion of combustible gases.
8	Type 8 Enclosures are designed for use in indoor areas classified as Hazardous Class 1, Division 1, Groups A, B, or C; this enclosure meets the requirements of Type 7 enclosures, except that the equipment is oil immersed.

SIDEBAR D NEMA Enclosure Ratings Continued

Type	Description
9	Type 9 Enclosures are designed for use in indoor areas classified as Hazardous Class 2, Division 1, Groups E, F, or G. Type 9 enclosures are designed to contain heat generated by internal components so that such heat will not cause accumulated dust to ignite. These enclosures are also designed to withstand the ingress of combustible dust and fibers.
10	Type 10 Enclosures are designed to conform to Schedule 2G of the Mining Enforcement Safety Administration for equipment to be used in mines where the atmosphere is expected to contain methane or natural gas, with or without coal dust.
11	Type 11 Enclosures are designed to protect enclosed equipment from corrosion due to the dripping, splashing or seepage of corrosive liquids, gases, or fumes. Internal equipment is further protected by immersion in oil.
12	Type 12 Enclosures are intended for use indoors to provide protection against falling dirt; airborne or blown dust; fibers, lint, or flyings; splashing, seepage, dripping or external condensation of noncorrosive liquids. Conduit knockouts are not permitted in Type 12 enclosures except where provided with oil-resistant gaskets. Doors on these enclosures must have a full hinge, with oil-resistant gaskets and requiring a tool to open. Doors must swing horizontally and closing hardware must be captive.
12K	Type 12K Enclosures conform to those specified in Type 12 above, but permit the use of conduit knockouts.
13	Type 13 Enclosures are designed for use indoors to enclose devices such as limit switches, foot switches, push buttons, pilot lights, etc. No conduit knockouts are permitted except where oil-resistant gaskets are installed. These enclosures provide protection from dirt, dust, lint seepage, external condensation or spraying, dripping, or splashing of water, oil, or coolant.

save the contractor days if not weeks of headaches. In contracts with multiple primes, spotting this ahead of time will go a long way in simplifying coordination and start-up.

Pulling and bending. It is important to review the characteristics of the cable that is being pulled. Most cable is supplied with a data sheet that states all of the information needed, such as maximum pull and bend radius. It is interesting to note that glass fibers (FO cable) have low tensile strength but can be tied in knots without any degradation in performance. Coaxial cables, particularly the 10base5 variety have limited bend radius but have high tensile strength. The contractor must plan his pulls and conduit layout accordingly.

Maximum segment length. As described previously, the different protocols and topologies make use of different cables. The electrical characteristics of these cables limit the distance a signal can be broadcast to segments of fixed lengths. The contractor must take these segment lengths into consideration when laying out conduit and cable. Devices used to extend the range of the network have been presented. These devices also need power in order to operate. Provisions must be made to enclose these devices and typically these enclosures will be dedicated to the repeating device only. The only time this problem comes into play is with the use of bus topologies. Star topologies have a centralized router that is local to the server with the drops fanning out from that location. Fiber-optic cabling also uses these devices, but the distances are much greater.

Signal loss in FO cables. Even though FO is capable of carrying a greater bandwidth over greater distances, the signal will degrade each time it reaches a connector or splice. For a review of the characteristics of FO, please refer to Chapter 3. Following is a brief example of calculating loss in a fiber-optic network.

Fundamental to the proper operation of the optical links is the need for the transmitted power to reach the receiver. This sounds obvious, but is often overlooked. Every component in an optical communication system introduces loss. Before specifying modems, the designer must have a thorough understanding of his proposed system's present and future performance goals. The first step is to calculate the optical budget and the total system loss.

Optical fiber has a characteristic loss per unit distance. This is usually expressed in dB/Km and is dependent on the wavelength of the light source. For a typical 8/125 general purpose fiber, the attenuation is .8 dB/Km@1550 nm. Higher quality fiber of the same size can approach .4 dB/Km. A good attenuation for a 100/140 fiber is 3.5 dB/Km@1300 nm.

Each connector introduces a loss, the insertion loss, into the link. This is expressed in dB/connector or simply dB. Connector insertion losses range from approximately 1.5 dB for SMA-905 to .3 dB for a SC connector.

Splices also introduce an insertion loss, which varies widely depending upon type and quality. For the sake of calculating system loss, a figure of .75 dB/splice should be used unless the splice has been tested and the loss is known.

Assume a link with four connectors, two splices and a total of six kilometers of fiber. We are using a 62.5/125 fiber with a loss of 1.0 dB/Km. Transmitter output power into the above fiber is −12 dBm (dBm = decibels as referenced to 1 milliwatt). Our receiver's sensitivity is −27 dBm.

Subtracting the transmitter's output power from the receiver's sensitivity gives us an *optical budget* of 15 dB. We now summarize the system or link losses:

6 Km fiber @ 1.0 dB/Km = 6.0 dB

4 connectors (ceramic ST) @ .3 dB = 1.2 dB

2 splices @ .75 dB = 1.5 dB

Total system loss = 8.7 dB

This calculation assumes a source of 1300 nm.

Subtracting the system loss from the optical budget gives us a margin of 6.3 dB. This is a comfortable margin that would allow future expansion of the link. If no expansion is planned, a less powerful and costly modem can be used. Other factors that must be considered are changes in fiber size over the length of the link or a bypassed node. Also, a 1 dB loss should be allowed for single-mode fiber links to allow for dispersion and absorption. In all cases, a margin of at least 3 dB should be maintained. A margin of 1 dB is not acceptable, considering that one bad splice could render such a link susceptible to erratic performance or complete failure.

Ingress Protection (IP) Ratings describe the degree of protection an enclosure provides against penetration of objects or liquids into an enclosure. IP ratings follow the form: IP s 1, with s representing solids penetration protection and 1 representing liquid penetration protection. The following table describes the defined levels of protection as specified in IEC Standard 529.

SIDEBAR E IP Enclosure Ratings

First digit(s)	Degree of protection	Second digit (1)	Degree of protection
0	No protection.	0	No protection.
1	Protection against access by hand or object equal to or greater than 50 MM.	1	Protection against water dripping vertically.
2	Protection against contact by fingers or objects greater than 12 MM.	2	Protection against drops of liquid dripping on enclosure tilted to 15°.
3	Protection against contact by tools or objects greater than 2.5 MM.	3	Protection against rain or spraying liquids with enclosure tilted to 60°.

SIDEBAR E IP Enclosure Ratings Continued

First digit(s)	Degree of protection	Second digit (1)	Degree of protection
4	Protection against small tools or wires or objects greater than 1 mm.	4	Protection from splashing from any direction.
5	Complete protection against contact with live or moving parts; protected against deposits of dusts.	5	Protection against water jets from any direction.
6	Complete protection of live or moving parts; protected against penetration of dust.	6	Protection from conditions on ships' decks. Water will not enter under stated conditions of pressure and time.
		7	Protection against immersion in water. Water will not enter under stated conditions of pressure and time.
		8	Protection against indefinite immersion in water under a specified pressure.

Example: An enclosure rated for IP45 means that the enclosure is protected against the entry of objects greater than 1 mm and can withstand water jets from any direction.

SIDEBAR H IEEE Codes

- LAN/MAN Bridging & Management (802.1)
- Logical Link Control (802.2)
- CSMA/CD Access Method (802.3)
- Conformance Test Methodology for IEEE-802.3
- Token-Passing Bus Access Method (802.4)
- Token-Ring Access Method (802.5)
- DQDB Access Method (802.6)
- Broadband LAN (802.7)
- Integrated Services (802.9)
- LAN/MAN Security (802.10)
- Wireless (802.11)
- Demand Priority Access Method (802.12)
- Cable TV (802.14)

SIDEBAR F Instrument Symbol Nomenclature

	First letter		Succeeding letters		
Letter	Measured or initiating variable	Modifier	Readout or passive function	Output function	Modifier
A	Analysis		Alarm		
B	Burner, Combustion		User defined	User defined	User defined
C	User defined			Control	Close
D	User defined	Differential			
E	Voltage		Sensor (primary element)		
F	Flow	Ratio			
G	User defined		Glass, Viewing device		
H	Hand				High
I	Current		Indicate		
J	Power	Scan			
K	Time, Schedule	Time, Rate of change		Control station	
L	Level		Light		Low
M	User defined	Momentary			Middle, Intermediate
N	User defined		User defined	User defined	User defined
O	User defined		Orifice, Restriction		Open
P	Pressure, Vacuum		Point (test connection)		
Q	Quantity	Integrate, Totalize			

SIDEBAR F Instrument Symbol Nomenclature Continued

	First letter		Succeeding letters		
Letter	Measured or initiating variable	Modifier	Readout or passive function	Output function	Modifier
R	Radiation		Record		
S	Speed, Frequency	Safety		Switch	
T	Temperature			Transmit	
U	Multivariable		Multifunction	Multifunction	Multifunction
V	Vibration, Mechanical			Valve, Damper, Louver	
W	Weight, Force		Well		
X	Unclassified	X Axis	Unclassified	Unclassified	Unclassified
Y	Event, State, Presence	Y Axis		Relay, Compute, Convert	
Z	Position, Dimension	Z Axis		Driver, Actuator, Unclassified final control element	

Notes: User defined letters are intended to define undefined instruments which are to be used repetitively on a single project and can be defined for use only on that project. Unclassified letters are intended to cover undefined instruments on a specific project which are to be used on a limited basis. Multifunctional letters are intended for use on undefined instruments and are optional.

Examples: To define a flow instrument with a local display which transmits a signal to a remote location, the proper designation is FIT-xxx, where xxx is the specific loop number. For a switch indicating open position, the proper designation is ZSO-xxx. A switch used to indicate a high level (of liquid, solid, or position) would be designated as LSH-xxx. Any primary measurement element in direct contact with the measured process uses the succeeding letter “E” (FE, LE, TE).

Testing and commissioning

Testing of fiber-optic cables is accomplished by several methods. The most basic is the pocket flashlight. This is a quick and dirty method for checking continuity or for "ringing out" a single fiber. A note of caution: if you are unaware of what is happening at the other end of the fiber, NEVER look into the end of a fiber-optic cable! One or more fibers may be carrying potentially harmful levels of light energy. Point the cable at the wall while testing.

Somewhat more sophisticated is the OTDR, or Optical Time Domain Reflectometer. The OTDR measures backscatter, or reflections, caused by a break or some other fault in the optical fiber. A laser pulse is injected into the cable being "short," and when a fault is detected, a portion of the pulse is reflected back to the OTDR. On well documented systems, the fault can pinpoint the location of any problems with the cable.

For periodic checkouts, the optical loss test set is used. It is basically a light source of specific or selectable frequency and power output and a receiver that gives an indication of optical power. The test set is inexpensive compared to the OTDR and is used to periodically check the performance of installed cables.

It is good practice to test optical cables before and after installation. Testing before will spot any problems developed in manufacture or shipping and will avoid pulling and then pulling back out in the event of a major failure. It also provides a "baseline" for later comparison. The cable should be tested after installation in all cases. This also provides a baseline and can spot a problem developed during installation.

Appendix A

RFC 791: Internet Protocol

The following document was prepared by the Defense Advanced Research Projects Agency (DARPA) for use in the development of software and equipment used on networks supporting the Internet. This document describes the Internet Protocol (IP) explained in Chapter 2. This document presents detailed information designed to allow full compliance across the various networks that support the Internet. The information is very technical and introduces many new concepts. The Internet Protocol is used to move information across interconnected networks and is responsible for virtually connecting nodes together to effect the transfer of data. The IP operates primarily at the Network layer of the OSI model.

Preface

This document specifies the DOD Standard Internet Protocol. This document is based on six earlier editions of the ARPA Internet Protocol Specification, and the present text draws heavily from them. There have been many contributors to this work both in terms of concepts and in terms of text. This edition revises aspects of addressing, error handling, option codes, and the security, precedence, compartments, and handling restriction features of the internet protocol.

Jon Postel
Editor
September 1981

1 Introduction

1.1 Motivation

The Internet Protocol is designed for use in interconnected systems of packet-switched computer communication networks. Such a system has been called a "catenet" [1]. The internet protocol provides for transmitting blocks of data called datagrams from sources to destinations, where sources and destinations are hosts identified by fixed length addresses. The internet protocol also provides for fragmentation and reassembly of long datagrams, if necessary, for transmission through "small packet" networks.

1.2 Scope

The internet protocol is specifically limited in scope to provide the functions necessary to deliver a package of bits (an internet datagram) from a source to a destination over an interconnected system of networks. There are no mechanisms to augment end-to-end data reliability, flow control, sequencing, or other services commonly found in host-to-host protocols. The internet protocol can capitalize on the services of its supporting networks to provide various types and qualities of service.

1.3 Interfaces

This protocol is called on by host-to-host protocols in an internet environment. This protocol calls on local network protocols to carry the internet datagram to the next gateway or destination host. For example, a TCP module would call on the internet module to take a TCP segment (including the TCP header and user data) as the data portion of an internet datagram. The TCP module would provide the addresses and other parameters in the internet header to the internet module as arguments of the call. The internet module would then create an internet datagram and call on the local network interface to transmit the internet datagram.

In the ARPANET case, for example, the internet module would call on a local net module which would add the 1822 leader [2] to the internet datagram creating an ARPANET message to transmit to the IMP. The ARPANET address would be derived from the internet address by the local network interface and would be the address of some host in the ARPANET, that host might be a gateway to other networks.

1.4 Operation

The internet protocol implements two basic functions: addressing and fragmentation.

The internet modules use the addresses carried in the internet header to transmit internet datagrams toward their destinations. The selection of a path for transmission is called routing.

The internet modules use fields in the internet header to fragment and reassemble internet datagrams when necessary for transmission through "small packet" networks.

The model of operation is that an internet module resides in each host engaged in internet communication and in each gateway that interconnects networks. These modules share common rules for interpreting address fields and for fragmenting and assembling internet datagrams. In addition, these modules (especially in gateways) have procedures for making routing decisions and other functions.

The internet protocol treats each internet datagram as an independent entity unrelated to any other internet datagram. There are no connections or logical circuits (virtual or otherwise). The internet protocol uses four key mechanisms in providing its service: Type of Service, Time to Live, Options, and Header Checksum.

The Type of Service is used to indicate the quality of the service desired. The type of service is an abstract or generalized set of parameters which characterize the service choices provided in the networks that make up the internet. This type of service indication is to be used by gateways to select the actual transmission parameters for a particular network, the network to be used for the next hop, or the next gateway when routing an internet datagram.

The Time to Live is an indication of an upper bound on the lifetime of an internet datagram. It is set by the sender of the datagram and reduced at the points along the route where it is processed. If the time to live reaches zero before the internet datagram reaches its destination, the internet datagram is destroyed. The time to live can be thought of as a self-destruct time limit.

The Options provide for control functions needed or useful in some situations but unnecessary for the most common communications. The options include provisions for timestamps, security, and special routing.

The Header Checksum provides a verification that the information used in processing internet datagram has been transmitted correctly. The data may contain errors. If the header checksum fails, the internet datagram is discarded at once by the entity which detects the error. The internet protocol does not provide a reliable communication facility. There are no acknowledgments either end-to-end or hop-by-hop. There is no error control for data, only a header checksum. There are no retransmissions. There is no flow control. Errors detected may be reported via the Internet Control Message Protocol (ICMP) [3] which is implemented in the internet protocol module.

2 Overview

2.1 Relation to Other Protocols

The following diagram illustrates the place of the internet protocol in the protocol hierarchy:

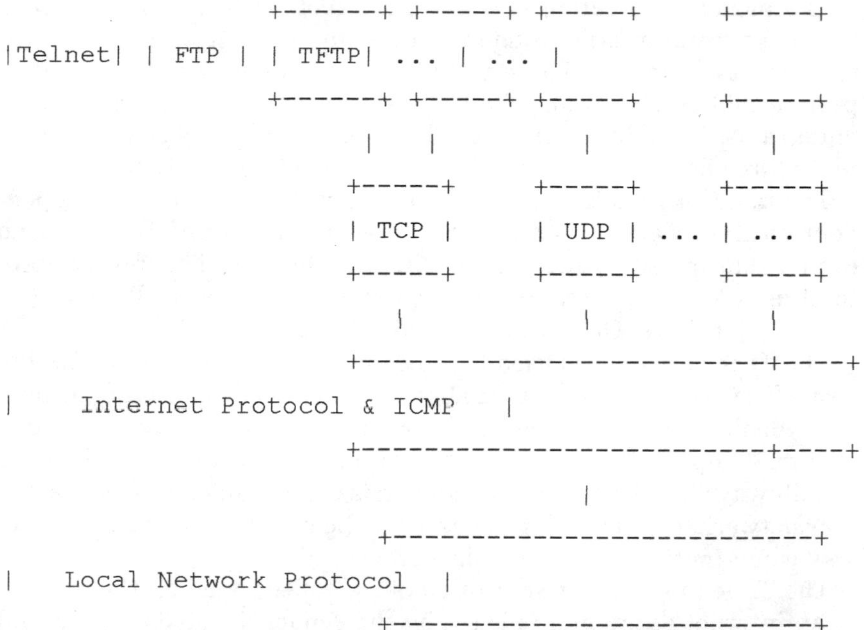

Figure 1 Protocol Relationships.

Internet protocol interfaces on one side to the higher level host-to-host protocols and on the other side to the local network protocol. In this context a "local network" may be a small network in a building or a large network such as the ARPANET.

2.2 Model of Operation

The model of operation for transmitting a datagram from one application program to another is illustrated by the following scenario:

We suppose that this transmission will involve one intermediate gateway.

The sending application program prepares its data and calls on its local internet module to send that data as a datagram and passes the destination address and other parameters as arguments of the call.

The internet module prepares a datagram header and attaches the data to it. The internet module determines a local network address for this internet address, in this case it is the address of a gateway. It sends this datagram and the local network address to the local network interface.

The local network interface creates a local network header, and attaches the datagram to it, then sends the result via the local network.

The datagram arrives at a gateway host wrapped in the local network header, the local network interface strips off this header, and turns the datagram over to the internet module. The internet module determines from the internet address that the datagram is to be forwarded to another host in a second network. The internet module determines a local net address for the destination host. It calls on the local network interface for that network to send the datagram.

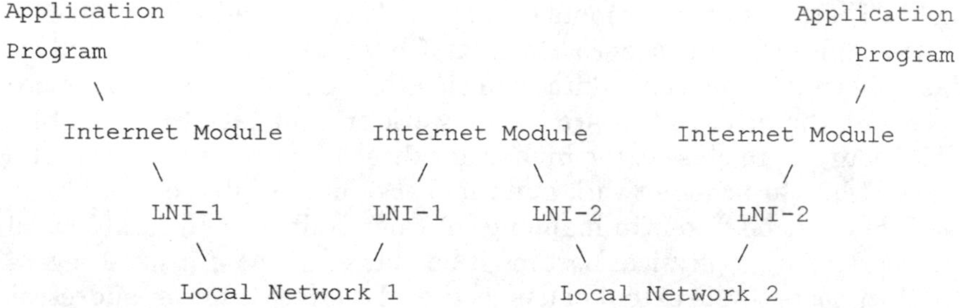

Figure 2 Transmission Path.

This local network interface creates a local network header and attaches the datagram sending the result to the destination host. At this destination host the datagram is stripped of the local net header by the local network interface and handed to the internet module.

The internet module determines that the datagram is for an application program in this host. It passes the data to the application program in response to a system call, passing the source address and other parameters as results of the call.

2.3 Function Description

The function or purpose of Internet Protocol is to move datagrams through an interconnected set of networks. This is done by passing the datagrams from one internet module to another until the destination is reached. The internet modules reside in hosts and gateways in the internet system. The datagrams are routed from one internet module to another through individual networks based on the interpretation of an internet address. Thus, one important mechanism of the internet protocol is the internet address.

In the routing of messages from one internet module to another, datagrams may need to traverse a network whose maximum packet size is smaller than the size of the datagram. To overcome this difficulty, a fragmentation mechanism is provided in the internet protocol.

Addressing. A distinction is made between names, addresses, and routes [4]. A name indicates what we seek. An address indicates where it is. A route indicates how to get there. The internet protocol deals primarily with addresses. It is the task of higher level (i.e., host-to-host or application) protocols to make the mapping from names to addresses. The internet module maps internet addresses to local net addresses. It is the task of lower level (i.e., local net or gateways) procedures to make the mapping from local net addresses to routes.

Addresses are fixed length of four octets (32 bits). An address begins with a network number, followed by local address (called the "rest" field). There are three formats or classes of internet addresses: in class a, the high-order bit is zero, the next 7 bits are the network, and the last 24 bits are the local address; in class b, the high-order two bits are one-zero, the next 14 bits are the network and the last 16 bits are the local address; in class c, the high-order three bits are one-one-zero, the next 21 bits are the network and the last 8 bits are the local address.

Care must be taken in mapping internet addresses to local net addresses; a single physical host must be able to act as if it were several distinct hosts to the extent of using several distinct internet addresses. Some hosts will also have several physical interfaces (multi-homing).

That is, provision must be made for a host to have several physical interfaces to the network with each having several logical internet addresses.

Examples of address mappings may be found in "Address Mappings" [5].

Fragmentation. Fragmentation of an internet datagram is necessary when it originates in a local net that allows a large packet size and must traverse a local net that limits packets to a smaller size to reach its destination.

An internet datagram can be marked "don't fragment." Any internet datagram so marked is not to be internet fragmented under any circumstances. If internet datagram marked "don't fragment" cannot be delivered to its destination without fragmenting it, it is to be discarded instead.

Fragmentation, transmission and reassembly across a local network which is invisible to the internet protocol module is called intranet fragmentation and may be used [6].

The internet fragmentation and reassembly procedure needs to be able to break a datagram into an almost arbitrary number of pieces that can be later reassembled. The receiver of the fragments uses the identification field to ensure that fragments of different datagrams are not mixed. The fragment offset field tells the receiver the position of a fragment in the original datagram. The fragment offset and length determine the portion of the original datagram covered by this fragment. The more-fragments flag indicates (by being reset) the last fragment. These fields provide sufficient information to reassemble datagrams.

The identification field is used to distinguish the fragments of one datagram from those of another. The originating protocol module of an internet datagram sets the identification field to a value that must be unique for that source-destination pair and protocol for the time the datagram will be active in the internet system. The originating protocol module of a complete datagram sets the more-fragments flag to zero and the fragment offset to zero.

To fragment a long internet datagram, an internet protocol module (for example, in a gateway), creates two new internet datagrams and copies the contents of the internet header fields from the long datagram into both new internet headers. The data of the long datagram is divided into two portions on an 8-octet (64 bit) boundary (the second portion might not be an integral multiple of 8 octets, but the first must be). Call the number of 8 octet blocks in the first portion NFB (for Number of Fragment Blocks). The first portion of the data is placed in the first new internet datagram, and the total length field is set to the length of the first datagram. The more-fragments flag is set to one. The second portion of the data is placed in the second new internet datagram, and the total length field is set to the length of the second datagram. The more-fragments flag carries the same value as the long datagram. The fragment offset field of the second new internet datagram is set to the value of that field in the long datagram plus NFB.

This procedure can be generalized for an n-way split, rather than the two-way split described.

To assemble the fragments of an internet datagram, an internet protocol module (for example at a destination host) combines internet datagrams that all have the same value for the four fields: identification, source, destination, and protocol. The combination is done by placing the data portion of each fragment in the relative position indicated by the fragment offset in that fragment's internet header. The first fragment will have the fragment offset zero, and the last fragment will have the more-fragments flag reset to zero.

2.4 Gateways

Gateways implement internet protocol to forward datagrams between networks. Gateways also implement the Gateway to Gateway Protocol (GGP) [7] to coordinate routing and other internet control information.

In a gateway the higher level protocols need not be implemented and the GGP functions are added to the IP module.

3 Specification

3.1 Internet Header Format

A summary of the contents of the internet header is shown below:

Note that each tick mark represents one bit position.

Version: 4 bits. The Version field indicates the format of the internet header. This document describes version 4.

IHL: 4 bits. Internet Header Length is the length of the internet header in 32-bit words, and thus points to the beginning of the data. Note that the minimum vlaue for a correct header is 5.

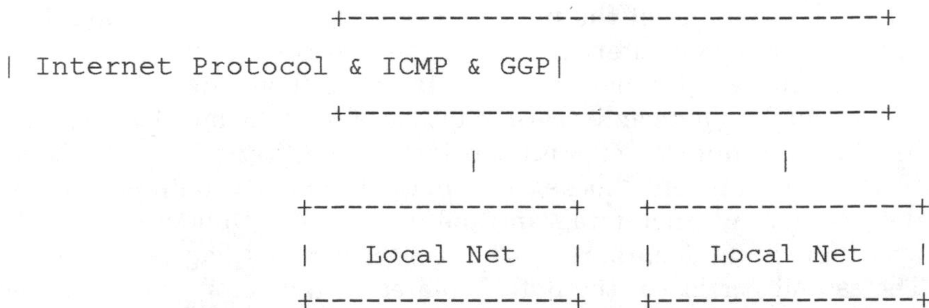

Figure 3 Gateway Protocols.

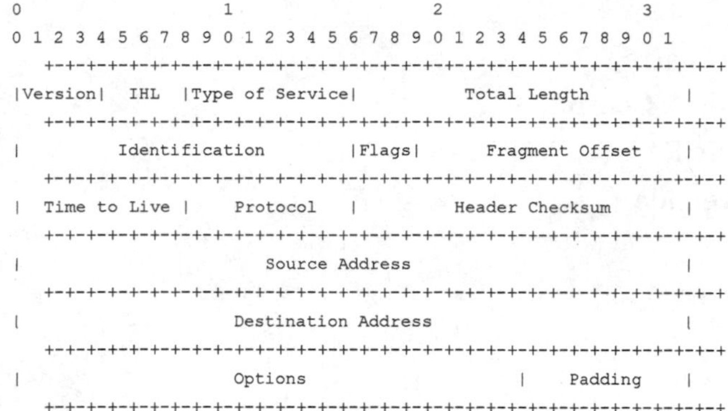

Figure 4 Example Internet Datagram Header.

Type of Service: 8 bits. The Type of Service provides an indication of the abstract parameters of the quality of service desired. These parameters are to be used to guide the selection of the actual service parameters when transmitting a datagram through a particular network. Several networks offer service precedence, which somehow treats high precedence traffic as more important than other traffic (generally by accepting only traffic above a certain precedence at time of high load). The major choice is a three-way tradeoff between low-delay, high-reliability, and high-throughput.

Bits 0-2: Precedence.

Bit 3: 0 = Normal Delay, 1 = Low Delay.

Bits 4: 0 = Normal Throughput, 1 = High Throughput.

Bits 5: 0 = Normal Reliability, 1 = High Reliability.

Bit 6–7: Reserved for Future Use.

```
0     1     2     3     4     5     6     7
      +-----+-----+-----+-----+-----+-----+-----+-----+
      |                 |     |     |     |     |     |
      |   PRECEDENCE    |  D  |  T  |  R  |  0  |  0  |
      |                 |     |     |     |     |     |
      +-----+-----+-----+-----+-----+-----+-----+-----+
```

Precedence

111—Network Control

110—Internetwork Control

101—CRITIC/ECP

100—Flash Override

011—Flash

010—Immediate

001—Priority

000—Routine

The use of the Delay, Throughput, and Reliability indications may increase the cost (in some sense) of the service. In many networks better performance for one of these parameters is coupled with worse performance on another. Except for very unusual cases at most two of these three indications should be set.

The type of service is used to specify the treatment of the datagram during its transmission through the internet system. Example mappings of the internet type of service to the actual service provided on networks such as AUTODIN II, ARPANET, SATNET, and PRNET is given in "Service Mappings" [8].

The Network Control precedence designation is intended to be used within a network only. The actual use and control of that designation is up to each network. The Internetwork Control designation is intended for use by gateway control originators only. If the actual use of these precedence designations is of concern to a particular network, it is the responsibility of that network to control the access to, and use of, those precedence designations.

Total Length: 16 bits. Total Length is the length of the datagram, measured in octets, including internet header and data. This field allows the length of a datagram to be up to 65,535 octets. Such long datagrams are impractical for most hosts and networks. All hosts must be prepared to accept datagrams of up to 576 octets (whether they arrive whole or in fragments). It is recommended that hosts only send datagrams larger than 576 octets if they have assurance that the destination is prepared to accept the larger datagrams.

The number 576 is selected to allow a reasonable sized data block to be transmitted in addition to the required header information. For example, this size allows a data block of 512 octets plus 64 header octets to fit in a dagram. The maximal internet header is 60 octets, and a typical internet header is 20 octets, allowing a margin for headers of higher level protocols.

Identification: 16 bits. An identifying value assigned by the sender to aid in assembling the fragments of a datagram.

Flags: 3 bits. Various Control Flags.

Bit 0: reserved, must be zero.

Bit 1: (DF) 0 = May Fragment, 1 = Don't Fragment.

Bit 2: (MF) 0 = Last Fragment, 1 = More Fragments.

```
0   1   2
        +---+---+---+
|   | D | M |
| 0 | F | F |
        +---+---+---+
```

Fragment Offset: 13 bits. This field indicates where in the datagram this fragment belongs. The fragment offset is measured in units of 8 octets (64 bits). The first fragment has offset zero.

Time to Live: 8 bits. This field indicates the maximum time the datagram is allowed to remain in the internet system. If this field contains the value zero, then the datagram must be destroyed. This field is modified in internet header processing. The time is measured in units of seconds, but since every module that processes a datagram must decrease the TTL by at least one even if it processes the datagram in less than a second, the TTL must be thought of only as an upper bound on the time a datagram may exist. The intention is to cause undeliverable datagrams to be discarded, and to bound the maximum datagram lifetime.

Protocol: 8 bits. This field indicates the next level protocol used in the data portion of the internet datagram. The values for various protocols are specified in "Assigned Numbers" [9].

Header Checksum: 16 bits. A checksum on the header only. Since some header fields change (e.g., time to live), this is recomputed and verified at each point that the internet header is processed.

The checksum algorithm is as follows. The checksum field is the 16 bit one's complement of the one's complement sum of all 16-bit words in the header. For purposes of computing the checksum, the value of the checksum field is zero. This is a simple to compute checksum and experimental evidence indicates it is adequate, but it is provisional and may be replaced by a CRC procedure, depending on further experience.

Source Address: 32 bits. The source address. See section 3.2.

Destination Address: 32 bits. The destination address. See section 3.2.

Options: variable. The options may appear or not in datagrams. They must be implemented by all IP modules (host and gateways). What is optional is their transmission in any particular datagram, not their implementation.

In some environments the security option may be required in all datagrams.

The option field is variable in length. There may be zero or more options. There are two cases for the format of an option:

Case 1: A single octet of option-type.

Case 2: An option-type octet, an option-length octet, and the actual option-data octets.

The option-length octet counts the option-type octet and the option-length octet as well as the option-data octets.

The option-type octet is viewed as having 3 fields:

1 bit: copied flag,

2 bits: option class,

5 bits: option number.

The copied flag indicates that this option is copied into all fragments on fragmentation.

0 = not copied

1 = copied

The option classes are:

0 = control

1 = reserved for future use

2 = debugging and measurement

3 = reserved for future use

The following internet options are defined:

CLASS	NUMBER	LENGTH	DESCRIPTION
0	0	—	End of Option list. This option occupies only 1 octet; it has no length octet.
0	1	—	No Operation. This option occupies only 1 octet; it has no length octet.
0	2	11	Security. Used to carry Security, Compartmentation, User Group (TCC), and Handling Restriction Codes compatible with DOD requirements.
0	3	var.	Loose Source Routing. Used to route the internet datagram based on information supplied by the source.
0	9	var.	Strict Source Routing. Used to route the internet datagram based on information supplied by the source.
0	7	var.	Record Route. Used to trace the route an internet datagram takes.
0	8	4	Stream ID. Used to carry the stream identifier.
2	4	var.	Internet Timestamp.

Specific Option Definitions.

End of Option List. This option indicates the end of the option list. This might not coincide with the end of the internet header according to the internet header length. This is used at the end of all options, not the end of each option, and need only be used if the end of the options would not otherwise coincide with the end of the internet header. It may be copied, introduced, or deleted on fragmentation, or for any other reason.

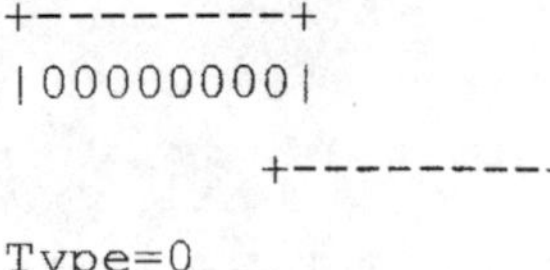
```
+--------+
|00000000|
+--------+
Type=0
```

No Operation. This option may be used between options, for example, to align the beginning of a subsequent option on a 32-bit boundary. May be copied, introduced, or deleted on fragmentation, or for any other reason.

```
+--------+
|00000001|
+--------+
Type=1
```

Security. This option provides a way for hosts to send security, compartmentation, handling restrictions, and TCC (closed user group) parameters. The format for this option is as follows:

```
        +--------+--------+---//---+---//---+---//---+---//---+
|10000010|00001011|SSS  SSS|CCC  CCC|HHH  HHH|  TCC   |
        +--------+--------+---//---+---//---+---//---+---//---+
Type=130 Length=11
```

Security (S field): 16 bits. Specifies one of 16 levels of security (eight of which are reserved for future use).

00000000 00000000—Unclassified

11110001 00110101—Confidential

01111000 10011010—EFTO

10111100 01001101—MMMM

01011110 00100110—PROG

10101111 00010011—Restricted

11010111 10001000—Secret

01101011 11000101—Top Secret

00110101 11100010—(Reserved for future use)

10011010 11110001—(Reserved for future use)

01001101 01111000—(Reserved for future use)

00100100 10111101—(Reserved for future use)

00010011 01011110—(Reserved for future use)

10001001 10101111—(Reserved for future use)

11000100 11010110—(Reserved for future use)

11100010 01101011—(Reserved for future use)

Compartments (C field): 16 bits. An all-zero value is used when the information transmitted is not compartmented. Other values for the compartments field may be obtained from the Defense Intelligence Agency.

Handling Restrictions (H field): 16 bits. The values for the control and release markings are alphanumeric digraphs and are defined in the Defense Intelligence Agency Manual DIAM 65–19, "Standard Security Markings."

Transmission Control Code (TCC field): 24 bits. Provides a means to segregate traffic and define controlled communities of interest among subscribers. The TCC values are trigraphs, and are available from HQ DCA Code 530. Must be copied on fragmentation. This option appears at most once in a datagram.

Loose Source and Record Route. The loose source and record route (LSRR) option provides a means for the source of an internet datagram to supply routing information to be used by the gateways in forwarding the datagram to the destination, and to record the route information.

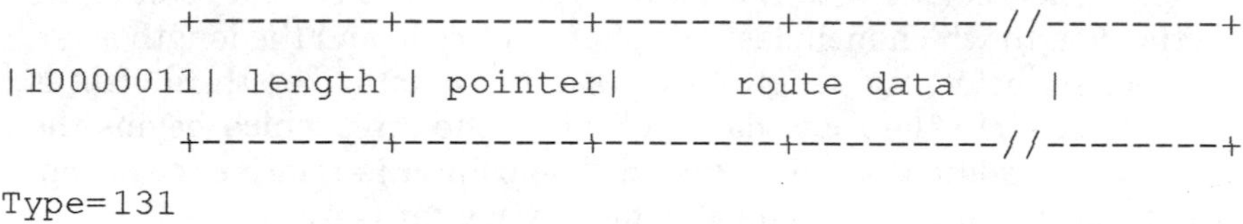

```
        +--------+--------+--------+---------//--------+
|10000011| length | pointer|     route data    |
        +--------+--------+--------+---------//--------+
Type=131
```

The option begins with the option type code. The second octet is the option length which includes the option type code and the length octet, the pointer octet, and length-3 octets of route data. The third octet is the pointer into the route data indicating the octet which begins the next source address to be processed. The pointer is relative to this option, and the smallest legal value for the pointer is 4.

A route data is composed of a series of internet addresses. Each internet address is 32 bits or 4 octets. If the pointer is greater than the length, the source route is empty (and the recorded route full) and the routing is to be based on the destination address field.

If the address in destination address field has been reached and the pointer is not greater than the length, the next address in the source route replaces the address in the destination address field, and the recorded route address replaces the source address just used, and pointer is increased by four.

The recorded route address is the internet module's own internet address as known in the environment into which this datagram is being forwarded.

This procedure of replacing the source route with the recorded route (though it is in the reverse of the order it must be in to be used as a source route) means the option (and the IP header as a whole) remains a constant length as the datagram progresses through the internet.

This option is a loose source route because the gateway or host IP is allowed to use any route of any number of other intermediate gateways to reach the next address in the route. Must be copied on fragmentation. Appears at most once in a datagram.

Strict Source and Record Route. The strict source and record route (SSRR) option provides a means for the source of an internet datagram to supply routing information to be used by the gateways in forwarding the datagram to the destination, and to record the route information.

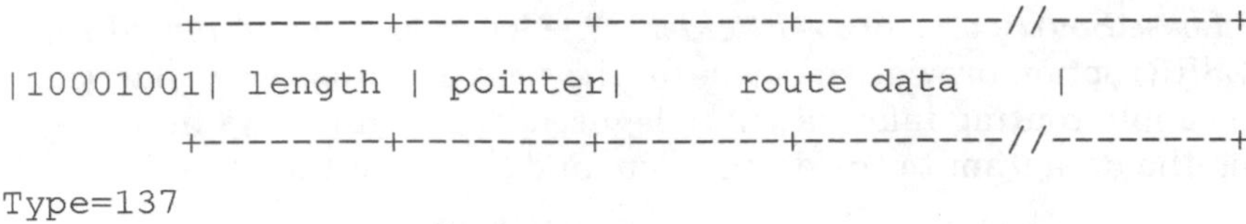

The option begins with the option type code. The second octet is the option length which includes the option type code and the length octet, the pointer octet, and length-3 octets of route data. The third octet is the pointer into the route data indicating the octet which begins the next source address to be processed. The pointer is relative to this option, and the smallest legal value for the pointer is 4.

A route data is composed of a series of internet addresses. Each internet address is 32 bits or 4 octets. If the pointer is greater than the length, the source route is empty (and the recorded route full) and the routing is to be based on the destination address field.

If the address in destination address field has been reached and the pointer is not greater than the length, the next address in the source route replaces the address in the destination address field, and the recorded route address replaces the source address just used, and pointer is increased by four.

The recorded route address is the internet module's own internet address as known in the environment into which this datagram is being forwarded.

This procedure of replacing the source route with the recorded route (though it is in the reverse of the order it must be in to be used as a source route) means the option (and the IP header as a whole) remains a constant length as the datagram progresses through the internet.

This option is a strict source route because the gateway or host IP must send the datagram directly to the next address in the source route through only the directly connected network indicated in the next address to reach the next gateway or host specified in the route.

Must be copied on fragmentation. Appears at most once in a datagram.

Record Route. The record route option provides a means to record the route of an internet datagram.

The option begins with the option type code. The second octet is the option length which includes the option type code and the length octet,

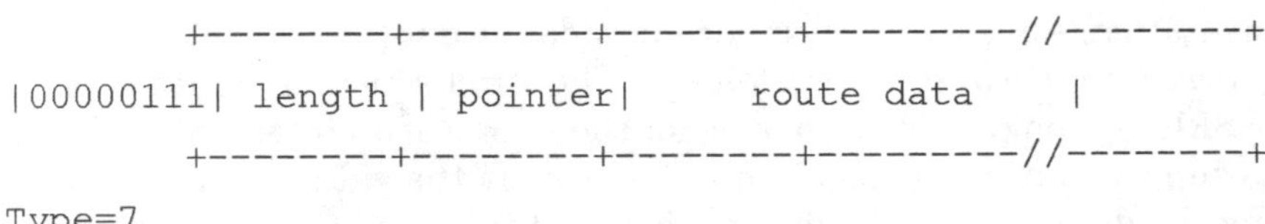

the pointer octet, and length-3 octets of route data. The third octet is the pointer into the route data indicating the octet which begins the next area to store a route address. The pointer is relative to this option, and the smallest legal value for the pointer is 4.

A recorded route is composed of a series of internet addresses. Each internet address is 32 bits or 4 octets. If the pointer is greater than the length, the recorded route data area is full. The originating host must compose this option with a large enough route data area to hold all the addresses expected. The size of the option does not change due to adding addresses. The initial contents of the route data area must be zero.

When an internet module routes a datagram it checks to see if the record route option is present. If it is, it inserts its own internet address as known in the environment into which this datagram is being forwarded into the recorded route beginning at the octet indicated by the pointer, and increments the pointer by four.

If the route data area is already full (the pointer exceeds the length) the datagram is forwarded without inserting the address into the recorded route. If there is some room but not enough room for a full address to be inserted, the original datagram is considered to be in error and is discarded. In either case an ICMP parameter problem message may be sent to the source host [3].

Not copied on fragmentation, goes in first fragment only. Appears at most once in a datagram.

Stream Identifier. This option provides a way for the 16-bit SATNET stream identifier to be carried through networks that do not support the stream concept.

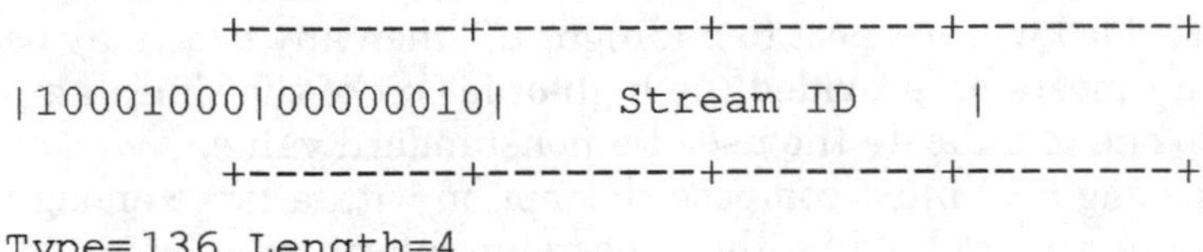

Must be copied on fragmentation. Appears at most once in a datagram.

Internet Timestamp. The Option Length is the number of octets in the option counting the type, length, pointer, and overflow/flag octets (maximum length 40).

The Pointer is the number of octets from the beginning of this option to the end of timestamps plus one (i.e., it points to the octet beginning the space for next timestamp). The smallest legal value is 5. The timestamp area is full when the pointer is greater than the length.

The Overflow (oflw) [4 bits] is the number of IP modules that cannot register timestamps due to lack of space.

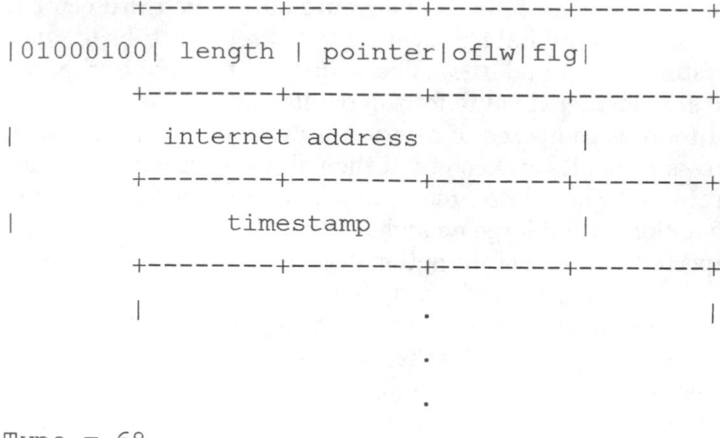

The Flag (flag) [4 bits] values are

0—time stamps only, stored in consecutive 32-bit words,

1—each timestamp is preceded with internet address of the registering entity,

3—the internet address fields are prespecified. An IP module only registers its timestamp if it matches its own address with the next specified internet address.

The Timestamp is a right-justified, 32-bit timestamp in milliseconds since midnight UT. If the time is not available in milliseconds or cannot be provided with respect to midnight UT then any time may be inserted as a timestamp provided the high-order bit of the timestamp field is set to one to indicate the use of a nonstandard value.

The originating host must compose this option with a large enough timestamp data area to hold all the timestamp information expected. The size of the option does not change due to adding timestamps. The initial contents of the timestamp data area must be zero or internet address/zero pairs.

If the timestamp data area is already full (the pointer exceeds the length) the datagram is forwarded without inserting the timestamp, but the overflow count is incremented by one. If there is some room but not enough room for a full timestamp to be inserted, or the overflow count itself overflows, the original datagram is considered to be in error and is discarded. In either case an ICMP parameter problem message may be sent to the source host [3].

The timestamp option is not copied upon fragmentation. It is carried in the first fragment. Appears at most once in a datagram.

Padding: variable. The internet header padding is used to ensure that the internet header ends on a 32-bit boundary. The padding is zero.

3.2 Discussion

The implementation of a protocol must be robust. Each implementation must expect to interoperate with others created by different individuals. While the goal of this specification is to be explicit about the protocol there is the possibility of differing interpretations. In general, an implementation must be conservative in its sending behavior, and liberal in its receiving behavior. That is, it must be careful to send well-formed datagrams, but must accept any datagram that it can interpret (e.g., not object to technical errors where the meaning is still clear).

The basic internet service is datagram oriented and provides for the fragmentation of datagrams at gateways, with reassembly taking place at the destination internet protocol module in the destination host. Of course, fragmentation and reassembly of datagrams within a network or by private agreement between the gateways of a network is also allowed since this is transparent to the internet protocols and the higher level protocols. This transparent type of fragmentation and reassembly is termed "network-dependent" (or intranet) fragmentation and is not discussed further here.

Internet addresses distinguish sources and destinations to the host level and provide a protocol field as well. It is assumed that each protocol will provide for whatever multiplexing is necessary within a host.

Addressing. To provide for flexibility in assigning addresses to networks and to allow for the large number of small- to intermediate-sized networks, the interpretation of the address field is coded to specify a small number of networks with a large number of hosts, a moderate number of networks with a moderate number of hosts, and a large number of networks with a small number of hosts. In addition there is an escape code for extended addressing mode.

Address Formats

High Order Bits	Format	Class
0	7 bits of net, 24 bits of host	a
10	14 bits of net, 16 bits of host	b
110	21 bits of net, 8 bits of host	c
111	escape to extended addressing mode	

A value of zero in the network field means this network. This is only used in certain ICMP messages. The extended addressing mode is undefined. Both of these features are reserved for future use. The actual values assigned for network addresses is given in “Assigned Numbers” [9].

The local address, assigned by the local network, must allow for a single physical host to act as several distinct internet hosts. That is, there must be a mapping between internet host addresses and network/host interfaces that allows several internet addresses to correspond to one interface. It must also be allowed for a host to have several physical interfaces and to treat the datagrams from several of them as if they were all addressed to a single host. Address mappings between internet addresses and addresses for ARPANET, SATNET, PRNET, and other networks are described in “Address Mappings” [5].

Fragmentation and Reassembly. The internet identification field (ID) is used together with the source and destination address, and the protocol fields, to identify datagram fragments for reassembly.

The More Fragments flag bit (MF) is set if the datagram is not the last fragment. The Fragment Offset field identifies the fragment location, relative to the beginning of the original unfragmented datagram. Fragments are counted in units of 8 octets. The fragmentation strategy is designed so that an unfragmented datagram has all zero fragmentation information (MF = 0, fragment offset = 0). If an internet datagram is fragmented, its data portion must be broken on 8-octet boundaries.

This format allows 2**13 = 8192 fragments of 8 octets each for a total of 65,536 octets. Note that this is consistent with the datagram total length field (of course, the header is counted in the total length and not in the fragments).

When fragmentation occurs, some options are copied, but others remain with the first fragment only.

Every internet module must be able to forward a datagram of 68 octets without further fragmentation. This is because an internet header may be up to 60 octets, and the minimum fragment is 8 octets. Every internet destination must be able to receive a datagram of 576 octets either in one piece or in fragments to be reassembled.

The fields which may be affected by fragmentation include:

1. options field
2. more fragments flag
3. fragment offset
4. internet header length field
5. total length field
6. header checksum

If the Don't Fragment flag (DF) bit is set, then internet fragmentation of this datagram is NOT permitted, although it may be discarded. This can be used to prohibit fragmentation in cases where the receiving host does not have sufficient resources to reassemble internet fragments.

One example of use of the Don't Fragment feature is to down line load a small host. A small host could have a boot strap program that accepts a datagram, stores it in memory, and then executes it. The fragmentation and reassembly procedures are most easily described by examples. The following procedures are example implementations.

General notation in the following pseudo programs: "=<" means "less than or equal", "#" means "not equal", "=" means "equal", "<–" means "is set to". Also, "x to y" includes x and excludes y; for example, "4 to 7" would include 4, 5, and 6 (but not 7).

AN EXAMPLE FRAGMENTATION PROCEDURE. The maximum sized datagram that can be transmitted through the next network is called the maximum transmission unit (MTU). If the total length is less than or equal the maximum transmission unit then submit this datagram to the next step in datagram processing; otherwise cut the datagram into two fragments, the first fragment being the maximum size, and the second fragment being the rest of the datagram. The first fragment is submitted to the next step in datagram processing, while the second fragment is submitted to this procedure in case it is still too large.

Notation:

FO—Fragment Offset

IHL—Internet Header Length

DF—Don't Fragment flag

MF—More Fragments flag

TL—Total Length

OFO—Old Fragment Offset

OIHL—Old Internet Header Length

OMF—Old More Fragments flag

OTL—Old Total Length

NFB—Number of Fragment Blocks

MTU—Maximum Transmission Unit

Procedure:

IF TL =< MTU THEN Submit this datagram to the next step in datagram processing ELSE IF DF = 1 THEN discard the datagram ELSE

To produce the first fragment:

1. Copy the original internet header;
2. OIHL <– IHL; OTL <– TL; OFO <– FO; OMF <– MF;
3. NFB <– (MTU-IHL*4)/8;
4. Attach the first NFB*8 data octets;
5. Correct the header:
   MF <– 1; TL <– (IHL*4) + (NFB*8);
   Recompute Checksum;
6. Submit this fragment to the next step in datagram processing;
   To produce the second fragment:
7. Selectively copy the internet header (some options are not copied, see option definitions);
8. Append the remaining data;
9. Correct the header;
   IHL <– (((OIHL*4) – (length of options not copied))+3)/4;
   TL <– OTL – NFB*8 – (OIHL-IHL)*4);
   FO <– OFO + NFB; MF <– OMF; Recompute Checksum;
10. Submit this fragment to the fragmentation test; DONE.

In the above procedure each fragment (except the last) was made the maximum allowable size. An alternative might produce less than the maximum size datagrams. For example, one could implement a fragmentation procedure that repeatedly divided large datagrams in half until the resulting fragments were less than the maximum transmission unit size.

AN EXAMPLE REASSEMBLY PROCEDURE. For each datagram the buffer identifier is computed as the concatenation of the source, destination, protocol, and identification fields. If this is a whole datagram (that is, both the fragment offset and the more fragments fields are zero), then any reassembly resources associated with this buffer identifier are released and the datagram is forwarded to the next step in datagram processing.

If no other fragment with this buffer identifier is on hand then reassembly resources are allocated. The reassembly resources consist of a data buffer, a header buffer, a fragment block bit table, a total data length field, and a timer. The data from the fragment is placed in the data buffer according to its fragment offset and length, and bits are set in the fragment block bit table corresponding to the fragment blocks received.

If this is the first fragment (that is the fragment offset is zero) this header is placed in the header buffer. If this is the last fragment (that is, the "more fragments" field is zero) the total data length is computed. If this fragment completes the datagram (tested by checking the bits set in the fragment block table), then the datagram is sent to the next step in datagram processing; otherwise the timer is set to the maximum of the current timer value and the value of the "time to live" field from this fragment; and the reassembly routine gives up control. If the timer runs out, then all reassembly resources for this buffer identifier are released. The initial setting of the timer is a lower bound on the reassembly waiting time. This is because the waiting time will be increased if the time to live in the arriving fragment is greater than the current timer value but will not be decreased if it is less. The maximum this timer value could reach is the maximum time to live (approximately 4.25 minutes). The current recommendation for the initial timer setting is 15 seconds. This may be changed as experience with this protocol accumulates. Note that the choice of this parameter value is related to the buffer capacity available and the data rate of the transmission medium; that is, data rate times timer value equals buffer size (e.g., 10kb/s × 15s = 150 Kb).

Notation:

FO—Fragment Offset

IHL—Internet Header Length

MF—More Fragments flag

TTL—Time To Live

NFB—Number of Fragment Blocks

TL—Total Length

TDL—Total Data Length

BUFID—Buffer Identifier

RCVBT—Fragment Received Bit Table

TLB—Timer Lower Bound

Procedure:

1. BUFID <− source | destination | protocol | identification;
2. IF FO = 0 AND MF = 0
3. THEN IF buffer with BUFID is allocated
4. THEN flush all reassembly for this BUFID;
5. Submit datagram to next step; DONE.
6. ELSE IF no buffer with BUFID is allocated
7. THEN allocate reassembly resources with BUFID;
   TIMER <− TLB; TDL <− 0;
8. put data from fragment into data buffer with BUFID from octet FO*8 to octet (TL-(IHL*4)) +FO*8;

9. set RCVBT bits from FO to FO+ ((TL- (IHL*4)+7)/8);
10. IF MF = 0 THEN TDL <– TL – (IHL*4)+(FO*8)
11. IF FO = 0 THEN put header in header buffer
12. IF TDL # 0
13. AND all RCVBT bits from 0 to (TDL+7)/8 are set
14. THEN TL <– TDL+(IHL*4)
15. Submit datagram to next step;
16. free all reassembly resources for this BUFID; DONE.
17. TIMER <– MAX (TIMER,TTL);
18. give up until next fragment or timer expires;
19. timer expires: flush all reassembly with this BUFID; DONE.

In the case that two or more fragments contain the same data either identically or through a partial overlap, this procedure will use the more recently arrived copy in the data buffer and datagram delivered.

Identification. The choice of the Identifier for a datagram is based on the need to provide a way to uniquely identify the fragments of a particular datagram. The protocol module assembling fragments judges fragments to belong to the same datagram if they have the same source, destination, protocol, and identifier. Thus, the sender must choose the identifier to be unique for this source, destination pair, and protocol for the time the datagram (or any fragment of it) could be alive in the internet.

It seems then that a sending protocol module needs to keep a table of identifiers, one entry for each destination it has communicated with in the last maximum packet lifetime for the internet.

However, since the Identifier field allows 65,536 different values, some host may be able to simply use unique identifiers independent of destination.

It is appropriate for some higher level protocols to choose the identifier. For example, TCP protocol modules may retransmit an identical TCP segment, and the probability for correct reception would be enhanced if the retransmission carried the same identifier as the original transmission since fragments of either datagram could be used to construct a correct TCP segment.

Type of Service. The type of service (TOS) is for internet service quality selection. The type of service is specified along the abstract parameters precedence, delay, throughput, and reliability. These abstract parameters are to be mapped into the actual service parameters of the particular networks the datagram traverses.

Precedence. An independent measure of the importance of this datagram.

Delay. Prompt delivery is important for datagrams with this indication.

Throughput. High data rate is important for datagrams with this indication.

Reliability. A higher level of effort to ensure delivery is important for datagrams with this indication.

For example, the ARPANET has a priority bit, and a choice between "standard" messages (type 0) and "uncontrolled" messages (type 3), (the choice between single packet and multipacket messages can also be considered a service parameter). The uncontrolled messages tend to be less reliably delivered and suffer less delay. Suppose an internet datagram is to be sent through the ARPANET. Let the internet type of service be given as:

Precedence: 5

Delay: 0

Throughput: 1

Reliability: 1

In this example, the mapping of these parameters to those available for the ARPANET would be to set the ARPANET priority bit on since the Internet precedence is in the upper half of its range, to select standard messages since the throughput and reliability requirements are indicated and delay is not. More details are given on service mappings in "Service Mappings" [8].

Time to Live. The time to live is set by the sender to the maximum time the datagram is allowed to be in the internet system. If the datagram is in the internet system longer than the time to live, then the datagram must be destroyed.

This field must be decreased at each point that the internet header is processed to reflect the time spent processing the datagram. Even if no local information is available on the time actually spent, the field must be decremented by 1. The time is measured in units of seconds (i.e., the value 1 means one second). Thus, the maximum time to live is 255 seconds or 4.25 minutes. Since every module that processes a datagram must decrease the TTL by at least one even if it processes the datagram in less than a second, the TTL must be thought of only as an upper bound on the time a datagram may exist. The intention is to cause undeliverable datagrams to be discarded, and to bound the maximum datagram lifetime.

Some higher level reliable connection protocols are based on assumptions that old duplicate datagrams will not arrive after a certain time elapses. The TTL is a way for such protocols to have an assurance that their assumption is met.

Options. The options are optional in each datagram, but required in implementations. That is, the presence or absence of an option is the choice of the sender, but each internet module must be able to parse every option. There can be several options present in the option field.

The options might not end on a 32-bit boundary. The internet header must be filled out with octets of zeros. The first of these would be interpreted as the end-of-options option, and the remainder as internet header padding.

Every internet module must be able to act on every option. The Security Option is required if classified, restricted, or compartmented traffic is to be passed.

Checksum. The internet header checksum is recomputed if the internet header is changed. For example, a reduction of the time to live, additions or changes to internet options, or due to fragmentation. This checksum at the internet level is intended to protect the internet header fields from transmission errors.

There are some applications where a few data bit errors are acceptable while retransmission delays are not. If the internet protocol enforced data correctness such applications could not be supported.

Errors. Internet protocol errors may be reported via the ICMP messages [3].

3.3 Interfaces

The functional description of user interfaces to the IP is, at best, fictional, since every operating system will have different facilities. Consequently, we must warn readers that different IP implementations may have different user interfaces. However, all IPs must provide a certain minimum set of services to guarantee that all IP implementations can support the same protocol hierarchy. This section specifies the functional interfaces required of all IP implementations.

Internet protocol interfaces on one side to the local network and on the other side to either a higher level protocol or an application program. In the following, the higher level protocol or application program (or even a gateway program) will be called the "user" since it is using the internet module. Since internet protocol is a datagram protocol, there is minimal memory or state maintained between datagram transmissions, and each call on the internet protocol module by the user supplies all information necessary for the IP to perform the service requested.

An Example Upper Level Interface

The following two example calls satisfy the requirements for the user to internet protocol module communication ("=>" means returns):

SEND (src, dst, prot, TOS, TTL, BufPTR, len, Id, DF, opt => result)

where:

src = source address
dst = destination address
prot = protocol
TOS = type of service
TTL = time to live
BufPTR = buffer pointer
len = length of buffer
Id = Identifier
DF = Don't Fragment
opt = option data
result = response
OK = datagram sent ok
Error = error in arguments or local network error

Note that the precedence is included in the TOS and the security/compartment is passed as an option.

RECV (BufPTR, prot, => result, src, dst, TOS, len, opt)

where:

BufPTR = buffer pointer
prot = protocol
result = response
OK = datagram received ok
Error = error in arguments
len = length of buffer
src = source address
dst = destination address
TOS = type of service
opt = option data

When the user sends a datagram, it executes the SEND call supplying all the arguments. The internet protocol module, on receiving this call, checks the arguments and prepares and sends the message. If the arguments are good and the datagram is accepted by the local network, the call returns successfully. If either the arguments are bad, or the datagram is not accepted by the local network, the call returns unsucessfully. On unsuccessful returns, a reasonable report must be made as to the cause of the problem, but the details of such reports are up to individual implementations.

When a datagram arrives at the internet protocol module from the local network, either there is a pending RECV call from the user addressed or there is not. In the first case, the pending call is satisfied by passing the information from the datagram to the user. In the second case, the user addressed is notified of a pending datagram. If the user addressed does not exist, an ICMP error message is returned to the sender, and the data is discarded.

The notification of a user may be via a pseudo interrupt or similar mechanism, as appropriate in the particular operating system environment of the implementation.

A user's RECV call may then either be immediately satisfied by a pending datagram, or the call may be pending until a datagram arrives. The source address is included in the send call in case the sending host has several addresses (multiple physical connections or logical addresses). The internet module must check to see that the source address is one of the legal addresses for this host.

An implementation may also allow or require a call to the internet module to indicate interest in or reserve exclusive use of a class of datagrams (e.g., all those with a certain value in the protocol field).

This section functionally characterizes a USER/IP interface. The notation used is similar to most procedure of function calls in high level languages, but this usage is not meant to rule out trap type service calls (e.g., SVCs, UUOs, EMTs), or any other form of interprocess communication.

APPENDIX A: Examples & Scenarios

Example 1.

This is an example of the minimal data carrying internet datagram (Figure 5).

Note that each tick mark represents one bit position.

This is a internet datagram in version 4 of internet protocol; the internet header consists of five 32-bit words, and the total length of the datagram is 21 octets. This datagram is a complete datagram (not a fragment).

Example 2.

In this example (Figure 6), we show first a moderate sized internet datagram (452 data octets), then two internet fragments that might result from the fragmentation of this datagram if the maximum sized transmission allowed were 280 octets.

```
 0                   1                   2                   3
 0 1 2 3 4 5 6 7 8 9 0 1 2 3 4 5 6 7 8 9 0 1 2 3 4 5 6 7 8 9 0 1
+-+-+-+-+-+-+-+-+-+-+-+-+-+-+-+-+-+-+-+-+-+-+-+-+-+-+-+-+-+-+-+-+
|Ver= 4 |IHL= 5 |Type of Service|        Total Length = 21      |
+-+-+-+-+-+-+-+-+-+-+-+-+-+-+-+-+-+-+-+-+-+-+-+-+-+-+-+-+-+-+-+-+
|      Identification = 111     |Flg=0|   Fragment Offset = 0   |
+-+-+-+-+-+-+-+-+-+-+-+-+-+-+-+-+-+-+-+-+-+-+-+-+-+-+-+-+-+-+-+-+
|   Time = 123  |  Protocol = 1 |        header checksum        |
+-+-+-+-+-+-+-+-+-+-+-+-+-+-+-+-+-+-+-+-+-+-+-+-+-+-+-+-+-+-+-+-+
|                         source address                        |
+-+-+-+-+-+-+-+-+-+-+-+-+-+-+-+-+-+-+-+-+-+-+-+-+-+-+-+-+-+-+-+-+
|                      destination address                      |
+-+-+-+-+-+-+-+-+-+-+-+-+-+-+-+-+-+-+-+-+-+-+-+-+-+-+-+-+-+-+-+-+
|     data      |
+-+-+-+-+-+-+-+-+
```

Figure 5 Example Internet Datagram.

```
 0                   1                   2                   3
 0 1 2 3 4 5 6 7 8 9 0 1 2 3 4 5 6 7 8 9 0 1 2 3 4 5 6 7 8 9 0 1
+-+-+-+-+-+-+-+-+-+-+-+-+-+-+-+-+-+-+-+-+-+-+-+-+-+-+-+-+-+-+-+-+
|Ver= 4 |IHL= 5 |Type of Service|       Total Length = 472      |
+-+-+-+-+-+-+-+-+-+-+-+-+-+-+-+-+-+-+-+-+-+-+-+-+-+-+-+-+-+-+-+-+
|     Identification = 111      |Flg=0|     Fragment Offset = 0 |
+-+-+-+-+-+-+-+-+-+-+-+-+-+-+-+-+-+-+-+-+-+-+-+-+-+-+-+-+-+-+-+-+
|   Time = 123  | Protocol = 6  |        header checksum        |
+-+-+-+-+-+-+-+-+-+-+-+-+-+-+-+-+-+-+-+-+-+-+-+-+-+-+-+-+-+-+-+-+
|                         source address                        |
+-+-+-+-+-+-+-+-+-+-+-+-+-+-+-+-+-+-+-+-+-+-+-+-+-+-+-+-+-+-+-+-+
|                      destination address                      |
+-+-+-+-+-+-+-+-+-+-+-+-+-+-+-+-+-+-+-+-+-+-+-+-+-+-+-+-+-+-+-+-+
|                             data                              |
+-+-+-+-+-+-+-+-+-+-+-+-+-+-+-+-+-+-+-+-+-+-+-+-+-+-+-+-+-+-+-+-+
|                             data                              |
\                                                               \
\                                                               \
|                             data                              |
+-+-+-+-+-+-+-+-+-+-+-+-+-+-+-+-+-+-+-+-+-+-+-+-+-+-+-+-+-+-+-+-+
|             data              |
+-+-+-+-+-+-+-+-+-+-+-+-+-+-+-+-+
```

Figure 6 Example Internet Datagram.

Now the first fragment that results from splitting the datagram after 256 data octets.

```
 0                   1                   2                   3
 0 1 2 3 4 5 6 7 8 9 0 1 2 3 4 5 6 7 8 9 0 1 2 3 4 5 6 7 8 9 0 1
+-+-+-+-+-+-+-+-+-+-+-+-+-+-+-+-+-+-+-+-+-+-+-+-+-+-+-+-+-+-+-+-+
|Ver= 4 |IHL= 5 |Type of Service|       Total Length = 276      |
+-+-+-+-+-+-+-+-+-+-+-+-+-+-+-+-+-+-+-+-+-+-+-+-+-+-+-+-+-+-+-+-+
|     Identification = 111      |Flg=1|     Fragment Offset = 0 |
+-+-+-+-+-+-+-+-+-+-+-+-+-+-+-+-+-+-+-+-+-+-+-+-+-+-+-+-+-+-+-+-+
|   Time = 119  | Protocol = 6  |        Header Checksum        |
+-+-+-+-+-+-+-+-+-+-+-+-+-+-+-+-+-+-+-+-+-+-+-+-+-+-+-+-+-+-+-+-+
|                        source address                         |
+-+-+-+-+-+-+-+-+-+-+-+-+-+-+-+-+-+-+-+-+-+-+-+-+-+-+-+-+-+-+-+-+
|                      destination address                      |
+-+-+-+-+-+-+-+-+-+-+-+-+-+-+-+-+-+-+-+-+-+-+-+-+-+-+-+-+-+-+-+-+
|                             data                              |
+-+-+-+-+-+-+-+-+-+-+-+-+-+-+-+-+-+-+-+-+-+-+-+-+-+-+-+-+-+-+-+-+
|                             data                              |
\                                                               \
\                                                               \
|                             data                              |
+-+-+-+-+-+-+-+-+-+-+-+-+-+-+-+-+-+-+-+-+-+-+-+-+-+-+-+-+-+-+-+-+
|                             data                              |
+-+-+-+-+-+-+-+-+-+-+-+-+-+-+-+-+-+-+-+-+-+-+-+-+-+-+-+-+-+-+-+-+
```

Figure 7 Example Internet Fragment.

And the second fragment.

```
 0                   1                   2                   3
 0 1 2 3 4 5 6 7 8 9 0 1 2 3 4 5 6 7 8 9 0 1 2 3 4 5 6 7 8 9 0 1
+-+-+-+-+-+-+-+-+-+-+-+-+-+-+-+-+-+-+-+-+-+-+-+-+-+-+-+-+-+-+-+-+
|Ver= 4 |IHL= 5 |Type of Service|       Total Length = 216      |
+-+-+-+-+-+-+-+-+-+-+-+-+-+-+-+-+-+-+-+-+-+-+-+-+-+-+-+-+-+-+-+-+
|     Identification = 111      |Flg=0|  Fragment Offset  =  32 |
+-+-+-+-+-+-+-+-+-+-+-+-+-+-+-+-+-+-+-+-+-+-+-+-+-+-+-+-+-+-+-+-+
|   Time = 119  | Protocol = 6  |        Header Checksum        |
+-+-+-+-+-+-+-+-+-+-+-+-+-+-+-+-+-+-+-+-+-+-+-+-+-+-+-+-+-+-+-+-+
|                        source address                         |
+-+-+-+-+-+-+-+-+-+-+-+-+-+-+-+-+-+-+-+-+-+-+-+-+-+-+-+-+-+-+-+-+
```

Figure 8 Example Internet Fragment.

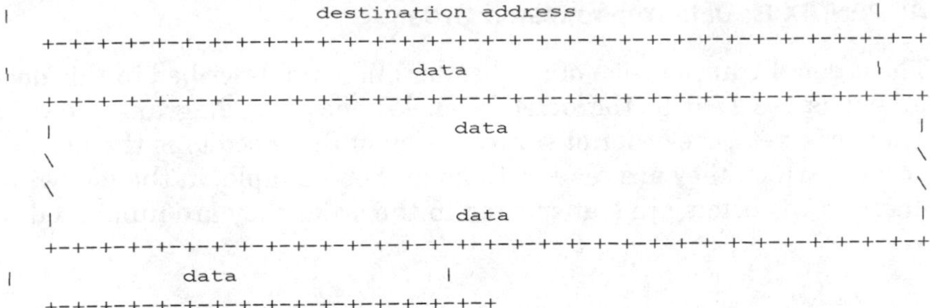

```
|                    destination address                        |
+-+-+-+-+-+-+-+-+-+-+-+-+-+-+-+-+-+-+-+-+-+-+-+-+-+-+-+-+-+-+-+-+
|                             data                              |
+-+-+-+-+-+-+-+-+-+-+-+-+-+-+-+-+-+-+-+-+-+-+-+-+-+-+-+-+-+-+-+-+
|                             data                              |
\                                                               \
\                                                               \
|                             data                              |
+-+-+-+-+-+-+-+-+-+-+-+-+-+-+-+-+-+-+-+-+-+-+-+-+-+-+-+-+-+-+-+-+
|            data               |
+-+-+-+-+-+-+-+-+-+-+-+-+-+-+-+-+
```

Figure 8 Example Internet Fragment. (*Continued*)

Example 3.

Here, we show an example of a datagram containing options:

```
 0                   1                   2                   3
 0 1 2 3 4 5 6 7 8 9 0 1 2 3 4 5 6 7 8 9 0 1 2 3 4 5 6 7 8 9 0 1
+-+-+-+-+-+-+-+-+-+-+-+-+-+-+-+-+-+-+-+-+-+-+-+-+-+-+-+-+-+-+-+-+
|Ver= 4 |IHL= 8 |Type of Service|       Total Length = 576      |
+-+-+-+-+-+-+-+-+-+-+-+-+-+-+-+-+-+-+-+-+-+-+-+-+-+-+-+-+-+-+-+-+
|       Identification = 111    |Flg=0|     Fragment Offset = 0 |
+-+-+-+-+-+-+-+-+-+-+-+-+-+-+-+-+-+-+-+-+-+-+-+-+-+-+-+-+-+-+-+-+
|   Time = 123  |  Protocol = 6 |       Header Checksum         |
+-+-+-+-+-+-+-+-+-+-+-+-+-+-+-+-+-+-+-+-+-+-+-+-+-+-+-+-+-+-+-+-+
|                        source address                         |
+-+-+-+-+-+-+-+-+-+-+-+-+-+-+-+-+-+-+-+-+-+-+-+-+-+-+-+-+-+-+-+-+
|                    destination address                        |
+-+-+-+-+-+-+-+-+-+-+-+-+-+-+-+-+-+-+-+-+-+-+-+-+-+-+-+-+-+-+-+-+
| Opt. Code = x | Opt.  Len.= 3 | option value  | Opt. Code = x |
+-+-+-+-+-+-+-+-+-+-+-+-+-+-+-+-+-+-+-+-+-+-+-+-+-+-+-+-+-+-+-+-+
| Opt. Len. = 4 |           option value        | Opt. Code = 1 |
+-+-+-+-+-+-+-+-+-+-+-+-+-+-+-+-+-+-+-+-+-+-+-+-+-+-+-+-+-+-+-+-+
| Opt. Code = y | Opt. Len. = 3 |  option value | Opt. Code = 0 |
+-+-+-+-+-+-+-+-+-+-+-+-+-+-+-+-+-+-+-+-+-+-+-+-+-+-+-+-+-+-+-+-+
|                             data                              |
\                                                               \
\                                                               \
|                             data                              |
+-+-+-+-+-+-+-+-+-+-+-+-+-+-+-+-+-+-+-+-+-+-+-+-+-+-+-+-+-+-+-+-+
|                             data                              |
+-+-+-+-+-+-+-+-+-+-+-+-+-+-+-+-+-+-+-+-+-+-+-+-+-+-+-+-+-+-+-+-+
```

Figure 9 Example Internet Datagram.

APPENDIX B: Data Transmission Order

The order of transmission of the header and data described in this document is resolved to the octet level. Whenever a diagram shows a group of octets, the order of transmission of those octets is the normal order in which they are read in English. For example, in the following diagram the octets are transmitted in the order they are numbered.

```
0                   1                   2                   3
0 1 2 3 4 5 6 7 8 9 0 1 2 3 4 5 6 7 8 9 0 1 2 3 4 5 6 7 8 9 0 1
  +-+-+-+-+-+-+-+-+-+-+-+-+-+-+-+-+-+-+-+-+-+-+-+-+-+-+-+-+-+-+-+-+
|       1       |       2       |       3       |       4       |
  +-+-+-+-+-+-+-+-+-+-+-+-+-+-+-+-+-+-+-+-+-+-+-+-+-+-+-+-+-+-+-+-+
|       5       |       6       |       7       |       8       |
  +-+-+-+-+-+-+-+-+-+-+-+-+-+-+-+-+-+-+-+-+-+-+-+-+-+-+-+-+-+-+-+-+
|       9       |      10       |      11       |      12       |
  +-+-+-+-+-+-+-+-+-+-+-+-+-+-+-+-+-+-+-+-+-+-+-+-+-+-+-+-+-+-+-+-+
```

Figure 10 Transmission Order of Bytes.

Whenever an octet represents a numeric quantity, the left most bit in the diagram is the high order or most significant bit. That is, the bit labeled 0 is the most significant bit. For example, the following diagram represents the value 170 (decimal).

Figure 11 Significance of Bits.

Similarly, whenever a multi-octet field represents a numeric quantity the left most bit of the whole field is the most significant bit. When a multi-octet quantity is transmitted the most significant octet is transmitted first.

Glossary

1822 BBN Report 1822, "The Specification of the Interconnection of a Host and an IMP." The specification of interface between a host and the ARPANET.

ARPANET leader The control information on an ARPANET message at the host-IMP interface.

ARPANET message The unit of transmission between a host and an IMP in the ARPANET. The maximum size is about 1012 octets (8096 bits).

ARPANET packet A unit of transmission used internally in the ARPANET between IMPs. The maximum size is about 126 octets (1008 bits).

Destination The destination address, an internet header field.

DF The Don't Fragment bit carried in the flags field.

Flags An internet header field carrying various control flags.

Fragment Offset This internet header field indicates where in the internet datagram a fragment belongs.

GGP Gateway to Gateway Protocol, the protocol used primarily between gateways to control routing and other gateway functions.

Header Control information at the beginning of a message, segment, datagram, packet, or block of data.

ICMP Internet Control Message Protocol; implemented in the internet module, the ICMP is used from gateways to hosts and between hosts to report errors and make routing suggestions.

Identification An internet header field carrying the identifying value assigned by the sender to aid in assembling the fragments of a datagram.

IHL The internet header field Internet Header Length is the length of the internet header measured in 32-bit words.

IMP The Interface Message Processor, the packet switch of the ARPANET.

Internet Address A four octet (32 bit) source or destination address consisting of a Network field and a Local Address field.

Internet Datagram The unit of data exchanged between a pair of internet modules (includes the internet header).

Internet Fragment A portion of the data of an internet datagram with an internet header.

Local Address The address of a host within a network. The actual mapping of an internet local address onto the host addresses in a network is quite general, allowing for many to one mappings.

MF The More-Fragments Flag carried in the internet header flags field.

Module An implementation, usually in software, of a protocol or other procedure.

More-Fragments flag A flag indicating whether or not this internet datagram contains the end of an internet datagram, carried in the internet header Flags field.

NFB The Number of Fragment Blocks in the data portion of an internet fragment. That is, the length of a portion of data measured in 8-octet units.

Octet An eight-bit byte.

Options The internet header Options field may contain several options, and each option may be several octets in length.

Padding The internet header Padding field is used to ensure that the data begins on a 32-bit word boundary. The padding is zero.

Protocol In this document, the next higher level protocol identifier, an internet header field.

Rest The local address portion of an Internet Address.

Source The source address, an internet header field.

TCP Transmission Control Protocol: A host-to-host protocol for reliable communication in internet environments.

TCP Segment The unit of data exchanged between TCP modules (including the TCP header).

TFTP Trivial File Transfer Protocol: A simple file transfer protocol built on UDP.

Time to Live An internet header field which indicates the upper bound on how long this internet datagram may exist.

TOS Type of Service.

Total Length The internet header field Total Length is the length of the datagram in octets including internet header and data.

TTL Time to Live.

Type of Service An internet header field which indicates the type (or quality) of service for this internet datagram.

UDP User Datagram Protocol: A user-level protocol for transaction-oriented applications.

User The user of the internet protocol. This may be a higher level protocol module, an application program, or a gateway program.

Version The Version field indicates the format of the internet header.

References

1. Cerf, V., "The Catenet Model for Internetworking," Information Processing Techniques Office, Defense Advanced Research Projects Agency, IEN *48,* July 1978.
2. Bolt Beranek and Newman, "Specification for the Interconnection of a Host and an IMP", BBN Technical Report 1822, Revised May 1978.
3. Postel, J., "Internet Control Message Protocol - DARPA Internet Program Protocol Specification," RFC *792*, USC/Information Sciences Institute, September 1981.
4. Shoch, J., "Inter-Network Naming, Addressing, and Routing," COMPCON, IEEE Computer Society, Fall 1978.
5. Postel, J., "Address Mappings," RFC *796*, USC/Information Sciences Institute, September 1981.
6. Shoch, J., "Packet Fragmentation in Inter-Network Protocols," Computer Networks, v. 3, n. 1, February 1979.
7. Strazisar, v., "How to Build a Gateway", IEN *109*, Bolt Beranek and Newman, August 1979.
8. Postel, J., "Service Mappings," RFC *795*, USC/Information Sciences Institute, September 1981.
9. Postel, J., "Assigned Numbers," RFC *790*, USC/Information Sciences Institute, September 1981.

Appendix

B

RFC 793: Transmission Control Protocol

This document was also developed by DARPA to ensure compatibility of networks and equipment used in internetworking of systems. Transmission Control Protocol (TCP) works in conjunction with the IP described in the previous document. TCP operates primarily at the Transport level of the OSI model and is responsible for setting up and controlling the connection between nodes. TCP and its relationship to the IP are described in Chapter 2.

Preface

This document describes the DOD Standard Transmission Control Protocol (TCP). There have been nine earlier editions of the ARPA TCP specification on which this standard is based, and the present text draws heavily from them. There have been many contributors to this work both in terms of concepts and in terms of text. This edition clarifies several details and removes the end-of-letter buffer-size adjustments, and redescribes the letter mechanism as a push function.

Jon Postel
Editor
September 1981

1. Introduction

The Transmission Control Protocol (TCP) is intended for use as a highly reliable host-to-host protocol between hosts in packet-switched computer communication networks, and in interconnected systems of such networks. This document describes the functions to be performed by the Transmission Control Protocol, the program that implements it, and its interface to programs or users that require its services.

1.1. Motivation

Computer communication systems are playing an increasingly important role in military, government, and civilian environments. This document focuses its attention primarily on military computer communication requirements, especially robustness in the presence of communication unreliability and availability in the presence of congestion, but many of these problems are found in the civilian and government sector as well.

As strategic and tactical computer communication networks are developed and deployed, it is essential to provide means of interconnecting them and to provide standard interprocess communication protocols which can support a broad range of applications. In anticipation of the need for such standards, the Deputy Undersecretary of Defense for Research and Engineering has declared the Transmission Control Protocol (TCP) described herein to be a basis for DOD-wide interprocess communication protocol standardization.

TCP is a connection-oriented, end-to-end reliable protocol designed to fit into a layered hierarchy of protocols which support multinetwork applications. The TCP provides for reliable interprocess communication between pairs of processes in host computers attached to distinct but interconnected computer communication networks. Very few assumptions are made as to the reliability of the communication protocols below the TCP layer. TCP assumes it can obtain a simple, potentially unreliable datagram service from the lower level protocols. In principle, the TCP should be able to operate above a wide spectrum of communication systems ranging from hard-wired connections to packet-switched or circuit-switched networks.

TCP is based on concepts first described by Cerf and Kahn in [1]. The TCP fits into a layered protocol architecture just above a basic Internet Protocol [2] which provides a way for the TCP to send and receive variable-length segments of information enclosed in internet datagram "envelopes." The internet datagram provides a means for addressing source and destination TCPs in different networks. The internet protocol also deals with any fragmentation or reassembly of the TCP segments required to achieve transport and delivery through multiple networks and interconnecting gateways. The internet protocol also carries information on the precedence, security classification and compartmentation of the TCP segments, so this information can be communicated end-to-end across multiple networks.

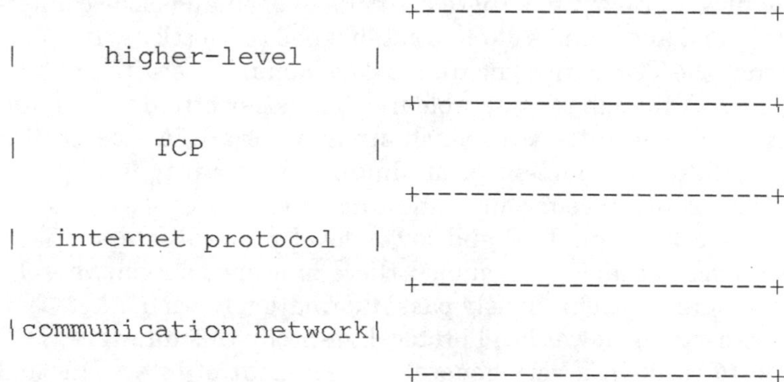

Figure 1 Protocol Layering.

Much of this document is written in the context of TCP implementations which are co-resident with higher level protocols in the host computer. Some computer systems will be connected to networks via front-end computers which house the TCP and internet protocol layers, as well as network specific software. The TCP specification describes an interface to the higher level protocols which appears to be implementable even for the front-end case, as long as a suitable host-to-front end protocol is implemented.

1.2. Scope

The TCP is intended to provide a reliable process-to-process communication service in a multinetwork environment. The TCP is intended to be a host-to-host protocol in common use in multiple networks.

1.3. About This Document

This document represents a specification of the behavior required of any TCP implementation, both in its interactions with higher level protocols and in its interactions with other TCPs. The rest of this section offers a very brief view of the protocol interfaces and operation. Section 2 summarizes the philosophical basis for the TCP design. Section 3 offers both a detailed description of the actions required of TCP when various events occur (arrival of new segments, user calls, errors, etc.) and the details of the formats of TCP segments.

1.4. Interfaces

The TCP interfaces on one side to user or application processes and on the other side to a lower level protocol such as Internet Protocol. The interface between an application process and the TCP is illustrated in reasonable detail. This interface consists of a set of calls much like the calls an operating system provides to an application process for manipulating files. For example, there are calls to open and close connections and to send and receive data on established connections. It is also expected that the TCP can asynchronously communicate with application programs. Although considerable freedom is permitted to TCP implementors to design interfaces which are appropriate to a particular operating system environment, a minimum functionality is required at the TCP/user interface for any valid implementation.

The interface between TCP and lower level protocol is essentially unspecified except that it is assumed there is a mechanism whereby the two levels can asynchronously pass information to each other. Typically, one expects the lower level protocol to specify this interface. TCP is designed to work in a very general environment of interconnected networks. The lower level protocol which is assumed throughout this document is the Internet Protocol [2].

1.5. Operation

As noted above, the primary purpose of the TCP is to provide reliable, securable logical circuit or connection service between pairs of processes. To provide this service on top of a less-reliable internet communication system requires facilities in the following areas:

Basic Data Transfer

Reliability

Flow Control

Multiplexing

Connections

Precedence and Security

The basic operation of the TCP in each of these areas is described in the following paragraphs.

Basic data transfer. The TCP is able to transfer a continuous stream of octets in each direction between its users by packaging some number of octets into segments for transmission through the internet system. In general, the TCPs decide when to block and forward data at their own convenience.

Sometimes users need to be sure that all the data they have submitted to the TCP has been transmitted. For this purpose a push function is defined. To assure that data submitted to a TCP is actually transmitted, the sending user indicates that it should be pushed through to the receiving user. A push causes the TCPs to promptly forward and deliver data up to that point to the receiver. The exact push point might not be visible to the receiving user and the push function does not supply a record boundary marker.

Reliability. The TCP must recover from data that is damaged, lost, duplicated, or delivered out of order by the internet communication system. This is achieved by assigning a sequence number to each octet transmitted, and requiring a positive acknowledgment (ACK) from the receiving TCP. If the ACK is not received within a timeout interval, the data is retransmitted. At the receiver, the sequence numbers are used to correctly order segments that may be received out of order and to eliminate duplicates. Damage is handled by adding a checksum to each segment transmitted, checking it at the receiver, and discarding damaged segments.

As long as the TCPs continue to function properly and the internet system does not become completely partitioned, no transmission errors will affect the correct delivery of data. TCP recovers from internet communication system errors.

Flow control. TCP provides a means for the receiver to govern the amount of data sent by the sender. This is achieved by returning a "window" with every ACK indicating a range of acceptable sequence numbers beyond the last segment successfully received. The window indicates an allowed number of octets that the sender may transmit before receiving further permission.

Multiplexing. To allow for many processes within a single host to use TCP communication facilities simultaneously, the TCP provides a set of addresses or ports within each host. Concatenated with the network and host addresses from the internet communication layer, this forms a socket. A pair of sockets uniquely identifies each connection. That is, a socket may be simultaneously used in multiple connections.

The binding of ports to processes is handled independently by each host. However, it proves useful to attach frequently used processes (e.g., a "logger" or timesharing service) to fixed sockets which are made known to the public. These services can then be accessed through the known addresses. Establishing and learning the port addresses of other processes may involve more dynamic mechanisms.

Connections. The reliability and flow control mechanisms described above require that TCPs initialize and maintain certain status information for each data stream. The combination of this information, including sockets, sequence numbers, and window sizes, is called a connection. Each connection is uniquely specified by a pair of sockets identifying its two sides.

When two processes wish to communicate, their TCPs must first establish a connection (initialize the status information on each side). When their communication is complete, the connection is terminated or closed to free the resources for other uses.

Since connections must be established between unreliable hosts and over the unreliable internet communication system, a handshake mechanism with clock-based sequence numbers is used to avoid erroneous initialization of connections.

Precedence and security. The users of TCP may indicate the security and precedence of their communication. Provision is made for default values to be used when these features are not needed.

2. Philosophy

2.1. Elements of the internetwork system

The internetwork environment consists of hosts connected to networks which are in turn interconnected via gateways. It is assumed here that the networks may be either local networks (e.g., the Ethernet) or large networks (e.g., the ARPANET), but in any case they are based on packet-switching technology. The active agents that produce and consume messages are processes. Various levels of protocols in the networks, the gateways, and the hosts support an interprocess communication system that provides two-way data flow on logical connections between process ports.

The term packet is used generically here to mean the data of one transaction between a host and its network. The format of data blocks exchanged within the network will generally not be of concern to us. Hosts are computers attached to a network, and from the communication network's point of view, are the sources and destinations of packets.

Processes are viewed as the active elements in host computers (in accordance with the fairly common definition of a process as a program in execution). Even terminals and files or other I/O devices are viewed as communicating with each other through the use of processes. Thus, all communication is viewed as interprocess communication. Since a process may need to distinguish among several communication streams between itself and another process (or processes), we imagine that each process may have a number of ports through which it communicates with the ports of other processes.

2.2. Model of operation

Processes transmit data by calling on the TCP and passing buffers of data as arguments. The TCP packages the data from these buffers into segments and calls on the internet module to transmit each segment to the destination TCP. The receiving TCP places the data from a segment into the receiving user's buffer and notifies the receiving user. The TCPs include control information in the segments which they use to ensure reliable ordered data transmission.

The model of internet communication is that there is an internet protocol module associated with each TCP which provides an interface to the local network. This internet module packages TCP segments inside internet datagrams and routes these datagrams to a destination internet module or intermediate gateway. To transmit the datagram through the local network, it is embedded in a local network packet. The packet switches may perform further packaging, fragmentation, or other operations to achieve the delivery of the local packet to the destination internet module.

At a gateway between networks, the internet datagram is "unwrapped" from its local packet and examined to determine through which network the internet datagram should travel next. The internet datagram is then "wrapped" in a local packet suitable to the next network and routed to the next gateway, or to the final destination.

A gateway is permitted to break up an internet datagram into smaller internet datagram fragments if this is necessary for transmission through the next network. To do this, the gateway produces a set of internet datagrams; each carrying a fragment. Fragments may be further broken into smaller fragments at subsequent gateways. The internet datagram fragment format is designed so that the destination internet module can reassemble fragments into internet datagrams. A destination internet module unwraps the segment from the datagram (after reassembling the datagram, if necessary) and passes it to the destination TCP.

This simple model of the operation glosses over many details. One important feature is the type of service. This provides information to the gateway (or internet module) to guide it in selecting the service parameters to be used in traversing the next network.

Included in the type of service information is the precedence of the datagram. Datagrams may also carry security information to permit host and gateways that operate in multilevel secure environments to properly segregate datagrams for security considerations.

2.3. The host environment

The TCP is assumed to be a module in an operating system. The users access the TCP much like they would access the file system. The TCP may call on other operating system functions, for example, to manage data structures. The actual interface to the network is assumed to be controlled by a device driver module. The TCP does not call on the network device driver directly, but rather calls on the internet datagram protocol module which may in turn call on the device driver. The mechanisms of TCP do not preclude implementation of the TCP in a front-end processor. However, in such an implementation, a host-to-front-end protocol must provide the functionality to support the type of TCP-user interface described in this document.

2.4. Interfaces

The TCP/user interface provides for calls made by the user on the TCP to OPEN or CLOSE a connection, to SEND or RECEIVE data, or to obtain STATUS about a connection. These calls are like other calls from user programs on the operating system, for example, the calls to open, read from, and close a file.

The TCP/internet interface provides calls to send and receive datagrams addressed to TCP modules in hosts anywhere in the internet system. These calls have parameters for passing the address, type of service, precedence, security, and other control information.

2.5. Relation to other protocols

The diagram below illustrates the place of the TCP in the protocol hierarchy.

It is expected that the TCP will be able to support higher level protocols efficiently. It should be easy to interface higher level protocols like the ARPANET Telnet or AUTODIN II THP to the TCP.

2.6. Reliable communication

A stream of data sent on a TCP connection is delivered reliably and in order at the destination.

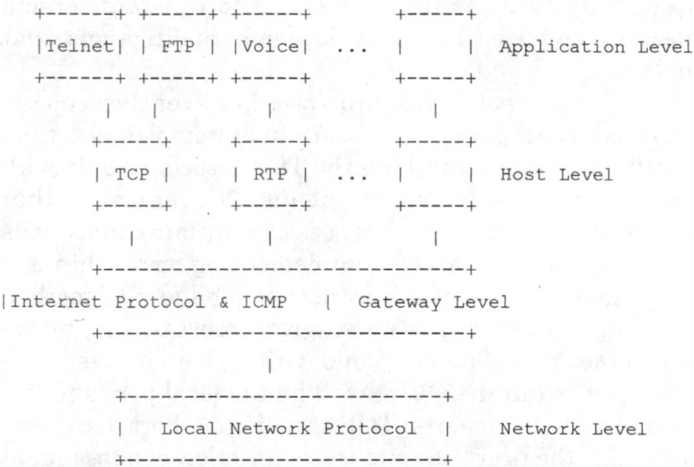

Figure 2 Protocol Relationships.

Transmission is made reliable via the use of sequence numbers and acknowledgments. Conceptually, each octet of data is assigned a sequence number. The sequence number of the first octet of data in a segment is transmitted with that segment and is called the segment sequence number. Segments also carry an acknowledgment number which is the sequence number of the next expected data octet of transmissions in the reverse direction. When the TCP transmits a segment containing data, it puts a copy on a retransmission queue and starts a timer; when the acknowledgment for that data is received, the segment is deleted from the queue. If the acknowledgment is not received before the timer runs out, the segment is retransmitted. An acknowledgment by TCP does not guarantee that the data has been delivered to the end user, but only that the receiving TCP has taken the responsibility to do so.

To govern the flow of data between TCPs, a flow control mechanism is employed. The receiving TCP reports a "window" to the sending TCP. This window specifies the number of octets, starting with the acknowledgment number, that the receiving TCP is currently prepared to receive.

2.7. Connection establishment and clearing

To identify the separate data streams that a TCP may handle, the TCP provides a port identifier. Since port identifiers are selected independently by each TCP, they might not be unique. To provide for unique addresses within each TCP, we concatenate an internet address identifying the TCP with a port identifier to create a socket which will be unique throughout all networks connected together.

A connection is fully specified by the pair of sockets at the ends. A local socket may participate in many connections to different foreign sockets. A connection can be used to carry data in both directions, that is, it is "full duplex."

TCPs are free to associate ports with processes however they choose. However, several basic concepts are necessary in any implementation. There must be well-known sockets which the TCP associates only with the "appropriate" processes by some means. We envision that processes may "own" ports, and that processes can initiate connections only on the ports they own. (Means for implementing ownership is a local issue, but we envision a Request Port user command, or a method of uniquely allocating a group of ports to a given process, e.g., by associating the high-order bits of a port name with a given process.)

A connection is specified in the OPEN call by the local port and foreign socket arguments. In return, the TCP supplies a (short) local connection name by which the user refers to the connection in subsequent calls. There are several things that must be remembered about a connection. To store this information we imagine that there is a data structure called a Transmission Control Block (TCB). One implementation strategy would have the local connection name be a pointer to the TCB for this connection. The OPEN call also specifies whether the connection establishment is to be actively pursued, or to be passively waited for.

A passive OPEN request means that the process wants to accept incoming connection requests rather than attempting to initiate a connection. Often the process requesting a passive OPEN will accept a connection request from any caller. In this case a foreign socket of all zeros is used to denote an unspecified socket. Unspecified foreign sockets are allowed only on passive OPENs.

A service process that wished to provide services for unknown other processes would issue a passive OPEN request with an unspecified foreign socket. Then a connection could be made with any process that requested a connection to this local socket. It would help if this local socket were known to be associated with this service. Well-known sockets are a convenient mechanism for a priori associating a socket address with a standard service. For instance, the "Telnet-Server" process is permanently assigned to a particular socket, and other sockets are reserved for File Transfer, Remote Job Entry, Text Generator, Echoer, and Sink processes (the last three being for test purposes). A socket address might be reserved for access to a "Look-Up" service which would return the specific socket at which a newly created service would be provided. The concept of a well-known socket is part of the TCP specification, but the assignment of sockets to services is outside this specification. (See [4].)

Processes can issue passive OPENs and wait for matching active OPENs from other processes and be informed by the TCP when connections have been established. Two processes which issue active OPENs to each other at the same time will be correctly connected. This flexibility is critical for the support of distributed computing in which components act asynchronously with respect to each other.

There are two principal cases for matching the sockets in the local passive OPENs and in foreign active OPENs. In the first case, the local passive OPENs have fully specified the foreign socket. In this case, the match must be exact. In the second case, the local passive OPENs have left the foreign socket unspecified. In this case, any foreign socket is acceptable as long as the local sockets match. Other possibilities include partially restricted matches. If there are several pending passive OPENs (recorded in TCBs) with the same local socket, a foreign active OPEN will be matched to a TCB with the specific foreign socket in the foreign active OPEN, if such a TCB exists, before selecting a TCB with an unspecified foreign socket.

The procedures for establishing connections utilize the synchronized (SYN) control flag and involve an exchange of three messages. This exchange has been termed a three-way handshake [3].

A connection is initiated by the rendezvous of an arriving segment containing a SYN and a waiting TCB entry each created by a user OPEN command. The matching of local and foreign sockets determines when a connection has been initiated. The connection becomes "established" when sequence numbers have been synchronized in both directions. The clearing of a connection also involves the exchange of segments, in this case carrying the FIN control flag.

2.8. Data communication

The data that flows on a connection may be thought of as a stream of octets. The sending user indicates in each SEND call whether the data in that call (and any preceeding calls) should be immediately pushed through to the receiving user by the setting of the PUSH flag.

A sending TCP is allowed to collect data from the sending user and to send that data in segments at its own convenience, until the push function is signaled, then it must send all unset data. When a receiving TCP sees the PUSH flag, it must not wait for more data from the sending TCP before passing the data to the receiving process. There is no necessary relationship between push functions and segment boundaries. The data in any particular segment may be the result of a single SEND call, in whole or in part, or of multiple SEND calls. The purpose of the push function and the PUSH flag is to push data through from the sending user to the receiving user. It does not provide a record service.

There is a coupling between the push function and the use of buffers of data that cross the TCP/user interface. Each time a PUSH flag is associated with data placed into the receiving user's buffer, the buffer is returned to the user for processing even if the buffer is not filled. If data arrives that fills the user's buffer before a PUSH is seen, the data is passed to the user in buffer-size units. TCP also provides a means to communicate to the receiver of data that at some point further along in the data stream than the receiver is currently reading there is urgent data. TCP does not attempt to define what the user specifically does upon being notified of pending urgent data, but the general notion is that the receiving process will take action to process the urgent data quickly.

2.9. Precedence and security

The TCP makes use of the internet protocol type of service field and security option to provide precedence and security on a per connection basis to TCP users. Not all TCP modules will necessarily function in a multilevel secure environment; some may be limited to unclassified use only, and others may operate at only one security level and compartment. Consequently, some TCP implementations and services to users may be limited to a subset of the multilevel secure case. TCP modules which operate in a multilevel secure environment must properly mark outgoing segments with the security, compartment, and precedence. Such TCP modules must also provide to their users or higher level protocols such as Telnet or THP an interface to allow them to specify the desired security level, compartment, and precedence of connections.

2.10. Robustness principle

TCP implementations will follow a general principle of robustness: be conservative in what you do, be liberal in what you accept from others.

3. Functional Specification

3.1. Header format

TCP segments are sent as internet datagrams. The Internet Protocol header carries several information fields, including the source and destination host addresses [2]. A TCP header follows the internet header, supplying information specific to the TCP protocol. This division allows for the existence of host-level protocols other than TCP.

Source port: 16 bits. The source port number.

Destination port: 16 bits. The destination port number.

Sequence number: 32 bits. The sequence number of the first data octet in this segment (except when SYN is present). If SYN is present the sequence number is the initial sequence number (ISN) and the first data octet is ISN+1.

Acknowledgment number: 32 bits. If the ACK control bit is set this field contains the value of the next sequence number the sender of the segment is expecting to receive. Once a connection is established this is always sent.

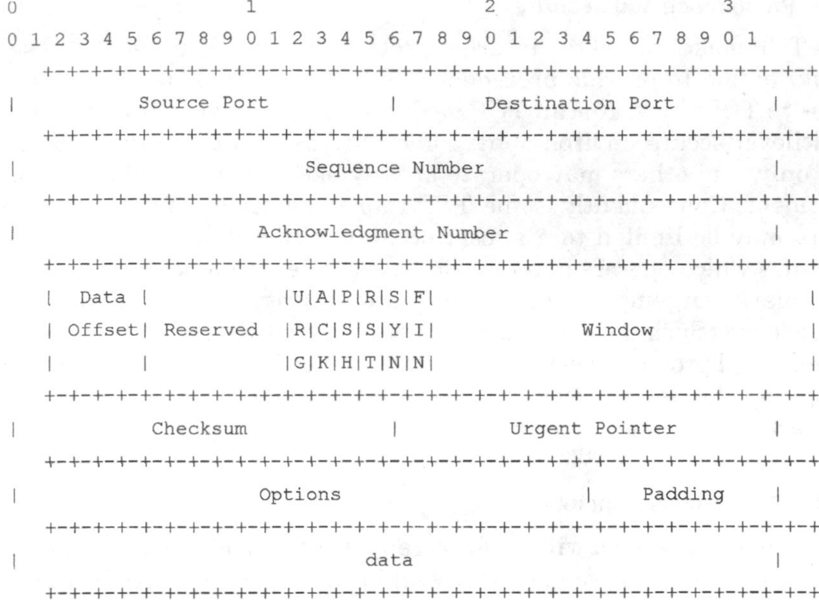

Figure 3 TCP Header Format. Note that one tick mark represents one bit position.

Data offset: 4 bits. The number of 32-bit words in the TCP Header. This indicates where the data begins. The TCP header (even one including options) is an integral number of 32-bits long.

Reserved: 6 bits. Reserved for future use. Must be zero.

Control bits: 6 bits (from left to right).

URG: Urgent Pointer field significant

ACK: Acknowledgment field significant

PSH: Push Function

RST: Reset the connection

SYN: Synchronize sequence numbers

FIN: No more data from sender

Window: 16 bits. The number of data octets beginning with the one indicated in the acknowledgment field which the sender of this segment is willing to accept.

Checksum: 16 bits. The checksum field is the 16 bit one's complement of the one's complement sum of all 16-bit words in the header and text. If a segment contains an odd number of header and text octets to be checksummed, the last octet is padded on the right with zeros to form a 16-bit word for checksum purposes. The pad is not transmitted as part of the segment. While computing the checksum, the checksum field itself is replaced with zeros.

The checksum also covers a 96-bit pseudo header conceptually prefixed to the TCP header. This pseudo header contains the Source Address, the Destination Address, the Protocol, and TCP length. This gives the TCP protection against misrouted segments. This information is carried in the Internet Protocol and is transferred across the TCP/Network interface in the arguments or results of calls by the TCP on the IP.

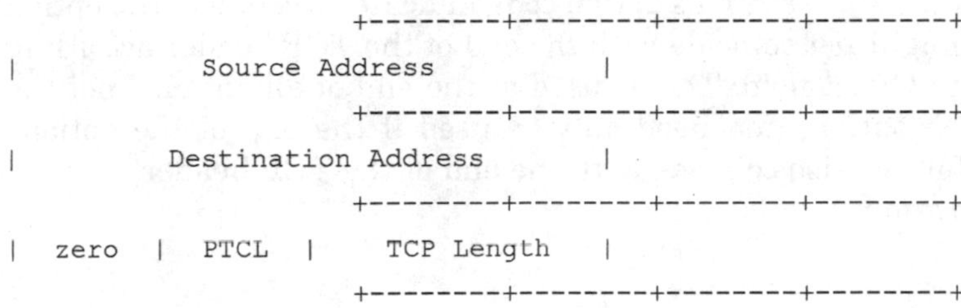

The TCP Length is the TCP header length plus the data length in octets (this is not an explicitly transmitted quantity, but is computed), and it does not count the 12 octets of the pseudo header.

Urgent pointer: 16 bits. This field communicates the current value of the urgent pointer as a positive offset from the sequence number in this segment. The urgent pointer points to the sequence number of the octet following the urgent data. This field is only to be interpreted in segments with the URG control bit set.

Options: variable. Options may occupy space at the end of the TCP header and are a multiple of 8 bits in length. All options are included in the checksum. An option may begin on any octet boundary. There are two cases for the format of an option:

Case 1: A single octet of option-kind.

Case 2: An octet of option-kind, an octet of option-length, and the actual option-data octets.

The option-length counts the two octets of option-kind and option-length as well as the option-data octets.

Note that the list of options may be shorter than the data offset field might imply. The content of the header beyond the End-of-Option option must be header padding (i.e., zero).

A TCP must implement all options. Currently defined options include (kind indicated in octal):

Kind	Length	Meaning
0	—	End of option list.
1	—	No-Operation.
2	4	Maximum Segment Size.

Specific option definitions.

End of Option List. This option code indicates the end of the option list. This might not coincide with the end of the TCP header according to the Data Offset field. This is used at the end of all options, not the end of each option, and need only be used if the end of the options would not otherwise coincide with the end of the TCP header.

No-Operation.

```
        +--------+
|00000000|
        +--------+
Kind=0
```

This option code may be used between options, for example, to align the beginning of a subsequent option on a word boundary. There is no guarantee that senders will use this option, so receivers must be prepared to process options even if they do not begin on a word boundary.

Maximum segment size.

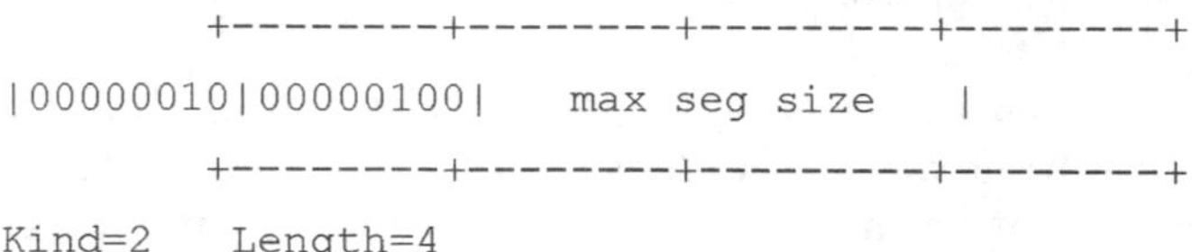

Maximum Segment Size Option Data: 16 bits.

If this option is present, then it communicates the maximum receive segment size at the TCP which sends this segment. This field must only be sent in the initial connection request (i.e., in segments with the SYN control bit set). If this option is not used, any segment size is allowed.

Padding: variable. The TCP header padding is used to ensure that the TCP header ends and data begins on a 32-bit boundary. The padding is composed of zeros.

3.2. Terminology

Before we can discuss very much about the operation of the TCP we need to introduce some detailed terminology. The maintenance of a TCP connection requires the remembering of several variables. We conceive of these variables being stored in a connection record called a Transmission Control Block or TCB. Among the variables stored in the TCB are the local and remote socket numbers, the security and precedence of the connection, pointers to the user's send and receive buffers, pointers to the retransmit queue and to the current segment. In addition several variables relating to the send and receive sequence numbers are stored in the TCB.

Send Sequence Variables:

SND.UNA—send unacknowledged

SND.NXT—send next

SND.WND—send window

SND.UP—send urgent pointer

SND.WL1—segment sequence number used for last window update

SND.WL2—segment acknowledgment number used for last window update

ISS—initial send sequence number

Receive Sequence Variables:

RCV.NXT—receive next

RCV.WND—receive window

RCV.UP—receive urgent pointer

IRS—initial receive sequence number

The following diagrams may help to relate some of these variables to the sequence space.

The send window is the portion of the sequence space labeled 3 in Figure 4.

The receive window is the portion of the sequence space labeled 2 in Figure 5.

There are also some variables used frequently in the discussion that take their values from the fields of the current segment.

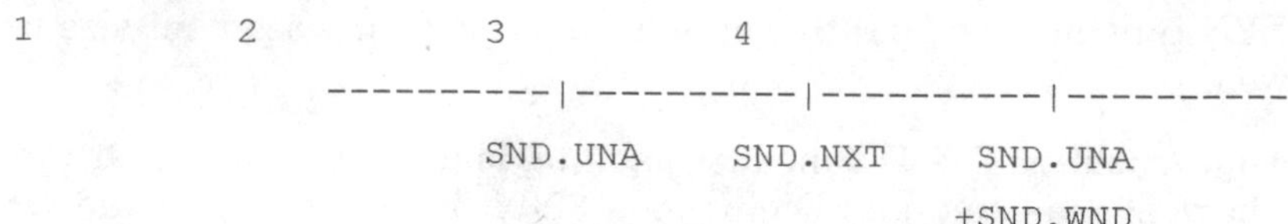

Figure 4 Send Sequence Space. (1) Old sequence numbers which have been acknowledged. (2) Sequence numbers of unacknowledged data. (3) Sequence numbers allowed for new data transmission. (4) Future sequence numbers which are not yet allowed.

Current Segment Variables

SEG.SEQ—segment sequence number

SEG.ACK—segment acknowledgment number

SEG.LEN—segment length

SEG.WND—segment window

SEG.UP—segment urgent pointer

SEG.PRC—segment precedence value

A connection progresses through a series of states during its lifetime. The states are: LISTEN, SYN-SENT, SYN-RECEIVED, ESTABLISHED, FIN-WAIT-1, FIN-WAIT-2, CLOSE-WAIT, CLOSING, LAST-ACK, TIME-WAIT, and the fictional state CLOSED. CLOSED is fictional because it represents the state when there is no TCB, and therefore, no connection. Briefly the meanings of the states are

LISTEN: represents waiting for a connection request from any remote TCP and port.

SYN-SENT: represents waiting for a matching connection request after having sent a connection request.

SYN-RECEIVED: represents waiting for a confirming connection request acknowledgment after having both received and sent a connection request.

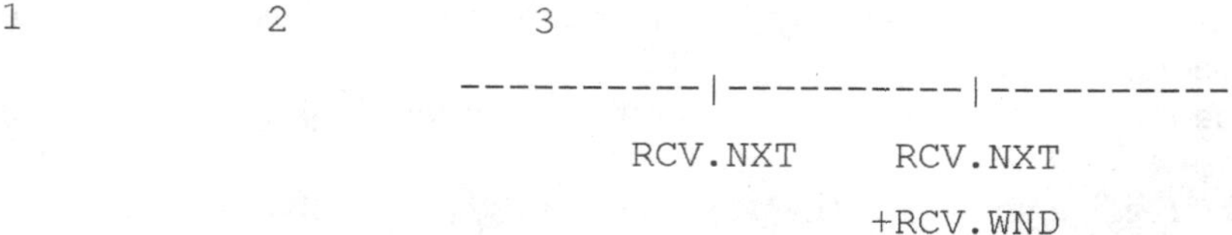

Figure 5 Receive Sequence Space. (1) Old sequence numbers which have been acknowledged. (2) Sequence numbers allowed for new reception. (3) Future sequence numbers which are not yet allowed.

ESTABLISHED: represents an open connection, data received can be delivered to the user. The normal state for the data transfer phase of the connection.

FIN-WAIT-1: represents waiting for a connection termination request from the remote TCP, or an acknowledgment of the connection termination request previously sent.

FIN-WAIT-2: represents waiting for a connection termination request from the remote TCP.

CLOSE-WAIT: represents waiting for a connection termination request from the local user.

CLOSING: represents waiting for a connection termination request acknowledgment from the remote TCP.

LAST-ACK: represents waiting for an acknowledgment of the connection termination request previously sent to the remote TCP (which includes an acknowledgment of its connection termination request).

TIME-WAIT: represents waiting for enough time to pass to be sure the remove TCP received the acknowledgment of its connection termination request.

CLOSED: represents no connection state at all.

A TCP connection progresses from one state to another in response to events. The events are the user calls, OPEN, SEND, RECEIVE, CLOSE, ABORT, and STATUS; the incoming segments, particularly those containing the SYN, ACK, RST and FIN flags; and timeouts. The state diagram in Figure 6 illustrates only state changes, together with

```
                    +---------+ ---------\      active OPEN
                    |  CLOSED |            \    -----------
                    +---------+<---------\   \   create TCB
                      |     ^              \   \  snd SYN
         passive OPEN |     |   CLOSE        \   \
         ------------ |     | ----------       \   \
          create TCB  |     | delete TCB         \   \
                      V     |                      \   \
                    +---------+            CLOSE    |    \
                    |  LISTEN |          ---------- |     |
                    +---------+          delete TCB |     |
         rcv SYN      |     |     SEND              |     |
        -----------   |     |    -------            |     V
-+      snd SYN,ACK  /       \   snd SYN          +---------+
```

```
|          |<-----------------           ----------------->|          |
|   SYN    |                    rcv SYN                     |   SYN    |
|   RCVD   |<-----------------------------------------------|   SENT   |
|          |                    snd ACK                     |          |
|          |------------------           -------------------|          |
+---------+   rcv ACK of SYN  \       /  rcv SYN,ACK       +---------+
  |           --------------   |     |   -----------
  |                  x         |     |     snd ACK
  |                            V     V
  |  CLOSE                   +---------+
  | -------                  |  ESTAB  |
  | snd FIN                  +---------+
  |                   CLOSE    |     |    rcv FIN
  V                  -------   |     |    -------
+---------+          snd FIN  /       \   snd ACK          +---------+
|  FIN    |<-----------------           ------------------>|  CLOSE  |
| WAIT-1  |------------------                              |   WAIT  |
+---------+          rcv FIN  \                            +---------+
  | rcv ACK of FIN   -------   |                            CLOSE  |
  | --------------   snd ACK   |                           ------- |
  V        x                   V                           snd FIN V
+---------+                  +---------+                   +---------+
|FINWAIT-2|                  | CLOSING |                   | LAST-ACK|
+---------+                  +---------+                   +---------+
  |                rcv ACK of FIN |                 rcv ACK of FIN |
  |  rcv FIN       -------------- |    Timeout=2MSL -------------- |
  |  -------              x       V    ------------        x       V
   \ snd ACK                 +---------+delete TCB         +---------+
------------------------>|TIME WAIT|------------------>| CLOSED  |
                             +---------+                   +---------+
```

Figure 6 TCP Connection State Diagram.

the causing events and resulting actions, but addresses neither error conditions nor actions which are not connected with state changes. In a later section, more detail is offered with respect to the reaction of the TCP to events.

Note Bene: This diagram is only a summary and must not be taken as the total specification.

3.3. Sequence numbers

A fundamental notion in the design is that every octet of data sent over a TCP connection has a sequence number. Since every octet is sequenced, each of them can be acknowledged. The acknowledgment mechanism employed is cumulative so that an acknowledgment of sequence number X indicates that all octets up to but not including X have been received. This mechanism allows for straight-forward duplicate detection in the presence of retransmission. Numbering of octets within a segment is that the first data octet immediately following the header is the lowest numbered, and the following octets are numbered consecutively.

It is essential to remember that the actual sequence number space is finite, though very large. This space ranges from 0 to 2**32–1. Since the space is finite, all arithmetic dealing with sequence numbers must be performed modulo 2**32. This unsigned arithmetic preserves the relationship of sequence numbers as they cycle from 2**32–1 to 0 again. There are some subtleties to computer modulo arithmetic, so great care should be taken in programming the comparison of such values. The symbol "=<" means "less than or equal" (modulo 2**32).

The typical kinds of sequence number comparisons which the TCP must perform include:

(a) Determining that an acknowledgment refers to some sequence number sent but not yet acknowledged.

(b) Determining that all sequence numbers occupied by a segment have been acknowledged (e.g., to remove the segment from a retransmission queue).

(c) Determining that an incoming segment contains sequence numbers which are expected (i.e., that the segment "overlaps" the receive window).

In response to sending data the TCP will receive acknowledgments. The following comparisons are needed to process the acknowledgments.

SND.UNA = oldest unacknowledged sequence number

SND.NXT = next sequence number to be sent

SEG.ACK = acknowledgment from the receiving TCP (next sequence number expected by the receiving TCP)

SEG.SEQ = first sequence number of a segment

SEG.LEN = the number of octets occupied by the data in the segment (counting SYN and FIN)

SEG.SEQ+SEG.LEN-1 = last sequence number of a segment

A new acknowledgment (called an "acceptable ack"), is one for which the inequality below holds:

SND.UNA < SEG.ACK =< SND.NXT

A segment on the retransmission queue is fully acknowledged if the sum of its sequence number and length is less than or equal to the acknowledgment value in the incoming segment.

When data is received the following comparisons are needed:

RCV.NXT = next sequence number expected on an incoming segment, and is the left or lower edge of the receive window

RCV.NXT+RCV.WND-1 = last sequence number expected on an incoming segment, and is the right or upper edge of the receive window

SEG.SEQ = first sequence number occupied by the incoming segment

SEG.SEQ+SEG.LEN-1 = last sequence number occupied by the incoming segment

A segment is judged to occupy a portion of valid receive sequence space if

RCV.NXT =< SEG.SEQ < RCV.NXT+RCV.WND

or

RCV.NXT =< SEG.SEQ+SEG.LEN-1 < RCV.NXT+RCV.WND

The first part of this test checks to see if the beginning of the segment falls in the window, the second part of the test checks to see if the end of the segment falls in the window; if the segment passes either part of the test it contains data in the window.

Actually, it is a little more complicated than this. Due to zero windows and zero-length segments, we have four cases for the acceptability of an incoming segment:

Segment Receive Test

Length	Window	
0	0	SEG.SEQ = RCV.NXT
0	>0	RCV.NXT =< SEG.SEQ < RCV.NXT+RCV.WND
>0	0	not acceptable
>0	>0	RCV.NXT =< SEG.SEQ < RCV.NXT+RCV.WND or RCV.NXT =< SEG.SEQ+SEG.LEN-1 < RCV.NXT+RCV.WND

Note that when the receive window is zero no segments should be acceptable except ACK segments. Thus, it is possible for a TCP to maintain a zero receive window while transmitting data and receiving ACKs. However, even when the receive window is zero, a TCP must process the RST and URG fields of all incoming segments.

We have taken advantage of the numbering scheme to protect certain control information as well. This is achieved by implicitly including some control flags in the sequence space so they can be retransmitted and acknowledged without confusion (i.e., one and only one copy of the control will be acted upon). Control information is not physically carried in the segment data space. Consequently, we must adopt rules for implicitly assigning sequence numbers to control. The SYN and FIN are the only controls requiring this protection, and these controls are used only at connection opening and closing. For sequence number purposes, the SYN is considered to occur before the first actual data octet of the segment in which it occurs, while the FIN is considered to occur after the last actual data octet in a segment in which it occurs. The segment length (SEG.LEN) includes both data and sequence space occupying controls. When a SYN is present then SEG.SEQ is the sequence number of the SYN.

Initial sequence number selection. The protocol places no restriction on a particular connection being used over and over again. A connection is defined by a pair of sockets. New instances of a connection will be referred to as incarnations of the connection. The problem that arises from this is—"how does the TCP identify duplicate segments from previous incarnations of the connection?" This problem becomes apparent if the connection is being opened and closed in quick succession, or if the connection breaks with loss of memory and is then reestablished.

To avoid confusion we must prevent segments from one incarnation of a connection from being used while the same sequence numbers may still be present in the network from an earlier incarnation. We want to assure this, even if a TCP crashes and loses all knowledge of the sequence numbers it has been using. When new connections are created, an initial sequence number (ISN) generator is employed which selects a new 32-bit ISN. The generator is bound to a (possibly fictitious) 32-bit clock whose low-order bit is incremented roughly every 4 microseconds. Thus, the ISN cycles approximately every 4.55 hours. Since we assume that segments will stay in the network no more than the Maximum Segment Lifetime (MSL) and that the MSL is less than 4.55 hours we can reasonably assume that ISNs will be unique.

For each connection there is a send sequence number and a receive sequence number. The initial send sequence number (ISS) is chosen by the data sending TCP, and the initial receive sequence number (IRS) is learned during the connection establishing procedure.

For a connection to be established or initialized, the two TCPs must synchronize on each other's initial sequence numbers. This is done in an exchange of connection establishing segments carrying a control bit called "SYN" (for synchronize) and the initial sequence numbers. As a shorthand, segments carrying the SYN bit are also called "SYNs." Hence, the solution requires a suitable mechanism for picking an initial sequence number and a slightly involved handshake to exchange the ISNs.

The synchronization requires each side to send its own initial sequence number and to receive a confirmation of it in acknowledgment from the other side. Each side must also receive the other side's initial sequence number and send a confirming acknowledgment.

(1) A → B SYN my sequence number is X

(2) A ← B ACK your sequence number is X

(3) A ← B SYN my sequence number is Y

(4) A → B ACK your sequence number is Y

Because steps 2 and 3 can be combined in a single message this is called the three-way (or three message) handshake.

A three-way handshake is necessary because sequence numbers are not tied to a global clock in the network, and TCPs may have different mechanisms for picking the ISNs. The receiver of the first SYN has no way of knowing whether the segment was an old delayed one or not, unless it remembers the last sequence number used on the connection (which is not always possible), and so it must ask the sender to verify this SYN. The three-way handsake and the advantages of a clock-driven scheme are discussed in [3].

Knowing when to keep quiet. To be sure that a TCP does not create a segment that carries a sequence number which may be duplicated by an old segment remaining in the network, the TCP must keep quiet for a maximum segment lifetime (MSL) before assigning any sequence numbers upon starting up or recovering from a crash in which memory of sequence numbers in use was lost. For this specification the MSL is taken to be 2 minutes. This is an engineering choice, and may be changed if experience indicates it is desirable to do so. Note that if a TCP is reinitialized in some sense, yet retains its memory of sequence numbers in use, then it need not wait at all; it must only be sure to use sequence numbers larger than those recently used.

The TCP quiet time concept. This specification provides that hosts which "crash" without retaining any knowledge of the last sequence numbers transmitted on each active (i.e., not closed) connection shall delay emitting any TCP segments for at least the agreed Maximum Segment Lifetime (MSL) in the internet system of which the host is a part. In the paragraphs below, an explanation for this specification is given. TCP implementors may violate the "quiet time" restriction, but only at the risk of causing some old data to be accepted as new or new data rejected as old duplicated by some receivers in the internet system.

TCPs consume sequence number space each time a segment is formed and entered into the network output queue at a source host. The duplicate detection and sequencing algorithm in the TCP protocol relies on the unique binding of segment data to sequence space to the extent that sequence numbers will not cycle through all 2**32 values before the segment data bound to those sequence numbers has been delivered and acknowledged by the receiver and all duplicate copies of the segments have "drained" from the internet. Without such an assumption, two distinct TCP segments could conceivably be assigned the same or overlapping sequence numbers, causing confusion at the receiver as to which data is new and which is old. Remember that each segment is bound to as many consecutive sequence numbers as there are octets of data in the segment.

Under normal conditions, TCPs keep track of the next sequence number to emit and the oldest awaiting acknowledgment so as to avoid mistakenly using a sequence number over before its first use has been acknowledged. This alone does not guarantee that old duplicate data is drained from the net, so the sequence space has been made very large to reduce the probability that a wandering duplicate will cause trouble upon arrival. At 2 megabits/sec. it takes 4.5 hours to use up 2**32 octets of sequence space. Since the maximum segment lifetime in the net is not likely to exceed a few tens of seconds, this is deemed ample protection for foreseeable nets, even if data rates escalate to 10s of megabits/sec. At 100 megabits/sec, the cycle time is 5.4 minutes which may be a little short, but still within reason.

The basic duplicate detection and sequencing algorithm in TCP can be defeated, however, if a source TCP does not have any memory of the sequence numbers it last used on a given connection. For example, if the TCP were to start all connections with sequence number 0, then upon crashing and restarting, a TCP might reform an earlier connection (possibly after half-open connection resolution) and emit packets with sequence numbers identical to or overlapping with packets still in the network which were emitted on an earlier incarnation of the same connection. In the absence of knowledge about the sequence numbers used on a particular connection, the TCP specification recommends that the source delay for MSL seconds before emitting segments on the connection, to allow time for segments from the earlier connection incarnation to drain from the system.

Even hosts which can remember the time of day and use it to select initial sequence number values are not immune from this problem (i.e., even if time of day is used to select an initial sequence number for each new connection incarnation).

Suppose, for example, that a connection is opened starting with sequence number S. Suppose that this connection is not used much and that eventually the initial sequence number function (ISN (t)) takes on a value equal to the sequence number, say S1, of the last segment sent by this TCP on a particular connection. Now suppose, at this instant, the host crashes, recovers, and establishes a new incarnation of the connection. The initial sequence number chosen is S1 = ISN(t)—last used sequence number on old incarnation of connection! If the recovery occurs quickly enough, any old duplicates in the net bearing sequence numbers in the neighborhood of S1 may arrive and be treated as new packets by the receiver of the new incarnation of the connection.

The problem is that the recovering host may not know for how long it crashed nor whether there are still old duplicates in the system from earlier connection incarnations.

One way to deal with this problem is to deliberately delay emitting segments for one MSL after recovery from a crash—this is the "quiet time" specification. Hosts which prefer to avoid waiting are willing to risk possible confusion of old and new packets at a given destination may choose not to wait for the "quiet time."

Implementors may provide TCP users with the ability to select on a connection by connection basis whether to wait after a crash, or may informally implement the "quiet time" for all connections. Obviously, even where a user selects to "wait," this is not necessary after the host has been "up" for at least MSL seconds.

To summarize: every segment emitted occupies one or more sequence numbers in the sequence space, the numbers occupied by a segment are "busy" or "in use" until MSL seconds have passed, upon crashing a block of space-time is occupied by the octets of the last emitted segment, if a new connection is started too soon and uses any of the sequence numbers in the space-time footprint of the last segment of the previous connection incarnation, there is a potential sequence number overlap area which could cause confusion at the receiver.

3.4. Establishing a connection

The "three-way handshake" is the procedure used to establish a connection. This procedure normally is initiated by one TCP and responded to by another TCP. The procedure also works if two TCP simultaneously initiate the procedure. When simultaneous attempt occurs, each TCP receives a "SYN" segment which carries no acknowledgment after it has sent a "SYN." Of course, the arrival of an old duplicate "SYN" segment can potentially make it appear, to the recipient, that a simultaneous connection initiation is in progress. Proper use of "reset" segments can disambiguate these cases.

Several examples of connection initiation follow. Although these examples do not show connection synchronization using data-carrying segments, this is perfectly legitimate, so long as the receiving TCP does not deliver the data to the user until it is clear the data is valid (i.e., the data must be buffered at the receiver until the connection reaches the ESTABLISHED state). The three-way handshake reduces the possibility of false connections. It is the implementation of a trade-off between memory and messages to provide information for this checking.

The simplest three-way handshake is shown in Figure 7 below. The figures should be interpreted in the following way. Each line is numbered for reference purposes. Right arrows (→) indicate departure of a TCP segment from TCP A to TCP B, or arrival of a segment at B from A. Left arrows (←) indicate the reverse. Ellipsis (. . .) indicates a segment which is still in the network (delayed). An "XXX" indicates a segment which is lost or rejected. Comments appear in parentheses. TCP states represent the state AFTER the departure or arrival of the segment (whose contents are shown in the center of each line). Segment contents are shown in abbreviated form, with sequence number, control flags, and ACK field. Other fields such as window, addresses, lengths, and text have been left out in the interest of clarity.

In line 2 of Figure 7, TCP A begins by sending a SYN segment indicating that it will use sequence numbers starting with sequence number 100. In line 3, TCP B sends a SYN and acknowledges the SYN it received from TCP A. Note that the acknowledgment field indicates TCP B is now expecting to hear sequence 101, acknowledging the SYN which occupied sequence 100.

At line 4, TCP A responds with an empty segment containing an ACK for TCP Bs SYN; and in line 5, TCP A sends some data. Note that

```
      TCP A                                                TCP B

  1.  CLOSED                                               LISTEN

  2.  SYN-SENT    --> <SEQ=100><CTL=SYN>               --> SYN-RECEIVED

  3.  ESTABLISHED <-- <SEQ=300><ACK=101><CTL=SYN,ACK>  <-- SYN-RECEIVED

  4.  ESTABLISHED --> <SEQ=101><ACK=301><CTL=ACK>       --> ESTABLISHED

  5.  ESTABLISHED --> <SEQ=101><ACK=301><CTL=ACK><DATA> --> ESTABLISHED
```

Figure 7 Basic 3-Way Handshake for Connection Synchronization.

the sequence number of the segment in line 5 is the same as in line 4 because the ACK does not occupy sequence number space (if it did, we would wind up ACKing ACKs!).

Simultaneous initiation is only slightly more complex, as is shown in Figure 8. Each TCP cycles from CLOSED to SYN-SENT to SYN-RECEIVED to ESTABLISHED.

The principle reason for the three-way handshake is to prevent old duplicate connection initiations from causing confusion. To deal with this, a special control message, reset, has been devised. If the receiving TCP is in a nonsynchronized state (i.e., SYN-SENT, SYN-RECEIVED), it returns to LISTEN on receiving an acceptable reset. If the TCP is in one of the synchronized states (ESTABLISHED, FIN-WAIT-1, FIN-WAIT-2, CLOSE-WAIT, CLOSING, LAST-ACK, TIME-WAIT), it aborts the connection and informs its user. We discuss this latter case under "half-open" connections below.

As a simple example of recovery from old duplicates, consider Figure 9. At line 3, an old duplicate SYN arrives at TCP B. TCP B cannot tell that this is an old duplicate, so it responds normally (line 4). TCP A detects that the ACK field is incorrect and returns a RST (reset) with its SEQ field selected to make the segment believable. TCP B, on receiving the RST, returns to the LISTEN state. When the original SYN (pun intended) finally arrives at line 6, the synchronization proceeds normally. If the SYN at line 6 had arrived before the RST, a more complex exchange might have occurred with RSTs sent in both directions.

Half-open connections and other anomalies. An established connection is said to be "half-open" if one of the TCPs has closed or aborted the

```
     TCP A                                                TCP B

1.   CLOSED                                               CLOSED

2.   SYN-SENT     --> <SEQ=100><CTL=SYN>                  ...

3.   SYN-RECEIVED <-- <SEQ=300><CTL=SYN>                  <-- SYN-SENT

4.                ... <SEQ=100><CTL=SYN>                  --> SYN-RECEIVED

5.  SYN-RECEIVED --> <SEQ=100><ACK=301><CTL=SYN,ACK> ...
6.  ESTABLISHED  <-- <SEQ=300><ACK=101><CTL=SYN,ACK> <-- SYN-RECEIVED
7.               ... <SEQ=101><ACK=301><CTL=ACK>      --> ESTABLISHED
```

Figure 8 Simultaneous Connection Synchronization.

```
    TCP A                                                TCP B

1.  CLOSED                                               LISTEN

2.  SYN-SENT    --> <SEQ=100><CTL=SYN>               ...

3.  (duplicate) ... <SEQ=90><CTL=SYN>               --> SYN-RECEIVED

4.  SYN-SENT    <-- <SEQ=300><ACK=91><CTL=SYN,ACK>  <-- SYN-RECEIVED

5.  SYN-SENT    --> <SEQ=91><CTL=RST>               --> LISTEN

6.              ... <SEQ=100><CTL=SYN>               --> SYN-RECEIVED

7.  SYN-SENT    <-- <SEQ=400><ACK=101><CTL=SYN,ACK>  <-- SYN-RECEIVED

8.  ESTABLISHED --> <SEQ=101><ACK=401><CTL=ACK>      --> ESTABLISHED
```

Figure 9 Recovery from Old Duplicate SYN.

connection at its end without the knowledge of the other, or if the two ends of the connection have become desynchronized owing to a crash that resulted in loss of memory. Such connections will automatically become reset if an attempt is made to send data in either direction. However, half-open connections are expected to be unusual, and the recovery procedure is mildly involved.

If at site A the connection no longer exists, then an attempt by the user at site B to send any data on it will result in the site B TCP receiving a reset control message. Such a message indicates to the site B TCP that something is wrong, and it is expected to abort the connection.

Assume that two user processes, A and B, are communicating with one another when a crash occurs causing loss of memory to A's TCP. Depending on the operating system supporting A's TCP, it is likely that some error recovery mechanism exists. When the TCP is up again, A is likely to start again from the beginning or from a recovery point. As a result, A will probably try to OPEN the connection again or try to SEND on the connection it believes open. In the latter case, it receives the error message "connection not open" from the local (A's) TCP. In an attempt to establish the connection, A's TCP will send a segment containing SYN. This scenario leads to the example shown in Figure 10. After TCP A crashes, the user attempts to re-open the connection. TCP B, in the meantime, thinks the connection is open.

When the SYN arrives at line 3, TCP B, being in a synchronized

```
    TCP A                                                  TCP B

1.  (CRASH)                                 (send 300,receive 100)

2.  CLOSED                                             ESTABLISHED

3.  SYN-SENT --> <SEQ=400><CTL=SYN>                    --> (??)

4.  (!!)     <-- <SEQ=300><ACK=100><CTL=ACK>      <-- ESTABLISHED

5.  SYN-SENT --> <SEQ=100><CTL=RST>               --> (Abort!!)

6.  SYN-SENT                                           CLOSED

7.  SYN-SENT --> <SEQ=400><CTL=SYN>               -->
```

Figure 10 Half-Open Connection Discovery.

state, and the incoming segment outside the window, responds with an acknowledgment indicating what sequence it next expects to hear (ACK 100). TCP A sees that this segment does not acknowledge anything it sent and, being unsynchronized, sends a reset (RST) because it has detected a half-open connection. TCP B aborts at line 5. TCP A will continue to try to establish the connection; the problem is now reduced to the basic 3-way handshake of Figure 7.

An interesting alternative case occurs when TCP A crashes and TCP B tries to send data on what it thinks is a synchronized connection. This is illustrated in Figure 11. In this case, the data arriving at TCP A from TCP B (line 2) is unacceptable because no such connection exists, so TCP A sends a RST. The RST is acceptable so TCP B processes it and aborts the connection.

```
    TCP A                                                    TCP B

1.  (CRASH)                                   (send 300,receive 100)

2.  (??)  <-- <SEQ=300><ACK=100><DATA=10><CTL=ACK> <-- ESTABLISHED

3.        --> <SEQ=100><CTL=RST>                   --> (ABORT!!)
```

Figure 11 Active Side Causes Half-Open Connection Discovery.

In Figure 12, we find the two TCPs, A and B, with passive connections waiting for SYN. An old duplicate arriving at TCP B (line 2) stirs B into action. A SYN-ACK is returned (line 3) and causes TCP A to generate a RST (the ACK in line 3 is not acceptable). TCP B accepts the reset and returns to its passive LISTEN state.

A variety of other cases are possible, all of which are accounted for by the following rules for RST generation and processing.

Reset Generation

As a general rule, reset (RST) must be sent whenever a segment arrives which apparently is not intended for the current connection. A reset must not be sent if it is not clear that this is the case.

There are three groups of states:

1. If the connection does not exist (CLOSED) then a reset is sent in response to any incoming segment except another reset. In particular, SYNs addressed to a nonexistent connection are rejected by this means.

   If the incoming segment has an ACK field, the reset takes its sequence number from the ACK field of the segment, otherwise the reset has sequence number zero and the ACK field is set to the sum of the sequence number and segment length of the incoming segment. The connection remains in the CLOSED state.

2. If the connection is in any nonsynchronized state (LISTEN, SYN-SENT, SYN-RECEIVED), and the incoming segment acknowledges something not yet sent (the segment carries an unacceptable ACK), or if an incoming segment has a security level or compartment which does not exactly match the level and compartment requested for the connection, a reset is sent.

   If our SYN has not been acknowledged and the precedence level of the incoming segment is higher than the precedence level requested, then either raise the local precedence level (if allowed by

```
    TCP A                                                TCP B

1.  LISTEN                                               LISTEN

2.       ... <SEQ=Z><CTL=SYN>                       -->  SYN-RECEIVED

3.  (??) <-- <SEQ=X><ACK=Z+1><CTL=SYN,ACK>          <--  SYN-RECEIVED

4.       --> <SEQ=Z+1><CTL=RST>                     -->  (return to LISTEN!)

5.  LISTEN                                               LISTEN
```

Figure 12 Old Duplicate SYN Initiates a Reset on Two Passive Sockets.

the user and the system) or send a reset; or if the precedence level of the incoming segment is lower than the precedence level requested, then continue as if the precedence matched exactly (if the remote TCP cannot raise the precedence level to match ours this will be detected in the next segment it sends, and the connection will be terminated then). If our SYN has been acknowledged (perhaps in this incoming segment) the precedence level of the incoming segment must match the local precedence level exactly, if it does not, a reset must be sent.

If the incoming segment has an ACK field, the reset takes its sequence number from the ACK field of the segment, otherwise the reset has sequence number zero and the ACK field is set to the sum of the sequence number and segment length of the incoming segment. The connection remains in the same state.

3. If the connection is in a synchronized state (ESTABLISHED, FIN-WAIT-1, FIN-WAIT-2, CLOSE-WAIT, CLOSING, LAST-ACK, TIME-WAIT), any unacceptable segment (out of window sequence number or unacceptable acknowledgement number) must elicit only an empty acknowledgment segment containing the current send-sequence number and an acknowledgment indicating the next sequence number expected to be received, and the connection remains in the same state. If an incoming segment has a security level, or compartment, or precedence which does not exactly match the level, compartment, and precedence requested for the connection, a reset is sent and the connection goes to the CLOSED state. The reset takes its sequence number from the ACK field of the incoming segment.

Reset Processing

In all states except SYN-SENT, all reset (RST) segments are validated by checking their SEQ-fields. A reset is valid if its sequence number is in the window. In the SYN-SENT state (a RST received in response to an initial SYN), the RST is acceptable if the ACK field acknowledges the SYN.

The receiver of a RST first validates it, then changes state. If the receiver was in the LISTEN state, it ignores it. If the receiver was in SYN-RECEIVED state and had previously been in the LISTEN state, then the receiver returns to the LISTEN state, otherwise the receiver aborts the connection and goes to the CLOSED state. If the receiver was in any other state, it aborts the connection and advises the user and goes to the CLOSED state.

3.5. Closing a connection

CLOSE is an operation meaning "I have no more data to send." The notion of closing a full-duplex connection is subject to ambiguous interpretation, of course, since it may not be obvious how to treat the receiving side of the connection. We have chosen to treat CLOSE in a simplex fashion. The user who CLOSEs may continue to RECEIVE until he is told that the other side has CLOSED also. Thus, a program could initiate several SENDs followed by a CLOSE, and then continue to RECEIVE until signaled that a RECEIVE failed because the other side has CLOSED. We assume that the TCP will signal a user, even if no RECEIVEs are outstanding, that the other side has closed, so the user can terminate his side gracefully. A TCP will reliably deliver all buffers SENT before the connection was CLOSED so a user who expects no data in return need only wait to hear the connection was CLOSED successfully to know that all his data was received at the destination TCP. Users must keep reading connections they close for sending until the TCP says no more data.

There are essentially three cases:

(1) The user initiates by telling the TCP to CLOSE the connection.

(2) The remote TCP initiates by sending a FIN control signal.

(3) Both users CLOSE simultaneously.

Case 1: local user initiates the close. In this case, a FIN segment can be constructed and placed on the outgoing segment queue. No further SENDs from the user will be accepted by the TCP, and it enters the FIN-WAIT-1 state. RECEIVEs are allowed in this state. All segments preceding and including FIN will be retransmitted until acknowledged. When the other TCP has both acknowledged the FIN and sent a FIN of its own, the first TCP can ACK this FIN. Note that a TCP receiving a FIN will ACK but not send its own FIN until its user has CLOSED the connection also.

Case 2: TCP receives a FIN from the network. If an unsolicited FIN arrives from the network, the receiving TCP can ACK it and tell the user that the connection is closing. The user will respond with a CLOSE, upon which the TCP can send a FIN to the other TCP after sending any remaining data. The TCP then waits until its own FIN is acknowledged whereupon it deletes the connection. If an ACK is not forthcoming, after the user timeout the connection is aborted and the user is told.

Case 3: both users close simultaneously. A simultaneous CLOSE by users at both ends of a connection causes FIN segments to be exchanged. When all segments preceding the FINs have been processed and acknowledged, each TCP can ACK the FIN it has received. Both will, upon receiving these ACKs, delete the connection.

```
     TCP A                                                  TCP B

 1.  ESTABLISHED                                            ESTABLISHED

 2.  (Close)
FIN-WAIT-1         --> <SEQ=100><ACK=300><CTL=FIN,ACK>  --> CLOSE-WAIT
 3.  FIN-WAIT-2    <-- <SEQ=300><ACK=101><CTL=ACK>      <-- CLOSE-WAIT
 4.                                                        (Close)
     TIME-WAIT     <-- <SEQ=300><ACK=101><CTL=FIN,ACK>  <-- LAST-ACK
 5.  TIME-WAIT     --> <SEQ=101><ACK=301><CTL=ACK>      --> CLOSED

 6.  (2 MSL)
     CLOSED
```

Figure 13 Normal Close Sequence.

3.6. Precedence and security

The intent is that connection be allowed only between ports operating with exactly the same security and compartment values and at the higher of the precedence level requested by the two ports. The precedence and security parameters used in TCP are exactly those defined in the Internet Protocol (IP) [2]. Throughout this TCP specification the term "security/compartment" is intended to indicate the security parameters used in IP including security, compartment, user group, and handling restriction.

A connection attempt with mismatched security/compartment val-

```
     TCP A                                                  TCP B

 1.  ESTABLISHED                                            ESTABLISHED

 2.  (Close)                                                (Close)
FIN-WAIT-1  --> <SEQ=100><ACK=300><CTL=FIN,ACK>  ... FIN-WAIT-1
<-- <SEQ=300><ACK=100><CTL=FIN,ACK>  <--
... <SEQ=100><ACK=300><CTL=FIN,ACK>  -->
 3.  CLOSING     --> <SEQ=101><ACK=301><CTL=ACK>      ... CLOSING
                 <-- <SEQ=301><ACK=101><CTL=ACK>      <--
                 ... <SEQ=101><ACK=301><CTL=ACK>      -->

 4.  TIME-WAIT                                              TIME-WAIT
     (2 MSL)                                                (2 MSL)
     CLOSED                                                 CLOSED
```

Figure 14 Simultaneous Close Sequence.

ues or a lower precedence value must be rejected by sending a reset. Rejecting a connection due to too low a precedence only occurs after an acknowledgment of the SYN has been received.

Note that TCP modules which operate only at the default value of precedence will still have to check the precedence of incoming segments and possibly raise the precedence level they use on the connection.

The security parameters may be used even in a nonsecure environment (the values would indicate unclassified data), thus hosts in nonsecure environments must be prepared to receive the security parameters, though they need not send them.

3.7. Data communication

Once the connection is established data is communicated by the exchange of segments. Because segments may be lost due to errors (checksum test failure), or network congestion, TCP uses retransmission (after a timeout) to ensure delivery of every segment. Duplicate segments may arrive due to network or TCP retransmission. As discussed in the section on sequence numbers the TCP performs certain tests on the sequence and acknowledgment numbers in the segments to verify their acceptability.

The sender of data keeps track of the next sequence number to use in the variable SND.NXT. The receiver of data keeps track of the next sequence number to expect in the variable RCV.NXT. The sender of data keeps track of the oldest unacknowledged sequence number in the variable SND.UNA. If the data flow is momentarily idle and all data sent has been acknowledged then the three variables will be equal. When the sender creates a segment and transmits it, the sender advances SND.NXT. When the receiver accepts a segment it advances RCV.NXT and sends an acknowledgment. When the data sender receives an acknowledgment it advances SND.UNA. The extent to which the values of these variables differ is a measure of the delay in the communication. The amount by which the variables are advanced is the length of the data in the segment. Note that once in the ESTABLISHED state all segments must carry current acknowledgment information.

The CLOSE user call implies a push function, as does the FIN control flag in an incoming segment.

Retransmission timeout. Because of the variability of the networks that compose an internetwork system and the wide range of uses of TCP connections the retransmission timeout must be dynamically determined. One procedure for determining a retransmission timeout is given here as an illustration.

An example retransmission timeout procedure. Measure the elapsed time between sending a data octet with a particular sequence number and receiving an acknowledgment that covers that sequence number (segments sent do not have to match segments received). This measured elapsed time is the Round Trip Time (RTT). Next compute a Smoothed Round Trip Time (SRTT) as

SRTT = (ALPHA * SRTT) + ((1-ALPHA) * RTT)

and based on this, compute the retransmission timeout (RTO) as

RTO = min[UBOUND, max[LBOUND, (BETA*SRTT)]],

where UBOUND is an upper bound on the timeout (e.g., 1 minute), LBOUND is a lower bound on the timeout (e.g., 1 second), ALPHA is a smoothing factor (e.g., .8 to .9), and BETA is a delay variance factor (e.g., 1.3 to 2.0).

The communication of urgent information. The objective of the TCP urgent mechanism is to allow the sending user to stimulate the receiving user to accept some urgent data and to permit the receiving TCP to indicate to the receiving user when all the currently known urgent data has been received by the user. This mechanism permits a point in the data stream to be designated as the end of urgent information. Whenever this point is in advance of the receive sequence number (RCV.NXT) at the receiving TCP, that TCP must tell the user to go into "urgent mode"; when the receive sequence number catches up to the urgent pointer, the TCP must tell the user to go into "normal mode." If the urgent pointer is updated while the user is in "urgent mode," the update will be invisible to the user.

The method employs an urgent field which is carried in all segments transmitted. The URG control flag indicates that the urgent field is meaningful and must be added to the segment sequence number to yield the urgent pointer. The absence of this flag indicates that there is no urgent data outstanding.

To send an urgent indication the user must also send at least one data octet. If the sending user also indicates a push, timely delivery of the urgent information to the destination process is enhanced.

Managing the window. The window sent in each segment indicates the range of sequence numbers the sender of the window (the data receiver) is currently prepared to accept. There is an assumption that this is related to the currently available data buffer space available for this connection.

Indicating a large window encourages transmissions. If more data arrives than can be accepted, it will be discarded. This will result in excessive retransmissions, adding unnecessarily to the load on the network and the TCPs. Indicating a small window may restrict the transmission of data to the point of introducing a round trip delay between each new segment transmitted.

The mechanisms provided allow a TCP to advertise a large window and to subsequently advertise a much smaller window without having accepted that much data. This so-called "shrinking the window," is strongly discouraged. The robustness principle dictates that TCPs will not shrink the window themselves, but will be prepared for such behavior on the part of other TCPs.

The sending TCP must be prepared to accept from the user and send at least one octet of new data even if the send window is zero. The sending TCP must regularly retransmit to the receiving TCP even when the window is zero. Two minutes is recommended for the retransmission interval when the window is zero. This retransmission is essential to guarantee that when either TCP has a zero window the reopening of the window will be reliably reported to the other.

When the receiving TCP has a zero window and a segment arrives it must still send an acknowledgment showing its next expected sequence number and current window (zero).

The sending TCP packages the data to be transmitted into segments which fit the current window, and may repackage segments on the retransmission queue. Such repackaging is not required, but may be helpful.

In a connection with a one-way data flow, the window information will be carried in acknowledgment segments that all have the same sequence number so there will be no way to reorder them if they arrive out of order. This is not a serious problem, but it will allow the window information to be on occasion temporarily based on old reports from the data receiver. A refinement to avoid this problem is to act on the window information from segments that carry the highest acknowledgment number (that is, segments with acknowledgment number equal or greater than the highest previously received).

The window management procedure has significant influence on the communication performance. The following comments are suggestions to implementers.

Window management suggestions. Allocating a very small window causes data to be transmitted in many small segments when better performance is achieved using fewer large segments.

One suggestion for avoiding small windows is for the receiver to defer updating a window until the additional allocation is at least × percent of the maximum allocation possible for the connection (where × might be 20 to 40).

Another suggestion is for the sender to avoid sending small segments by waiting until the window is large enough before sending data. If the user signals a push function then the data must be sent even if it is a small segment.

Note that the acknowledgments should not be delayed or unnecessary retransmissions will result. One strategy would be to send an acknowledgment when a small segment arrives (without updating the window information), and then to send another acknowledgment with new window information when the window is larger.

The segment sent to probe a zero window may also begin a break up of transmitted data into smaller and smaller segments. If a segment containing a single data octet sent to probe a zero window is accepted, it consumes one octet of the window now available. If the sending TCP simply sends as much as it can whenever the window is nonzero, the transmitted data will be broken into alternating big and small segments. As time goes on, occasional pauses in the receiver making window allocation available will result in breaking the big segments into a small and not quite so big pair. And after a while the data transmission will be in mostly small segments.

The suggestion here is that the TCP implementations need to actively attempt to combine small window allocations into larger windows, since the mechanisms for managing the window tend to lead to many small windows in the simplest minded implementations.

3.8. Interfaces

There are of course two interfaces of concern: the user/TCP interface and the TCP/lower-level interface. We have a fairly elaborate model of the user/TCP interface, but the interface to the lower level protocol module is left unspecified here, since it will be specified in detail by the specification of the lower level protocol. For the case that the lower level is IP we note some of the parameter values that TCPs might use.

User/TCP interface. The following functional description of user commands to the TCP is, at best, fictional, since every operating system will have different facilities. Consequently, we must warn readers that different TCP implementations may have different user interfaces. However, all TCPs must provide a certain minimum set of services to guarantee that all TCP implementations can support the same protocol hierarchy. This section specifies the functional interfaces required of all TCP implementations.

TCP user commands. The following sections functionally characterize a USER/TCP interface. The notation used is similar to most procedure or function calls in high-level languages, but this usage is not meant to rule out trap type service calls (e.g., SVCs, UUOs, EMTs).

The user commands described below specify the basic functions the TCP must perform to support interprocess communication. Individual implementations must define their own exact format, and may provide combinations or subsets of the basic functions in single calls. In particular, some implementations may wish to automatically OPEN a connection on the first SEND or RECEIVE issued by the user for a given connection.

In providing interprocess communication facilities, the TCP must not only accept commands, but must also return information to the processes it serves. The latter consists of: (a) general information about a connection (e.g., interrupts, remote close, binding of unspecified foreign socket), and (b) replies to specific user commands indicating success or various types of failure.

Open Format: OPEN (local port, foreign socket, active/passive [, timeout] [, precedence] [, security/compartment] [, options]). > local connection name

We assume that the local TCP is aware of the identity of the processes it serves and will check the authority of the process to use the connection specified. Depending upon the implementation of the TCP, the local network and TCP identifiers for the source address will either be supplied by the TCP or the lower level protocol (e.g., IP). These considerations are the result of concern about security, to the extent that no TCP is able to masquerade as another one, and so on. Similarly, no process can masquerade as another without the collusion of the TCP.

If the active/passive flag is set to passive, then this is a call to LISTEN for an incoming connection. A passive open may have either a fully specified foreign socket to wait for a particular connection or an unspecified foreign socket to wait for any call. A fully specified passive call can be made active by the subsequent execution of a SEND.

A transmission control block (TCB) is created and partially filled in with data from the OPEN command parameters. On an active OPEN command, the TCP will begin the procedure to synchronize (i.e., establish) the connection at once.

The timeout, if present, permits the caller to set up a timeout for all data submitted to TCP. If data is not successfully delivered to the destination within the timeout period, the TCP will abort the connection. The present global default is five minutes.

The TCP or some component of the operating system will verify the users authority to open a connection with the specified precedence or security/compartment. The absence of precedence or security/compartment specification in the OPEN call indicates the default values must be used.

TCP will accept incoming requests as matching only if the security/compartment information is exactly the same and only if the precedence is equal to or higher than the precedence requested in the OPEN call.

The precedence for the connection is the higher of the values requested in the OPEN call and received from the incoming request, and fixed at that value for the life of the connection. Implementers may want to give the user control of this precedence negotiation. For example, the user might be allowed to specify that the precedence must be exactly matched, or that any attempt to raise the precedence be confirmed by the user.

A local connection name will be returned to the user by the TCP. The local connection name can then be used as a short-hand term for the connection defined by the <local socket, foreign socket> pair.

Send Format: SEND (local connection name, buffer address, byte count, PUSH flag, URGENT flag [,timeout])

This call causes the data contained in the indicated user buffer to be sent on the indicated connection. If the connection has not been opened, the SEND is considered an error. Some implementations may allow users to SEND first; in which case, an automatic OPEN would be done. If the calling process is not authorized to use this connection, an error is returned. If the PUSH flag is set, the data must be transmitted promptly to the receiver, and the PUSH bit will be set in the last TCP segment created from the buffer. If the PUSH flag is not set, the data may be combined with data from subsequent SENDs for transmission efficiency.

If the URGENT flag is set, segments sent to the destination TCP will have the urgent pointer set. The receiving TCP will signal the urgent condition to the receiving process if the urgent pointer indicates that data preceding the urgent pointer has not been consumed by the receiving process. The purpose of urgent is to stimulate the receiver to process the urgent data and to indicate to the receiver when all the currently known urgent data has been received. The number of times the sending user's TCP signals urgent will not necessarily be equal to the number of times the receiving user will be notified of the presence of urgent data.

If no foreign socket was specified in the OPEN, but the connection is established (e.g., because a LISTENing connection has become specific due to a foreign segment arriving for the local socket), then the designated buffer is sent to the implied foreign socket. Users who make use of OPEN with an unspecified foreign socket can make use of SEND without ever explicitly knowing the foreign socket address.

However, if a SEND is attempted before the foreign socket becomes specified, an error will be returned. Users can use the STATUS call to determine the status of the connection. In some implementations the TCP may notify the user when an unspecified socket is bound.

If a timeout is specified, the current user timeout for this connection is changed to the new one.

In the simplest implementation, SEND would not return control to the sending process until either the transmission was complete or the timeout had been exceeded. However, this simple method is both subject to deadlocks (for example, both sides of the connection might try to do SENDs before doing any RECEIVEs) and offers poor performance, so it is not recommended. A more sophisticated implementation would return immediately to allow the process to run concurrently with network I/O, and, furthermore, to allow multiple SENDs to be in progress. Multiple SENDs are served in first come, first served order, so the TCP will queue those it cannot service immediately.

We have implicitly assumed an asynchronous user interface in which a SEND later elicits some kind of SIGNAL or pseudo-interrupt from the serving TCP. An alternative is to return a response immediately. For instance, SENDs might return immediate local acknowledgment, even if the segment sent had not been acknowledged by the distant TCP. We could optimistically assume eventual success. If we are wrong, the connection will close anyway due to the timeout. In implementations of this kind (synchronous), there will still be some asynchronous signals, but these will deal with the connection itself, and not with specific segments or buffers.

In order for the process to distinguish among error or success indications for different SENDs, it might be appropriate for the buffer address to be returned along with the coded response to the SEND request. TCP-to-user signals are discussed below, indicating the information which should be returned to the calling process.

Receive Format: RECEIVE (local connection name, buffer address, byte count) −> byte count, urgent flag, push flag

This command allocates a receiving buffer associated with the specified connection. If no OPEN precedes this command or the calling process is not authorized to use this connection, an error is returned.

In the simplest implementation, control would not return to the calling program until either the buffer was filled, or some error occurred, but this scheme is highly subject to deadlocks.

A more sophisticated implementation would permit several RECEIVEs to be outstanding at once. These would be filled as segments arrive. This strategy permits increased throughput at the cost of a more elaborate scheme (possibly asynchronous) to notify the calling program that a PUSH has been seen or a buffer filled.

If enough data arrive to fill the buffer before a PUSH is seen, the PUSH flag will not be set in the response to the RECEIVE. The buffer will be filled with as much data as it can hold. If a PUSH is seen before the buffer is filled, the buffer will be returned partially filled and PUSH indicated.

If there is urgent data the user will have been informed as soon as it arrived via a TCP-to-user signal. The receiving user should thus be in "urgent mode." If the URGENT flag is on, additional urgent data remains. If the URGENT flag is off, this call to RECEIVE has returned all the urgent data, and the user may now leave "urgent mode." Note that data following the urgent pointer (nonurgent data) cannot be delivered to the user in the same buffer with preceeding urgent data unless the boundary is clearly marked for the user.

To distinguish among several outstanding RECEIVEs and to take care of the case that a buffer is not completely filled, the return code is accompanied by both a buffer pointer and a byte count indicating the actual length of the data received. Alternative implementations of RECEIVE might have the TCP allocate buffer storage, or the TCP might share a ring buffer with the user.

Close Format: CLOSE (local connection name)

This command causes the connection specified to be closed. If the connection is not open or the calling process is not authorized to use this connection, an error is returned. Closing connections is intended to be a graceful operation in the sense that outstanding SENDs will be transmitted (and retransmitted), as flow control permits, until all have been serviced. Thus, it should be acceptable to make several SEND calls, followed by a CLOSE, and expect all the data to be sent to the destination. It should also be clear that users should continue to RECEIVE on CLOSING connections, since the other side may be trying to transmit the last of its data. Thus, CLOSE means "I have no more to send" but does not mean "I will not receive any more." It may happen (if the user level protocol is not well thought out) that the closing side is unable to get rid of all its data before timing out. In this event, CLOSE turns into ABORT, and the closing TCP gives up.

The user may CLOSE the connection at any time on his own initiative, or in response to various prompts from the TCP (e.g., remote close executed, transmission timeout exceeded, destination inaccessible).

Because closing a connection requires communication with the foreign TCP, connections may remain in the closing state for a short time. Attempts to reopen the connection before the TCP replies to the CLOSE command will result in error responses. Close also implies push function.

Status Format: STATUS (local connection name) −>status data

This is an implementation dependent user command and could be excluded without adverse effect. Information returned would typically come from the TCB associated with the connection. This command returns a data block containing the following information:

local socket,
foreign socket,
local connection name,
receive window,
send window,
connection state,
number of buffers awaiting acknowledgment,
number of buffers pending receipt,
urgent state,
precedence,
security/compartment,
and transmission timeout.

Depending on the state of the connection, or on the implementation itself, some of this information may not be available or meaningful. If the calling process is not authorized to use this connection, an error is returned. This prevents unauthorized processes from gaining information about a connection.

Abort Format: ABORT (local connection name)

This command causes all pending SENDs and RECEIVES to be aborted, the TCB to be removed, and a special RESET message to be sent to the TCP on the other side of the connection. Depending on the implementation, users may receive abort indications for each outstanding SEND or RECEIVE, or may simply receive an ABORT-acknowledgment.

TCP-to-user messages. It is assumed that the operating system environment provides a means for the TCP to asynchronously signal the user program. When the TCP does signal a user program, certain information is passed to the user. Often in the specification the information will be an error message. In other cases there will be information relating to the completion of processing a SEND or RECEIVE or other user call.

The following information is provided:

Local Connection Name	Always
Response String	Always
Buffer Address	Send & Receive
Byte count (counts bytes received)	Receive
Push flag	Receive
Urgent flag	Receive

TCP/lower-level interface. The TCP calls on a lower level protocol module to actually send and receive information over a network. One case is that of the ARPA internetwork system where the lower level module is the Internet Protocol (IP) [2].

If the lower level protocol is IP it provides arguments for a type of service and for a time to live. TCP uses the following settings for these parameters:

Type of Service = Precedence: routine,

Delay: normal,

Throughput: normal,

Reliability: normal, or 00000000,

Time to Live = one minute, or 00111100.

Note that the assumed maximum segment lifetime is two minutes. Here we explicitly ask that a segment be destroyed if it cannot be delivered by the internet system within one minute.

If the lower level is IP (or other protocol that provides this feature) and source routing is used, the interface must allow the route information to be communicated. This is especially important so that the source and destination addresses used in the TCP checksum be the originating source and ultimate destination. It is also important to preserve the return route to answer connection requests.

Any lower level protocol will have to provide the source address, destination address, and protocol fields, and some way of determining the "TCP length," both to provide the functional equivalent service of IP and to be used in the TCP checksum.

3.9. Event processing

The processing depicted in this section is an example of one possible implementation. Other implementations may have slightly different processing sequences, but they should differ from those in this section only in detail, not in substance.

The activity of the TCP can be characterized as responding to events. The events that occur can be cast into three categories: user calls, arriving segments, and timeouts. This section describes the processing the TCP does in response to each of the events. In many cases the processing required depends on the state of the connection. Events that occur:

User Calls

OPEN

SEND

RECEIVE

CLOSE

ABORT

STATUS

Arriving Segments

SEGMENT ARRIVES

Timeouts

USER TIMEOUT

RETRANSMISSION TIMEOUT

TIME-WAIT TIMEOUT

The model of the TCP/user interface is that user commands receive an immediate return and possibly a delayed response via an event or pseudointerrupt. In the following descriptions, the term "signal" means cause a delayed response.

Error responses are given as character strings. For example, user commands referencing connections that do not exist receive "error: connection not open."

Please note in the following that all arithmetic on sequence numbers, acknowledgment numbers, windows, etc., is modulo 2**32 the size of the sequence number space. Also note that "=<" means less than or equal to (modulo 2**32).

A natural way to think about processing incoming segments is to imagine that they are first tested for proper sequence number (i.e., that their contents lie in the range of the expected "receive window" in the sequence number space) and then that they are generally queued and processed in sequence number order.

When a segment overlaps other already received segments we reconstruct the segment to contain just the new data and adjust the header fields to be consistent.

Note that if no state change is mentioned the TCP stays in the same state.

OPEN Call

OPEN Call

CLOSED STATE (i.e., TCB does not exist)

Create a new transmission control block (TCB) to hold connection state information. Fill in local socket identifier, foreign socket, precedence, security/compartment, and user timeout information. Note that some parts of the foreign socket may be unspecified in a passive OPEN and are to be filled in by the parameters of the incoming SYN segment. Verify that the security and precedence requested are allowed for this user, and if not, return "error: precedence not allowed" or "error: security/compartment not allowed." If passive, enter the LISTEN state and return. If active and the foreign socket is unspecified, return "error: foreign socket unspecified"; if active and the foreign socket is specified, issue a SYN segment. An initial send sequence number (ISS) is selected. A SYN segment of the form <SEQ=ISS><CTL=SYN> is sent. Set SND.UNA to ISS, SND.NXT to ISS+1, enter SYN-SENT state, and return.

If the caller does not have access to the local socket specified, return "error: connection illegal for this process." If there is no room to create a new connection, return "error: insufficient resources."

LISTEN STATE

If active and the foreign socket is specified, then change the connection from passive to active, select an ISS. Send a SYN segment, set SND.UNA to ISS, SND.NXT to ISS+1. Enter SYN-SENT state. Data associated with SEND may be sent with SYN segment or queued for transmission after entering ESTABLISHED state. The urgent bit if requested in the command must be sent with the data segments sent as a result of this command. If there is no room to queue the request, respond with "error: insufficient resources." If foreign socket was not specified, then return "error: foreign socket unspecified."

OPEN Call

SYN-SENT STATE

SYN-RECEIVED STATE

ESTABLISHED STATE

FIN-WAIT-1 STATE

FIN-WAIT-2 STATE

CLOSE-WAIT STATE

CLOSING STATE

LAST-ACK STATE

TIME-WAIT STATE

Return "error: connection already exists."

SEND Call

SEND Call

CLOSED STATE (i.e., TCB does not exist)

If the user does not have access to such a connection, then return "error: connection illegal for this process." Otherwise, return "error: connection does not exist."

LISTEN STATE

If the foreign socket is specified, then change the connection from passive to active, select an ISS. Send a SYN segment, set SND.UNA to ISS, SND.NXT to ISS+1. Enter SYN-SENT state. Data associated with SEND may be sent with SYN segment or queued for transmission after entering ESTABLISHED state. The urgent bit if requested in the command must be sent with the data segments sent as a result of this command. If there is no room to queue the request, respond with "error: insufficient resources." If foreign socket was not specified, then return "error: foreign socket unspecified."

SYN-SENT STATE

SYN-RECEIVED STATE

Queue the data for transmission after entering ESTABLISHED state. If no space to queue, respond with "error: insufficient resources."

ESTABLISHED STATE

CLOSE-WAIT STATE

Segmentize the buffer and send it with a piggybacked acknowledgment (acknowledgment value=RCV.NXT). If there is insufficient space to remember this buffer, simply return "error: insufficient resources."

If the urgent flag is set, then SND.UP <−SND.NXT-1 and set the urgent pointer in the outgoing segments.

SEND Call

FIN-WAIT-1 STATE

FIN-WAIT-2 STATE

CLOSING STATE

LAST-ACK STATE

TIME-WAIT STATE

Return "error: connection closing" and do not service request.

RECEIVE Call

RECEIVE Call

CLOSED STATE (i.e., TCB does not exist)

If the user does not have access to such a connection, return "error: connection illegal for this process."

Otherwise return "error: connection does not exist."

LISTEN STATE

SYN-SENT STATE

SYN-RECEIVED STATE

Queue for processing after entering ESTABLISHED state. If there is no room to queue this request, respond with "error: insufficient resources."

ESTABLISHED STATE

FIN-WAIT-1 STATE

FIN-WAIT-2 STATE

If insufficient incoming segments are queued to satisfy the request, queue the request. If there is no queue space to remember the RECEIVE, respond with "error: insufficient resources."

Reassemble queued incoming segments into receive buffer and return to user. Mark "push seen" (PUSH) if this is the case. If RCV.UP is in advance of the data currently being passed to the user, notify the user of the presence of urgent data. When the TCP takes responsibility for delivering data to the user that fact must be communicated to the sender via an acknowledgment. The formation of such an acknowledgment is described below in the discussion of processing an incoming segment.

RECEIVE Call

CLOSE-WAIT STATE

Since the remote side has already sent FIN, RECEIVEs must be satisfied by text already on hand, but not yet delivered to the user. If no text is awaiting delivery, the RECEIVE will get an "error: connection closing" response. Otherwise, any remaining text can be used to satisfy the RECEIVE.

CLOSING STATE

LAST-ACK STATE

TIME-WAIT STATE

Return "error: connection closing."

CLOSE Call

CLOSE Call

CLOSED STATE (i.e., TCB does not exist)

If the user does not have access to such a connection, return "error: connection illegal for this process." Otherwise, return "error: connection does not exist."

LISTEN STATE

Any outstanding RECEIVEs are returned with "error: closing" responses. Delete TCB, enter CLOSED state, and return.

SYN-SENT STATE

Delete the TCB and return "error: closing" responses to any queued SENDs, or RECEIVEs.

SYN-RECEIVED STATE

If no SENDs have been issued and there is no pending data to send, then form a FIN segment and send it, and enter FIN-WAIT-1 state; otherwise queue for processing after entering ESTABLISHED state.

ESTABLISHED STATE

Queue this until all preceding SENDs have been segmentized, then form a FIN segment and send it. In any case, enter FIN-WAIT-1 state.

FIN-WAIT-1 STATE

FIN-WAIT-2 STATE

Strictly speaking, this is an error and should receive an "error: connection closing" response. An "ok" response would be acceptable too, as long as a second FIN is not emitted (the first FIN may be retransmitted though).

CLOSE Call

CLOSE-WAIT STATE

Queue this request until all preceding SENDs have been segmentized; then send a FIN segment, enter CLOSING state.

CLOSING STATE

LAST-ACK STATE

TIME-WAIT STATE

Respond with "error: connection closing."

ABORT Call

ABORT Call

CLOSED STATE (i.e., TCB does not exist)

If the user should not have access to such a connection, return "error: connection illegal for this process." Otherwise return "error: connection does not exist."

LISTEN STATE

Any outstanding RECEIVEs should be returned with "error: connection reset" responses. Delete TCB, enter CLOSED state, and return.

SYN-SENT STATE

All queued SENDs and RECEIVEs should be given "connection reset" notification, delete the TCB, enter CLOSED state, and return.

SYN-RECEIVED STATE

ESTABLISHED STATE

FIN-WAIT-1 STATE

FIN-WAIT-2 STATE

CLOSE-WAIT STATE

Send a reset segment:

<SEQ=SND.NXT> <CTL=RST>

All queued SENDs and RECEIVEs should be given "connection reset" notification; all segments queued for transmission (except for the RST formed above) or retransmission should be flushed, delete the TCB, enter CLOSED state, and return.

CLOSING STATE

LAST-ACK STATE

TIME-WAIT STATE

Respond with "ok" and delete the TCB, enter CLOSED state, and return.

STATUS Call

STATUS Call

CLOSED STATE (i.e., TCB does not exist)

If the user should not have access to such a connection, return "error: connection illegal for this process." Otherwise return "error: connection does not exist."

LISTEN STATE

Return "state = LISTEN," and the TCB pointer.

SYN-SENT STATE

Return "state = SYN-SENT," and the TCB pointer.

SYN-RECEIVED STATE

Return "state = SYN-RECEIVED," and the TCB pointer.

ESTABLISHED STATE

Return "state = ESTABLISHED," and the TCB pointer.

FIN-WAIT-1 STATE

Return "state = FIN-WAIT-1," and the TCB pointer.

FIN-WAIT-2 STATE

Return "state = FIN-WAIT-2," and the TCB pointer.

CLOSE-WAIT STATE

Return "state = CLOSE-WAIT," and the TCB pointer.

CLOSING STATE

Return "state = CLOSING," and the TCB pointer.

LAST-ACK STATE

Return "state = LAST-ACK," and the TCB pointer.

STATUS Call

TIME-WAIT STATE

Return "state = TIME-WAIT", and the TCB pointer.

SEGMENT ARRIVES

SEGMENT ARRIVES

If the state is CLOSED (i.e., TCB does not exist) then all data in the incoming segment is discarded. An incoming segment containing an RST is discarded. An incoming segment not containing an RST causes an RST to be sent in response. The acknowledgment and sequence field values are selected to make the reset sequence acceptable to the TCP that sent the offending segment.
If the ACK bit is off, sequence number zero is used,

<SEQ=0><ACK=SEG.SEQ+SEG.LEN> <CTL=RST,ACK>

If the ACK bit is on,

<SEQ=SEG.ACK><CTL=RST>

Return.
If the state is LISTEN then first check for an RST. An incoming RST should be ignored. Return. Second check for an ACK.

Any acknowledgment is bad if it arrives on a connection still in the LISTEN state. An acceptable reset segment should be formed for any arriving ACK-bearing segment. The RST should be formatted as follows:

<SEQ=SEG.ACK><CTL=RST>

Return.
Third check for a SYN.
If the SYN bit is set, check the security. If the security/compartment on the incoming segment does not exactly match the security/compartment in the TCB then send a reset and return.

<SEQ=SEG.ACK><CTL=RST>

SEGMENT ARRIVES

If the SEG.PRC is greater than the TCB.PRC then if allowed by the user and the system set TCB.PRC<−SEG.PRC, if not allowed send a reset and return.

<SEQ=SEG.ACK><CTL=RST>

If the SEG.PRC is less than the TCB.PRC then continue.
Set RCV.NXT to SEG.SEQ+1, IRS is set to SEG.SEQ and any other control or text should be queued for processing later. ISS should be selected and a SYN segment sent of the form:

<SEQ=ISS><ACK=RCV.NXT><CTL=SYN,ACK>

SND.NXT is set to ISS+1 and SND.UNA to ISS. The connection state should be changed to SYN-RECEIVED. Note that any other incoming control or data (combined with SYN) will be processed in the SYN-RECEIVED state, but processing of SYN and ACK should not be repeated. If the listen was not fully specified (i.e., the foreign socket was not fully specified), then the unspecified fields should be filled in now.
Fourth other text or control.
Any other control or text-bearing segment (not containing SYN) must have an ACK and thus would be discarded by the ACK processing. An incoming RST segment could not be valid, since it could not have been sent in response to anything sent by this incarnation of the connection. So you are unlikely to get here, but if you do, drop the segment, and return.
If the state is SYN-SENT then first check the ACK bit
If the ACK bit is set
If SEG.ACK =< ISS, or SEG.ACK > SND.NXT, send a reset (unless the RST bit is set, if so drop the segment and return)
<SEQ=SEG.ACK><CTL=RST> and discard the segment. Return.

If SND.UNA =< SEG.ACK =< SND.NXT then the ACK is acceptable. second check the RST bit

SEGMENT ARRIVES
If the RST bit is set
If the ACK was acceptable then signal the user "error: connection reset", drop the segment, enter CLOSED state, delete TCB, and return. Otherwise (no ACK) drop the segment and return.
third check the security and precedence
If the security/compartment in the segment does not exactly match the security/compartment in the TCB, send a reset
If there is an ACK

<SEQ=SEG.ACK><CTL=RST>

Otherwise

<SEQ=0><ACK=SEG.SEQ+SEG.LEN><CTL=RST,ACK>

If there is an ACK
The precedence in the segment must match the precedence in the TCB, if not, send a reset

<SEQ=SEG.ACK><CTL=RST>

If there is no ACK
If the precedence in the segment is higher than the precedence in the TCB then, if allowed by the user and the system, raise the precedence in the TCB to that in the segment, if not allowed to raise the prec then send a reset.

<SEQ=0 ><ACK=SEG.SEQ+SEG.LEN><CTL=RST,ACK>

If the precedence in the segment is lower than the precedence in the TCB continue.
If a reset was sent, discard the segment and return.
fourth check the SYN bit
This step should be reached only if the ACK is ok, or there is no ACK, and the segment did not contain an RST.
If the SYN bit is on and the security/compartment and precedence

SEGMENT ARRIVES

are acceptable, then RCV.NXT is set to SEG.SEQ+1, IRS is set to SEG.SEQ. SND.UNA should be advanced to equal SEG.ACK (if there is an ACK), and any segments on the retransmission queue which are thereby acknowledged should be removed.
If SND.UNA > ISS (our SYN has been ACKed), change the connection state to ESTABLISHED, form an ACK segment

<SEQ=SND.NXT><ACK=RCV.NXT><CTL=ACK>

and send it. Data or controls which were queued for transmission may be included. If there are other controls or text in the segment then continue processing at the sixth step below where the URG bit is checked, otherwise return.

Otherwise enter SYN-RECEIVED, form a SYN,ACK segment

<SEQ=ISS><ACK=RCV.NXT><CTL=SYN,ACK>

and send it. If there are other controls or text in the segment, queue them for processing after the ESTABLISHED state has been reached, return.

fifth, if neither of the SYN or RST bits is set then drop the segment and return.

SEGMENT ARRIVES

Otherwise,
first check sequence number

SYN-RECEIVED STATE

ESTABLISHED STATE

FIN-WAIT-1 STATE

FIN-WAIT-2 STATE

CLOSE-WAIT STATE

CLOSING STATE

LAST-ACK STATE

TIME-WAIT STATE

Segments are processed in sequence. Initial tests on arrival are used to discard old duplicates, but further processing is done in SEG.SEQ order. If a segment's contents straddle the boundary between old and new, only the new parts should be processed.

There are four cases for the acceptability test for an incoming segment:

Segment Receive Test

Length	Window	
0	0	SEG.SEQ = RCV.NXT
0	>0	RCV.NXT =< SEG.SEQ < RCV.NXT+RCV.WND
>0	0	not acceptable
>0	>0	RCV.NXT =< SEG.SEQ < RCV.NXT+RCV.WND

or RCV.NXT =< SEG.SEQ+SEG.LEN-1 < RCV.NXT+RCV.WND
If the RCV.WND is zero, no segments will be acceptable, but special allowance should be made to accept valid ACKs, URGs, and RSTs.

If an incoming segment is not acceptable, an acknowledgment should be sent in reply (unless the RST bit is set, if so drop the segment and return):

<SEQ=SND.NXT> <ACK=RCV.NXT> <CTL=ACK>

After sending the acknowledgment, drop the unacceptable segment and return.

SEGMENT ARRIVES

In the following it is assumed that the segment is the idealized segment that begins at RCV.NXT and does not exceed the window. One could tailor actual segments to fit this assumption by trimming off any portions that lie outside the window (including SYN and FIN), and only processing further if the segment then begins at RCV.NXT. Segments with higher beginning sequence numbers may be held for later processing.
second check the RST bit,

SYN-RECEIVED STATE

If the RST bit is set
If this connection was initiated with a passive OPEN (i.e., came from the LISTEN state), then return this connection to LISTEN state and return. The user need not be informed. If this connection was initiated with an active OPEN (i.e., came from SYN-SENT state) then the connection was refused, signal the user "connection refused." In either case, all segments on the retransmission queue should be removed. And in the active OPEN case, enter the CLOSED state and delete the TCB, and return.

ESTABLISHED

FIN-WAIT-1

FIN-WAIT-2

CLOSE-WAIT

If the RST bit is set then, any outstanding RECEIVEs and SEND should receive "reset" responses. All segment queues should be flushed. Users should also receive an unsolicited general "connection reset" signal. Enter the CLOSED state, delete the TCB, and return.

CLOSING STATE

LAST-ACK STATE

TIME-WAIT

If the RST bit is set then, enter the CLOSED state, delete the TCB, and return.

SEGMENT ARRIVES

third check security and precedence

SYN-RECEIVED

If the security/compartment and precedence in the segment do not exactly match the security/compartment and precedence in the TCB then send a reset, and return.

ESTABLISHED STATE

If the security/compartment and precedence in the segment do not exactly match the security/compartment and precedence in the TCB then send a reset, any outstanding RECEIVEs and SEND should receive "reset" responses. All segment queues should be flushed. Users should also receive an unsolicited general "connection reset" signal. Enter the CLOSED state, delete the TCB, and return.

Note this check is placed following the sequence check to prevent a segment from an old connection between these ports with a different security or precedence from causing an abort of the current connection.

fourth, check the SYN bit,

SYN-RECEIVED

ESTABLISHED STATE

FIN-WAIT STATE-1

FIN-WAIT STATE-2

CLOSE-WAIT STATE

CLOSING STATE

LAST-ACK STATE

TIME-WAIT STATE

If the SYN is in the window it is an error, send a reset, any outstanding RECEIVEs and SEND should receive "reset" responses, all segment queues should be flushed, the user should also receive an unsolicited general "connection reset" signal, enter the CLOSED state, delete the TCB, and return.

If the SYN is not in the window this step would not be reached and an ACK would have been sent in the first step (sequence number check).

SEGMENT ARRIVES

fifth check the ACK field,

if the ACK bit is off drop the segment and return

if the ACK bit is on

SYN-RECEIVED STATE

If SND.UNA =< SEG.ACK =< SND.NXT then enter ESTABLISHED state and continue processing.

If the segment acknowledgment is not acceptable, form a reset segment,

<SEQ=SEG.ACK> <CTL=RST>

and send it.

ESTABLISHED STATE

If SND.UNA < SEG.ACK =< SND.NXT then, set SND.UNA <- SEG.ACK.

Any segments on the retransmission queue which are thereby entirely acknowledged are removed. Users should receive positive acknowledgments for buffers which have been SENT and fully acknowledged (i.e., SEND buffer should be returned with "ok" response). If the ACK is a duplicate

(SEG.ACK < SND.UNA), it can be ignored. If the ACK acknowledges something not yet sent (SEG.ACK > SND.NXT) then send an ACK, drop the segment, and return.

If SND.UNA < SEG.ACK =< SND.NXT, the send window should be updated. If (SND.WL1 < SEG.SEQ or (SND.WL1 = SEG.SEQ and SND.WL2 =< SEG.ACK)), set SND.WND <- SEG.WND, set SND.WL1 <- SEG.SEQ, and set SND.WL2 <- SEG.ACK.

Note that SND.WND is an offset from SND.UNA, that SND.WL1 records the sequence number of the last segment used to update SND.WND, and that SND.WL2 records the acknowledgment number of the last segment used to update SND.WND. The check here prevents using old segments to update the window.

SEGMENT ARRIVES

FIN-WAIT-1 STATE

In addition to the processing for the ESTABLISHED state, if our FIN is now acknowledged then enter FIN-WAIT-2 and continue processing in that state.

FIN-WAIT-2 STATE

In addition to the processing for the ESTABLISHED state, if the retransmission queue is empty, the user's CLOSE can be acknowledged ("ok") but do not delete the TCB.

CLOSE-WAIT STATE

Do the same processing as for the ESTABLISHED state.

CLOSING STATE

In addition to the processing for the ESTABLISHED state, if the ACK acknowledges our FIN then enter the TIME-WAIT state, otherwise ignore the segment.

LAST-ACK STATE

The only thing that can arrive in this state is an acknowledgment of our FIN. If our FIN is now acknowledged, delete the TCB, enter the CLOSED state, and return.

TIME-WAIT STATE

The only thing that can arrive in this state is a retransmission of the remote FIN. Acknowledge it, and restart the 2 MSL timeout.

sixth, check the URG bit,

ESTABLISHED STATE

FIN-WAIT-1 STATE

FIN-WAIT-2 STATE

If the URG bit is set, RCV.UP <− max (RCV.UP,SEG.UP), and signal the user that the remote side has urgent data if the urgent pointer (RCV.UP) is in advance of the data consumed. If the user has already been signaled (or is still in the "urgent mode") for this continuous sequence of urgent data, do not signal the user again.

SEGMENT ARRIVES

CLOSE-WAIT STATE

CLOSING STATE

LAST-ACK STATE

TIME-WAIT

This should not occur, since a FIN has been received from the remote side. Ignore the URG.

seventh, process the segment text,

ESTABLISHED STATE

FIN-WAIT-1 STATE

FIN-WAIT-2 STATE

Once in the ESTABLISHED state, it is possible to deliver segment text to user RECEIVE buffers. Text from segments can be moved into buffers until either the buffer is full or the segment is empty. If the segment empties and carries a PUSH flag, then the user is informed when the buffer is returned that a PUSH has been received.

When the TCP takes responsibility for delivering the data to the user it must also acknowledge the receipt of the data. Once the TCP takes responsibility for the data it advances RCV.NXT over the data accepted, and adjusts RCV.WND as appropriate to the current buffer availability. The total of RCV.NXT and RCV.WND should not be reduced.

Please note the window management suggestions in section 3.7. Send an acknowledgment of the form:

<SEQ=SND.NXT> <ACK=RCV.NXT> <CTL=ACK>

This acknowledgment should be piggybacked on a segment being transmitted if possible without incurring undue delay.

SEGMENT ARRIVES

CLOSE-WAIT STATE

CLOSING STATE

LAST-ACK STATE

TIME-WAIT STATE

This should not occur, since a FIN has been received from the remote side. Ignore the segment text.

eighth, check the FIN bit,

Do not process the FIN if the state is CLOSED, LISTEN, or SYN-SENT since the SEG.SEQ cannot be validated; drop the segment and return.

If the FIN bit is set, signal the user "connection closing" and return any pending RECEIVEs with same message, advance RCV.NXT over the FIN, and send an acknowledgment for the FIN. Note that FIN implies PUSH for any segment text not yet delivered to the user.

SYN-RECEIVED STATE

ESTABLISHED STATE

Enter the CLOSE-WAIT state.

FIN-WAIT-1 STATE

If our FIN has been ACKed (perhaps in this segment), then enter TIME-WAIT, start the time-wait timer, turn off the other timers; otherwise enter the CLOSING state.

FIN-WAIT-2 STATE

Enter the TIME-WAIT state. Start the time-wait timer, turn off the other timers.

CLOSE-WAIT STATE

Remain in the CLOSE-WAIT state.

CLOSING STATE

Remain in the CLOSING state.

LAST-ACK STATE

Remain in the LAST-ACK state.

SEGMENT ARRIVES

TIME-WAIT STATE

Remain in the TIME-WAIT state. Restart the 2 MSL time-wait timeout.

and return.

USER TIMEOUT

USER TIMEOUT

For any state if the user timeout expires, flush all queues, signal the user "error: connection aborted due to user timeout" in general and for any outstanding calls, delete the TCB, enter the CLOSED state and return.

RETRANSMISSION TIMEOUT

For any state if the retransmission timeout expires on a segment in the retransmission queue, send the segment at the front of the retransmission queue again, reinitialize the retransmission timer, and return.

TIME-WAIT TIMEOUT

If the time-wait timeout expires on a connection delete the TCB, enter the CLOSED state and return.

Glossary

1822 BBN Report 1822, "The Specification of the Interconnection of a Host and an IMP." The specification of interface between a host and the ARPANET.

ACK A control bit (acknowledge) occupying no sequence space, which indicates that the acknowledgment field of this segment specifies the next sequence number the sender of this segment is expecting to receive, hence acknowledging receipt of all previous sequence numbers.

ARPANET message The unit of transmission between a host and an IMP in the ARPANET. The maximum size is about 1012 octets (8096 bits).

ARPANET packet A unit of transmission used internally in the ARPANET between IMPs. The maximum size is about 126 octets (1008 bits).

Connection A logical communication path identified by a pair of sockets.

Datagram A message sent in a packet-switched computer communications network.

Destination Address The destination address, usually the network and host identifiers.

FIN A control bit (finis) occupying one sequence number, which indicates that the sender will send no more data or control occupying sequence space.

Fragment A portion of a logical unit of data, in particular, an internet fragment is a portion of an internet datagram.

FTP A file transfer protocol.

Header Control information at the beginning of a message, segment, fragment, packet, or block of data.

Host A computer. In particular a source or destination of messages from the point of view of the communication network.

Identification An Internet Protocol field. This identifying value assigned by the sender aids in assembling the fragments of a datagram.

IMP The Interface Message Processor, the packet switch of the ARPANET.

Internet address A source or destination address specific to the host level.

Internet datagram The unit of data exchanged between an internet module and the higher level protocol together with the internet header.

Internet fragment A portion of the data of an internet datagram with an internet header.

IP Internet Protocol.

IRS The Initial Receive Sequence number. The first sequence number used by the sender on a connection.

ISN The Initial Sequence Number. The first sequence number used on a connection (either ISS or IRS). Selected on a clock-based procedure.

ISS The Initial Send Sequence number. The first sequence number used by the sender on a connection.

Leader Control information at the beginning of a message or block of data. In particular, in the ARPANET, the control information on an ARPANET message at the host-IMP interface.

Left sequence This is the next sequence number to be acknowledged by the data receiving TCP (or the lowest currently unacknowledged sequence number) and is sometimes referred to as the left edge of the send window.

Local packet The unit of transmission within a local network.

Module An implementation, usually in software, of a protocol or other procedure.

MSL Maximum Segment Lifetime, the time a TCP segment can exist in the internetwork system. Arbitrarily defined to be 2 minutes.

Octet An eight-bit byte.

Options An Option field may contain several options, and each option may be several octets in length. The options are used primarily in testing situations; for example, to carry timestamps. Both the Internet Protocol and TCP provide for options fields.

Packet A package of data with a header which may or may not be logically complete. More often a physical packaging than a logical packaging of data.

Port The portion of a socket that specifies which logical input or output channel of a process is associated with the data.

Process A program in execution. A source or destination of data from the point of view of the TCP or other host-to-host protocol.

PUSH A control bit occupying no sequence space, indicating that this segment contains data that must be pushed through to the receiving user.

RCV.NXT receive next sequence number

RCV.UP receive urgent pointer

RCV.WND receive window

Receive next sequence number This is the next sequence number the local TCP is expecting to receive.

Receive window This represents the sequence numbers the local (receiving) TCP is willing to receive. Thus, the local TCP considers that segments overlapping the range RCV.NXT to RCV.NXT + RCV.WND − 1 carry acceptable data or control. Segments containing sequence numbers entirely outside of this range are considered duplicates and discarded.

RST A control bit (reset), occupying no sequence space, indicating that the receiver should delete the connection without further interaction. The receiver can determine, based on the sequence number and acknowledgment fields of the incoming segment, whether it should honor the reset command or ignore it. In no case does receipt of a segment containing RST give rise to an RST in response.

RTP Real Time Protocol: A host-to-host protocol for communication of time critical information.

SEG.ACK segment acknowledgment

SEG.LEN segment length

SEG.PRC segment precedence value

SEG.SEQ segment sequence

SEG.UP segment urgent pointer field

SEG.WND segment window field

Segment A logical unit of data, in particular a TCP segment is the unit of data transferred between a pair of TCP modules.

Segment acknowledgment The sequence number in the acknowledgment field of the arriving segment.

Segment length The amount of sequence number space occupied by a segment, including any controls which occupy sequence space.

Segment sequence The number in the sequence field of the arriving segment.

Send sequence This is the next sequence number the local (sending) TCP will use on the connection. It is initially selected from an initial sequence number curve (ISN) and is incremented for each octet of data or sequence control transmitted.

Send window This represents the sequence numbers which the remote (receiving) TCP is willing to receive. It is the value of the window field specified in segments from the remote (data receiving) TCP. The range of new sequence numbers which may be emitted by a TCP lies between SND.NXT and SND.UNA + SND.WND − 1. (Retransmissions of sequence numbers between SND.UNA and SND.NXT are expected, of course.)

SND.NXT send sequence

SND.UNA left sequence

SND.UP send urgent pointer

SND.WL1 segment sequence number at last window update

SND.WL2 segment acknowledgment number at last window update

SND.WND send window

Socket An address which specifically includes a port identifier, that is, the concatenation of an Internet Address with a TCP port.

Source Address The source address, usually the network and host identifiers.

SYN A control bit in the incoming segment, occupying one sequence number, used at the initiation of a connection, to indicate where the sequence numbering will start.

TCB Transmission control block, the data structure that records the state of a connection.

TCB.PRC The precedence of the connection.

TCP Transmission Control Protocol: A host-to-host protocol for reliable communication in internetwork environments.

TOS Type of Service, an Internet Protocol field.

Type of Service An Internet Protocol field which indicates the type of service for this internet fragment.

URG A control bit (urgent), occupying no sequence space, used to indicate that the receiving user should be notified to do urgent processing as long as there is data to be consumed with sequence numbers less than the value indicated in the urgent pointer.

Urgent pointer A control field meaningful only when the URG bit is on. This field communicates the value of the urgent pointer which indicates the data octet associated with the sending user's urgent call.

References

1. Cerf, V., and R. Kahn, "A Protocol for Packet Network Intercommunication," IEEE Transactions on Communications, Vol. COM-22, No. 5, pp. 637–648, May 1974.
2. Postel, J. (ed.), "Internet Protocol—DARPA Internet Program Protocol Specification," *RFC 791,* USC/Information Sciences Institute, September 1981.
3. Dalal, Y., and C. Sunshine, "Connection Management in Transport Protocols," Computer Networks, Vol. 2, No. 6, pp. 454–473, December 1978.
4. Postel, J., "Assigned Numbers," *RFC 790,* USC/Information Sciences Institute, September 1981.

Appendix C

RFC 959: File Transfer Protocol (FTP)

File Transfer Protocol (FTP) is a commonly used protocol in the Internet and networking community. Websites are accessed and modified primarily through the use of FTP. Whenever a file is downloaded from a website, FTP is used to control and complete the transfer. This RFC is included as both a reference and enhancement to the body text. Reference is made in this document to the Telnet protocol. Telnet is a terminal emulation protocol which is used to effect communication between two computers over the Internet. Because Telnet deals with ASCII-based information and not HTML, it is used only in specialized communication sessions between dedicated computers. FTP operates on the Application layer of the OSI Model and is described in Chapter 2.

J. Postel
J. Reynolds

Status of this Memo

This memo is the official specification of the File Transfer Protocol (FTP). Distribution of this memo is unlimited. The following new optional commands are included in this edition of the specification: CDUP (Change to Parent Directory), SMNT (Structure Mount), STOU (Store Unique), RMD (Remove Directory), MKD (Make Directory), PWD (Print Directory), and SYST (System).

Note that this specification is compatible with the previous edition. (October 1985)

1 Introduction

The objectives of FTP are (1) to promote sharing of files (computer programs and/or data), (2) to encourage indirect or implicit (via programs) use of remote computers, (3) to shield a user from variations in file storage systems among hosts, and (4) to transfer data reliably and efficiently. FTP, though usable directly by a user at a terminal, is designed mainly for use by programs.

The attempt in this specification is to satisfy the diverse needs of users of maxi-hosts, mini-hosts, personal workstations, and TACs, with a simple, and easily implemented protocol design.

This paper assumes knowledge of the Transmission Control Protocol (TCP) [2] and the Telnet Protocol [3]. These documents are contained in the ARPA-Internet protocol handbook [1].

2 Overview

In this section, the history, the terminology, and the FTP model are discussed. The terms defined in this section are only those that have special significance in FTP. Some of the terminology is very specific to the FTP model; some readers may wish to turn to the section on the FTP model while reviewing the terminology.

2.1 History

FTP has had a long evolution over the years. Appendix C is a chronological compilation of Request for Comments documents relating to FTP. These include the first proposed file transfer mechanisms in 1971 that were developed for implementation on hosts at M.I.T. (RFC *114*), plus comments and discussion in RFC *141*. RFC *172* provided a user-level oriented protocol for file transfer between host computers (including terminal IMPs). A revision of this as RFC *265*, restated FTP for additional review, while RFC *281* suggested further changes. The use of a "Set Data Type" transaction was proposed in RFC *294* in January 1982.

RFC *354* obsoleted RFCs 264 and 265. The File Transfer Protocol was now defined as a protocol for file transfer between HOSTs on the ARPANET, with the primary function of FTP defined as transfering files efficiently and reliably among hosts and allowing the convenient use of remote file storage capabilities. RFC *385* further commented on errors, emphasis points, and additions to the protocol, while RFC *414* provided a status report on the working server and user FTPs. RFC *430*, issued in 1973, (among other RFCs too numerous to mention) presented further comments on FTP. Finally, an "official" FTP document was published as RFC *454*.

By July 1973, considerable changes from the last versions of FTP were made, but the general structure remained the same. RFC *542* was published as a new "official" specification to reflect these changes. However, many implementations based on the older specification were not updated.

In 1974, RFCs 607 and 614 continued comments on FTP. RFC *624* proposed further design changes and minor modifications. In 1975, RFC *686* entitled, "Leaving Well Enough Alone," discussed the differences between all of the early and later versions of FTP. RFC *691* presented a minor revision of RFC *686,* regarding the subject of print files.

Motivated by the transition from the NCP to the TCP as the underlying protocol, a phoenix was born out of all of the above efforts in RFC *765* as the specification of FTP for use on TCP. This current edition of the FTP specification is intended to correct some minor documentation errors, to improve the explanation of some protocol features, and to add some new optional commands.

In particular, the following new optional commands are included in this edition of the specification:

CDUP—Change to Parent Directory

SMNT—Structure Mount

STOU—Store Unique

RMD—Remove Directory

MKD—Make Directory

PWD—Print Directory

SYST—System

This specification is compatible with the previous edition. A program implemented in conformance to the previous specification should automatically be in conformance to this specification.

2.2 Terminology

ASCII The ASCII character set is as defined in the ARPA-Internet Protocol Handbook. In FTP, ASCII characters are defined to be the lower half of an eight-bit code set (i.e., the most significant bit is zero).

Access controls Access controls define users' access privileges to the use of a system, and to the files in that system. Access controls are necessary to prevent unauthorized or accidental use of files. It is the prerogative of a server-FTP process to invoke access controls.

Byte size There are two byte sizes of interest in FTP: the logical byte size of the file, and the transfer byte size used for the transmission of the data. The transfer byte size is always 8 bits. The transfer byte size is not necessarily the byte size in which data is to be stored in a system, nor the logical byte size for interpretation of the structure of the data.

Control connection The communication path between the USER-PI and SERVER-PI for the exchange of commands and replies. This connection follows the Telnet Protocol.

Data connection A full-duplex connection over which data is transferred, in a specified mode and type. The data transferred may be a part of a file, an entire file, or a number of files. The path may be between a server-DTP and a user-DTP, or between two server-DTPs.

Data port The passive data transfer process "listens" on the data port for a connection from the active transfer process in order to open the data connection.

DTP The data transfer process establishes and manages the data connection. The DTP can be passive or active.

End-of-Line The end-of-line sequence defines the separation of printing lines. The sequence is Carriage Return, followed by Line Feed.

EOF The end-of-file condition that defines the end of a file being transferred.

EOR The end-of-record condition that defines the end of a record being transferred.

Error recovery A procedure that allows a user to recover from certain errors such as failure of either host system or transfer process. In FTP, error recovery may involve restarting a file transfer at a given checkpoint.

FTP commands A set of commands that comprise the control information flowing from the user-FTP to the server-FTP process.

File An ordered set of computer data (including programs), of arbitrary length, uniquely identified by a pathname.

Mode The mode in which data is to be transferred via the data connection. The mode defines the data format during transfer including EOR and EOF. The transfer modes defined in FTP are described in the section on Transmission Modes.

NVT The Network Virtual Terminal as defined in the Telnet Protocol.

NVFS The Network Virtual File System. A concept which defines a standard network file system with standard commands and pathname conventions.

Page A file may be structured as a set of independent parts called pages. FTP supports the transmission of discontinuous files as independent indexed pages.

Pathname Pathname is defined to be the character string which must be input to a file system by a user in order to identify a file. Pathname normally contains device and/or directory names, and file name specification. FTP does not yet specify a standard pathname convention. Each user must follow the file naming conventions of the file systems involved in the transfer.

PI The protocol interpreter. The user and server sides of the protocol have distinct roles implemented in a user-PI and a server-PI.

Record A sequential file may be structured as a number of contiguous parts called records. Record structures are supported by FTP but a file need not have record structure.

Reply A reply is an acknowledgment (positive or negative) sent from server to user via the control connection in response to FTP commands. The general form of a reply is a completion code (including error codes) followed by a text string. The codes are for use by programs and the text is usually intended for human users.

Server-DTP The data transfer process, in its normal "active" state, establishes the data connection with the "listening" data port. It sets up parameters for transfer and storage, and transfers data on command from its PI. The DTP can be placed in a "passive" state to listen for, rather than initiate, a connection on the data port.

Server-FTP process A process or set of processes which perform the function of file transfer in cooperation with a user-FTP process and, possibly, another server. The functions consist of a protocol interpreter (PI) and a data transfer process (DTP).

Server-PI The server protocol interpreter "listens" on Port L for a connection from a user-PI and establishes a control communication connection. It receives standard FTP commands from the user-PI, sends replies, and governs the server-DTP.

Type The data representation type used for data transfer and storage. Type implies certain transformations between the time of data storage and data transfer. The representation types defined in FTP are described in the section on Establishing Data Connections.

User A person or a process on behalf of a person wishing to obtain file transfer service. The human user may interact directly with a server-FTP process, but use of a user-FTP process is preferred since the protocol design is weighted towards automata.

User-DTP The data transfer process "listens" on the data port for a connection from a server-FTP process. If two servers are transferring data between them, the user-DTP is inactive.

User-FTP process A set of functions including a protocol interpreter, a data transfer process and a user interface which together perform the function of file transfer in cooperation with one or more server-FTP processes. The user interface allows a local language to be used in the command-reply dialogue with the user.

User-PI The user protocol interpreter initiates the control connection from its port U to the server-FTP process, initiates FTP commands, and governs the user-DTP if that process is part of the file transfer.

2.3 The FTP Model

With the above definitions in mind, the following model (shown in Figure 1) may be diagrammed for an FTP service.

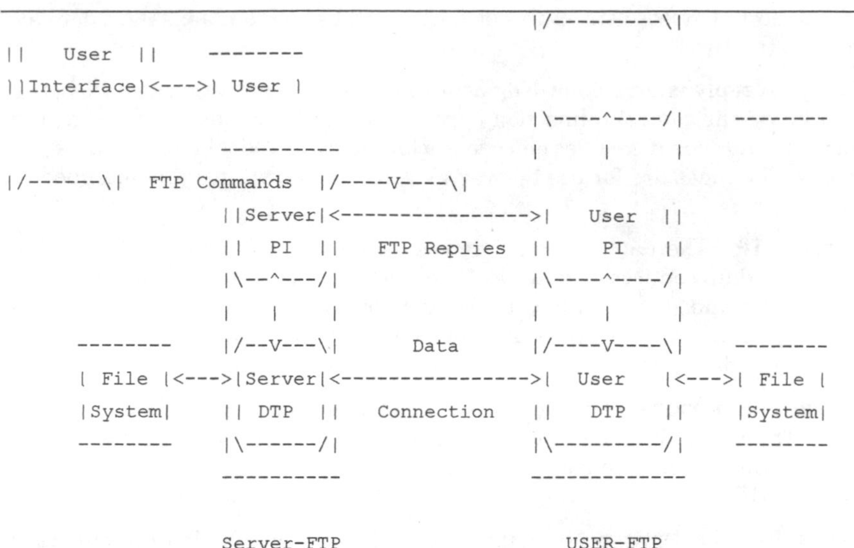

Figure 1 Model for FTP Use. NOTES: (1) The data connection may be used in either direction. (2) The data connection need not exist all of the time.

In the model described in Figure 1, the user-protocol interpreter initiates the control connection. The control connection follows the Telnet protocol. At the initiation of the user, standard FTP commands are generated by the user-PI and transmitted to the server process via the control connection. (The user may establish a direct control connection to the server-FTP, from a TAC terminal for example, and generate standard FTP commands independently, bypassing the user-FTP process.) Standard replies are sent from the server-PI to the user-PI over the control connection in response to the commands.

The FTP commands specify the parameters for the data connection (data port, transfer mode, representation type, and structure) and the nature of file system operation (store, retrieve, append, delete, etc.). The user-DTP or its designate should "listen" on the specified data port, and the server initiate the data connection and data transfer in accordance with the specified parameters. It should be noted that the data port need not be in the same host that initiates the FTP commands via the control connection, but the user or the user-FTP process must ensure a "listen" on the specified data port. It ought to also be noted that the data connection may be used for simultaneous sending and receiving.

In another situation a user might wish to transfer files between two hosts, neither of which is a local host. The user sets up control connections to the two servers and then arranges for a data connection between them. In this manner, control information is passed to the user-PI but data is transferred between the server data transfer processes. Following is a model of this server-server interaction.

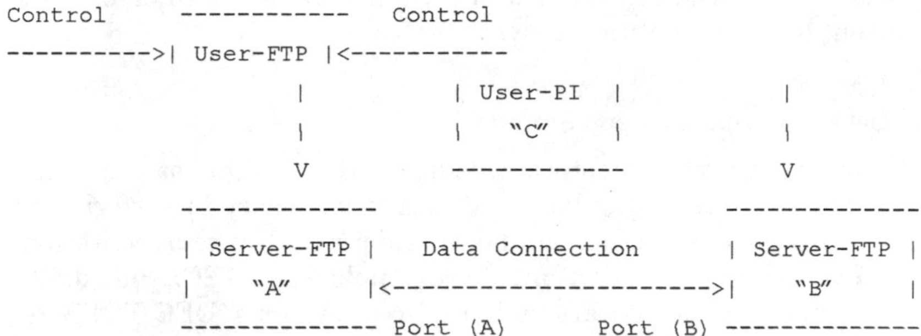

Figure 2

The protocol requires that the control connections be open while data transfer is in progress. It is the responsibility of the user to request the closing of the control connections when finished using the FTP service, while it is the server who takes the action. The server may abort data transfer if the control connections are closed without command.

The Relationship between FTP and Telnet. The FTP uses the Telnet protocol on the control connection. This can be achieved in two ways: first, the user-PI or the server-PI may implement the rules of the Telnet protocol directly in their own procedures; or, second, the user-PI or the server-PI may make use of the existing Telnet module in the system.

Ease of implementation, sharing code, and modular programming argue for the second approach. Efficiency and independence argue for the first approach. In practice, FTP relies on very little of the Telnet protocol, so the first approach does not necessarily involve a large amount of code.

3 Data Transfer Functions

Files are transferred only via the data connection. The control connection is used for the transfer of commands, which describe the functions to be performed, and the replies to these commands (see the section on FTP Replies). Several commands are concerned with the transfer of data between hosts. These data transfer commands include the MODE command which specifies how the bits of the data are to be transmitted, and the STRUcture and TYPE commands, which are used to define the way in which the data are to be represented. The transmission and representation are basically independent but the "Stream" transmission mode is dependent on the file structure attribute and if "Compressed" transmission mode is used, the nature of the filler byte depends on the representation type.

3.1 Data Representation and Storage

Data is transferred from a storage device in the sending host to a storage device in the receiving host. Often it is necessary to perform certain transformations on the data because data storage representations in the two systems are different. For example, NVT-ASCII has different data storage representations in different systems. DEC TOPS-20s generally store NVT-ASCII as five 7-bit ASCII characters, left-justified in a 36-bit word. IBM Mainframes store NVT-ASCII as 8-bit EBCDIC codes. Multics store NVT-ASCII as four 9-bit characters in a 36-bit word. It is desirable to convert characters into the standard NVT-ASCII representation when transmitting text between dissimilar systems. The sending and receiving sites would have to perform the necessary transformations between the standard representation and their internal representations.

A different problem in representation arises when transmitting binary data (not character codes) between host systems with different word lengths. It is not always clear how the sender should send data, and the receiver store it. For example, when transmitting 32-bit bytes from a 32-bit word-length system to a 36-bit word-length system, it may be desirable (for reasons of efficiency and usefulness) to store the 32-bit bytes right-justified in a 36-bit word in the latter system. In any case, the user should have the option of specifying data representation and transformation functions. It should be noted that FTP provides for very limited data type representations. Transformations desired beyond this limited capability should be performed by the user directly.

3.1.1 Data types. Data representations are handled in FTP by a user specifying a representation type. This type may implicitly (as in ASCII or EBCDIC) or explicitly (as in Local byte) define a byte size for interpretation which is referred to as the "logical byte size." Note that this has nothing to do with the byte size used for transmission over the data connection, called the "transfer byte size," and the two should not be confused. For example, NVT-ASCII has a logical byte size of 8 bits. If the type is Local byte, then the TYPE command has an obligatory second parameter specifying the logical byte size. The transfer byte size is always 8 bits.

3.1.1.1 ASCII type. This is the default type and must be accepted by all FTP implementations. It is intended primarily for the transfer of text files, except when both hosts would find the EBCDIC type more convenient.

The sender converts the data from an internal character representation to the standard 8-bit NVT-ASCII representation (see the Telnet specification). The receiver will convert the data from the standard form to his own internal form.

In accordance with the NVT standard, the <CRLF> sequence should be used where necessary to denote the end of a line of text. (See the discussion of file structure at the end of the section on Data Representation and Storage.)

Using the standard NVT-ASCII representation means that data must be interpreted as 8-bit bytes.

The Format parameter for ASCII and EBCDIC types is discussed below.

3.1.1.2 EBCDIC type. This type is intended for efficient transfer between hosts which use EBCDIC for their internal character representation.

For transmission, the data are represented as 8-bit EBCDIC characters. The character code is the only difference between the functional specifications of EBCDIC and ASCII types.

End-of-line (as opposed to end-of-record—see the discussion of structure) will probably be rarely used with EBCDIC type for purposes of denoting structure, but where it is necessary the <NL> character should be used.

3.1.1.3 Image type. The data are sent as contiguous bits which, for transfer, are packed into the 8-bit transfer bytes. The receiving site must store the data as contiguous bits. The structure of the storage system might necessitate the padding of the file (or of each record, for a record-structured file) to some convenient boundary (byte, word, or block). This padding, which must be all zeros, may occur only at the end of the file (or at the end of each record) and there must be a way of identifying the padding bits so that they may be stripped off if the file is retrieved. The padding transformation should be well publicized to enable a user to process a file at the storage site.

Image type is intended for the efficient storage and retrieval of files and for the transfer of binary data. It is recommended that this type be accepted by all FTP implementations.

3.1.1.4 Local type. The data is transferred in logical bytes of the size specified by the obligatory second parameter, byte size. The value of Byte size must be a decimal integer; there is no default value. The logical byte size is not necessarily the same as the transfer byte size. If there is a difference in byte sizes, then the logical bytes should be packed contiguously, disregarding transfer byte boundaries and with any necessary padding at the end.

When the data reaches the receiving host, it will be transformed in a manner dependent on the logical byte size and the particular host. This transformation must be invertible (i.e., an identical file can be retrieved if the same parameters are used) and should be well publicized by the FTP implementors.

For example, a user sending 36-bit floating-point numbers to a host with a 32-bit word could send that data as Local byte with a logical byte size of 36. The receiving host would then be expected to store the logical bytes so that they could be easily manipulated; in this example putting the 36-bit logical bytes into 64-bit double words should suffice.

In another example, a pair of hosts with a 36-bit word size may send data to one another in words by using TYPE L 36. The data would be sent in the 8-bit transmission bytes packed so that 9 transmission bytes carried two host words.

3.1.1.5 Format control. The types ASCII and EBCDIC also take a second (optional) parameter; this is to indicate what kind of vertical format control, if any, is associated with a file. The following data representation types are defined in FTP.

A character file may be transferred to a host for one of three purposes: for printing, for storage and later retrieval, or for processing. If a file is sent for printing, the receiving host must know how the vertical format control is represented. In the second case, it must be possible to store a file at a host and then retrieve it later in exactly the same form. Finally, it should be possible to move a file from one host to another and process the file at the second host without undue trouble. A single ASCII or EBCDIC format does not satisfy all these conditions. Therefore, these types have a second parameter specifying one of the following three formats.

3.1.1.5.1 Nonprint. This is the default format to be used if the second (format) parameter is omitted. Nonprint format must be accepted by all FTP implementations.

The file need contain no vertical format information. If it is passed to a printer process, this process may assume standard values for spacing and margins.

Normally, this format will be used with files destined for processing or just storage.

3.1.1.5.2 Telnet format controls. The file contains ASCII/EBCDIC vertical format controls (i.e., <CR>, <LF>, <NL>, <VT>, <FF>) which the printer process will interpret appropriately. <CRLF>, in exactly this sequence, also denotes end-of-line.

3.1.1.5.3 Carriage Control (ASA). The file contains ASA (FORTRAN) vertical format control characters. (See RFC *740* Appendix C;

and Communications of the ACM, Vol. 7, No. 10, p. 606, October 1964.) In a line or a record formatted according to the ASA Standard, the first character is not to be printed. Instead, it should be used to determine the vertical movement of the paper which should take place before the rest of the record is printed.

The ASA Standard specifies the following control characters:

CHARACTER	VERTICAL SPACING
blank	Move paper up one line
0	Move paper up two lines
1	Move paper to top of next page
•	No movement, i.e., overprint

Clearly there must be some way for a printer process to distinguish the end of the structural entity. If a file has record structure (see later) this is no problem; records will be explicitly marked during transfer and storage. If the file has no record structure, the <CRLF> end-of-line sequence is used to separate printing lines, but these format effectors are overridden by the ASA controls.

3.1.2 Data structures. In addition to different representation types, FTP allows the structure of a file to be specified. Three file structures are defined in FTP:

file-structure, where there is no internal structure and the file is considered to be a continuous sequence of data bytes,

record-structure, where the file is made up of sequential records,

and

page-structure, where the file is made up of independent indexed pages.

File-structure is the default to be assumed if the STRUcture command has not been used but both file and record structures must be accepted for "text" files (i.e., files with TYPE ASCII or EBCDIC) by all FTP implementations. The structure of a file will affect both the transfer mode of a file (see the section on Transmission Modes) and the interpretation and storage of the file.

The "natural" structure of a file will depend on which host stores the file. A source-code file will usually be stored on an IBM Mainframe in fixed length records but on a DEC TOPS-20 as a stream of characters partitioned into lines, for example by <CRLF>. If the transfer of files between such disparate sites is to be useful, there must be some way for one site to recognize the other's assumptions about the file.

With some sites being naturally file-oriented and others naturally record-oriented there may be problems if a file with one structure is sent to a host oriented to the other. If a text file is sent with record-structure to a host which is file oriented, then that host should apply an internal transformation to the file based on the record structure. Obviously, this transformation should be useful, but it must also be invertible so that an identical file may be retrieved using record structure.

In the case of a file being sent with file-structure to a record-oriented host, there exists the question of what criteria the host should use to divide the file into records which can be processed locally. If this division is necessary, the FTP implementation should use the end-of-line sequence, <CRLF> for ASCII, or <NL> for EBCDIC text files, as the delimiter. If an FTP implementation adopts this technique, it must be prepared to reverse the transformation if the file is retrieved with file-structure.

3.1.2.1 File-structure. File-structure is the default to be assumed if the STRUcture command has not been used.

In file-structure there is no internal structure and the file is considered to be a continuous sequence of data bytes.

3.1.2.2 Record-structure. Record-structures must be accepted for "text" files (i.e., files with TYPE ASCII or EBCDIC) by all FTP implementations. In record-structure the file is made up of sequential records.

3.1.2.3 Page-structure. To transmit files that are discontinuous, FTP defines a page structure. Files of this type are sometimes known as "random access files" or even as "holey files." In these files there is sometimes other information associated with the file as a whole (e.g., a file descriptor), or with a section of the file (e.g., page access controls), or both. In FTP, the sections of the file are called pages.

To provide for various page sizes and associated information, each page is sent with a page header. The page header has the following defined fields:

Header Length: The number of logical bytes in the page header including this byte. The minimum header length is 4.

Page Index: The logical page number of this section of the file. This is not the transmission sequence number of this page, but the index used to identify this page of the file.

Data Length: The number of logical bytes in the page data. The minimum data length is 0.

Page Type: The type of page this is. The following page types are defined:

0 = Last page: This is used to indicate the end of a paged structured transmission. The header length must be 4, and the data length must be 0.

1 = Simple Page: This is the normal type for simple paged files with no page level associated control information. The header length must be 4.

2 = Descriptor Page: This type is used to transmit the descriptive information for the file as a whole.

3 = Access Controlled Page: This type includes an additional header field for paged files with page level access control information. The header length must be 5.

Optional Fields: Further header fields may be used to supply per page control information, for example, per page access control.

All fields are one logical byte in length. The logical byte size is specified by the TYPE command. See Appendix I for further details and a specific case at the page structure. A note of caution about parameters: a file must be stored and retrieved with the same parameters if the retrieved version is to be identical to the version originally transmitted. Conversely, FTP implementations must return a file identical to the original if the parameters used to store and retrieve a file are the same.

3.2 Establishing Data Connections

The mechanics of transferring data consists of setting up the data connection to the appropriate ports and choosing the parameters for transfer. Both the user and the server-DTPs have a default data port. The user-process default data port is the same as the control connection port (i.e., U). The server-process default data port is the port adjacent to the control connection port (i.e., L-1).

The transfer byte size is 8-bit bytes. This byte size is relevant only for the actual transfer of the data; it has no bearing on representation of the data within a host's file system. The passive data transfer process (this may be a user-DTP or a second server-DTP) shall "listen" on the data port prior to sending a transfer request command. The FTP request command determines the direction of the data transfer. The server, upon receiving the transfer request, will initiate the data connection to the port. When the connection is established, the data transfer begins between DTPs, and the server-PI sends a confirming reply to the user-PI.

Every FTP implementation must support the use of the default data ports, and only the USER-PI can initiate a change to nondefault ports.

It is possible for the user to specify an alternate data port by use of the PORT command. The user may want a file dumped on a TAC line printer or retrieved from a third-party host. In the latter case, the user-PI sets up control connections with both server-PIs. One server is then told (by an FTP command) to "listen" for a connection which the other will initiate. The user-PI sends one server-PI a PORT command indicating the data port of the other. Finally, both are sent the appropriate transfer commands. The exact sequence of commands and replies sent between the user-controller and the servers is defined in the section on FTP Replies.

In general, it is the server's responsibility to maintain the data connection—to initiate it and to close it. The exception to this is when the user-DTP is sending the data in a transfer mode that requires the connection to be closed to indicate EOF. The server MUST close the data connection under the following conditions:

1. The server has completed sending data in a transfer mode that requires a close to indicate EOF.
2. The server receives an ABORT command from the user.
3. The port specification is changed by a command from the user.
4. The control connection is closed legally or otherwise.
5. An irrecoverable error condition occurs.

Otherwise the close is a server option, the exercise of which the server must indicate to the user-process by either a 250 or 226 reply only.

3.3 Data Connection Management

Default Data Connection Ports: All FTP implementations must support use of the default data connection ports, and only the User-PI may initiate the use of nondefault ports.

Negotiating nondefault data ports: The User-PI may specify a nondefault user side data port with the PORT command. The User-PI may request the server side to identify a nondefault server side data port with the PASV command. Since a connection is defined by the pair of addresses, either of these actions is enough to get a different data connection, still it is permitted to do both commands to use new ports on both ends of the data connection.

Reuse of the data connection: When using the stream mode of data transfer the end of the file must be indicated by closing the connection. This causes a problem if multiple files are to be transferred in the session, due to need for TCP to hold the connection record for a time out period to guarantee the reliable communication. Thus the connection cannot be reopened at once. There are two solutions to this problem. The first is to negotiate a nondefault port. The second is to use another transfer mode.

A comment on transfer modes: The stream transfer mode is inherently unreliable, since one cannot determine if the connection closed prematurely or not. The other transfer modes (Block, Compressed) do not close the connection to indicate the end of file. They have enough FTP encoding that the data connection can be parsed to determine the end of the file. Thus using these modes one can leave the data connection open for multiple file transfers.

3.4 Transmission Modes

The next consideration in transferring data is choosing the appropriate transmission mode. There are three modes: one which formats the data and allows for restart procedures; one which also compresses the data for efficient transfer; and one which passes the data with little or no processing. In this last case the mode interacts with the structure attribute to determine the type of processing. In the compressed mode, the representation type determines the filler byte.

All data transfers must be completed with an end-of-file (EOF) which may be explicitly stated or implied by the closing of the data connection. For files with record-structure, all the end-of-record markers (EOR) are explicit, including the final one. For files transmitted in page-structure, a "last-page" page type is used.

NOTE: In the rest of this section, byte means "transfer byte" except where explicitly stated otherwise.

For the purpose of standardized transfer, the sending host will translate its internal end-of-line or end-of-record denotation into the representation prescribed by the transfer mode and file structure, and the receiving host will perform the inverse translation to its internal denotation. An IBM Mainframe record count field may not be recognized at another host, so the end-of-record information may be transferred as a two-byte control code in Stream mode or as a flagged bit in a Block or Compressed mode descriptor. End-of-line in an ASCII or EBCDIC file with no record structure should be indicated by <CRLF> or <NL>, respectively. Since these transformations imply extra work for some systems, identical systems transferring nonrecord-structured text files might wish to use a binary representation and stream mode for the transfer.

The following transmission modes are defined in FTP.

3.4.1 Stream Mode. The data is transmitted as a stream of bytes. There is no restriction on the representation type used; record-structures are allowed.

In a record-structured file EOR and EOF will each be indicated by a two-byte control code. The first byte of the control code will be all ones, the escape character. The second byte will have the low-order bit on and zeros elsewhere for EOR and the second low-order bit on for EOF; that is, the byte will have value 1 for EOR and value 2 for EOF. EOR and EOF may be indicated together on the last byte transmitted by turning both low-order bits on (i.e., the value 3). If a byte of all ones was intended to be sent as data, it should be repeated in the second byte of the control code.

If the structure is a file-structure, the EOF is indicated by the sending host closing the data connection and all bytes are data bytes.

3.4.2 Block Mode. The file is transmitted as a series of data blocks preceded by one or more header bytes. The header bytes contain a count field and descriptor code. The count field indicates the total length of the data block in bytes, thus marking the beginning of the next data block (there are no filler bits).

The descriptor code defines: last block in the file (EOF), last block in the record (EOR), restart marker (see the section on Error Recovery and Restart) or suspect data (i.e., the data being transferred is suspected of errors and is not reliable). This last code is NOT intended for error control within FTP. It is motivated by the desire of sites exchanging certain types of data (e.g., seismic or weather data) to send and receive all the data despite local errors (such as "magnetic tape read errors"), but to indicate in the transmission that certain portions are suspect). Record-structures are allowed in this mode, and any representation type may be used.

The header consists of the three bytes. Of the 24 bits of header information, the 16 low-order bits shall represent byte count, and the 8 high-order bits shall represent descriptor codes as shown below.
Block Header

```
+----------------+----------------+----------------+
| Descriptor     |    Byte Count                   |
|         8 bits |                      16 bits    |
+----------------+----------------+----------------+
```

The descriptor codes are indicated by bit flags in the descriptor byte. Four codes have been assigned, where each code number is the decimal value of the corresponding bit in the byte.

CODE	MEANING
128	End of data block is EOR
64	End of data block is EOF
32	Suspected errors in data block
16	Data block is a restart marker

With this encoding, more than one descriptor coded condition may exist for a particular block. As many bits as necessary may be flagged.

The restart marker is embedded in the data stream as an integral number of 8-bit bytes representing printable characters in the language being used over the control connection (e.g., default-NVT-ASCII). <SP> (Space, in the appropriate language) must not be used WITHIN a restart marker. For example, to transmit a six-character marker, the following would be sent:

```
               +--------+--------+--------+
               |Descrptr|   Byte count    |
               |code= 16|             = 6 |
               +--------+--------+--------+

               +--------+--------+--------+
| Marker | Marker | Marker |
| 8 bits | 8 bits | 8 bits |
               +--------+--------+--------+

               +--------+--------+--------+
| Marker | Marker | Marker |
| 8 bits | 8 bits | 8 bits |
               +--------+--------+--------+
```

3.4.3 Compressed Mode. There are three kinds of information to be sent: regular data, sent in a byte string; compressed data, consisting of replications or filler; and control information, sent in a two-byte escape sequence. If $n > 0$ bytes (up to 127) of regular data are sent, these n bytes are preceded by a byte with the left-most bit set to 0 and the right-most 7 bits containing the number n.

Byte string:

```
          1     7              8                  8
          +-+-+-+-+-+-+-+-+ +-+-+-+-+-+-+-+-+   +-+-+-+-+-+-+-+-+
          |0|      n      | |     d(1)      | ... |     d(n)      |
          +-+-+-+-+-+-+-+-+ +-+-+-+-+-+-+-+-+   +-+-+-+-+-+-+-+-+
                                    ^            ^
|---n bytes---|
```

of data
string of *n* data bytes d(1), . . . , d(*n*)
Count *n* must be positive.
To compress a string of *n* replications of the data byte d, the following 2 bytes are sent:

Replicated Byte:

```
 2          6                  8
          +-+-+-+-+-+-+-+-+ +-+-+-+-+-+-+-+-+
 |1 0|      n      | |       d       |
          +-+-+-+-+-+-+-+-+ +-+-+-+-+-+-+-+-+
```

A string of *n* filler bytes can be compressed into a single byte, where the filler byte varies with the representation type. If the type is ASCII or EBCDIC the filler byte is <SP> (Space, ASCII code 32, EBCDIC code 64). If the type is Image or Local byte the filler is a zero byte.

Filler String:

```
 2          6
          +-+-+-+-+-+-+-+-+
 |1 1|      n      |
          +-+-+-+-+-+-+-+-+
```

The escape sequence is a double byte, the first of which is the escape byte (all zeros) and the second of which contains descriptor codes as defined in Block mode. The descriptor codes have the same meaning as in Block mode and apply to the succeeding string of bytes.

Compressed mode is useful for obtaining increased bandwidth on very large network transmissions at a little extra CPU cost. It can be most effectively used to reduce the size of printer files such as those generated by RJE hosts.

3.5 Error Recovery and Restart

There is no provision for detecting bits lost or scrambled in data transfer; this level of error control is handled by the TCP. However, a restart procedure is provided to protect users from gross system failures (including failures of a host, an FTP-process, or the underlying network).

The restart procedure is defined only for the block and compressed modes of data transfer. It requires the sender of data to insert a special marker code in the data stream with some marker information. The marker information has meaning only to the sender, but must consist of printable characters in the default or negotiated language of the control connection (ASCII or EBCDIC). The marker could represent a bit-count, a record-count, or any other information by which a system may identify a data checkpoint. The receiver of data, if it implements the restart procedure, would then mark the corresponding position of this marker in the receiving system, and return this information to the user.

In the event of a system failure, the user can restart the data transfer by identifying the marker point with the FTP restart procedure. The following example illustrates the use of the restart procedure.

The sender of the data inserts an appropriate marker block in the data stream at a convenient point. The receiving host marks the corresponding data point in its file system and conveys the last known sender and receiver marker information to the user, either directly or over the control connection in a 110 reply (depending on who is the sender). In the event of a system failure, the user or controller process restarts the server at the last server marker by sending a restart command with server's marker code as its argument. The restart command is transmitted over the control connection and is immediately followed by the command (such as RETR, STOR or LIST) which was being executed when the system failure occurred.

4 File Transfer Functions

The communication channel from the user-PI to the server-PI is established as a TCP connection from the user to the standard server port. The user protocol interpreter is responsible for sending FTP commands and interpreting the replies received; the server-PI interprets commands, sends replies and directs its DTP to set up the data connection and transfer the data. If the second party to the data transfer (the passive transfer process) is the user-DTP, then it is governed through the internal protocol of the user-FTP host; if it is a second server-DTP, then it is governed by its PI on command from the user-PI. The FTP replies are discussed in the next section. In the description of a few of the commands in this section, it is helpful to be explicit about the possible replies.

4.1 FTP Commands

4.1.1 Access control commands. The following commands specify access control identifiers (command codes are shown in parentheses).

USER NAME (USER) The argument field is a Telnet string identifying the user. The user identification is that which is required by the server for access to its file system. This command will normally be the first command transmitted by the user after the control connections are made (some servers may require this). Additional identification information in the form of a password and/or an account command may also be required by some servers. Servers may allow a new USER command to be entered at any point in order to change the access control and/or accounting information. This has the effect of flushing any user, password, and account information already supplied and beginning the login sequence again. All transfer parameters are unchanged and any file transfer in progress is completed under the old access control parameters.

PASSWORD (PASS) The argument field is a Telnet string specifying the user's password. This command must be immediately preceded by the user name command, and, for some sites, completes the user's identification for access control. Since password information is quite sensitive, it is desirable in general to "mask" it or suppress typeout. It appears that the server has no foolproof way to achieve this. It is therefore the responsibility of the user-FTP process to hide the sensitive password information.

ACCOUNT (ACCT) The argument field is a Telnet string identifying the user's account. The command is not necessarily related to the USER command, as some sites may require an account for login and others only for specific access, such as storing files. In the latter case the command may arrive at any time.

There are reply codes to differentiate these cases for the automation: when account information is required for login, the response to a successful PASSword command is reply code 332. On the other hand, if account information is NOT required for login, the reply to a successful PASSword command is 230; and if the account information is needed for a command issued later in the dialogue, the server should return a 332 or 532 reply depending on whether it stores (pending receipt of the ACCounT command) or discards the command, respectively.

CHANGE WORKING DIRECTORY (CWD) This command allows the user to work with a different directory or dataset for file storage or retrieval without altering his login or accounting information. Transfer parameters are similarly unchanged. The argument is a pathname specifying a directory or other system dependent file group designator.

CHANGE TO PARENT DIRECTORY (CDUP) This command is a special case of CWD, and is included to simplify the implementation of programs for transferring directory trees between operating systems having different syntaxes for naming the parent directory. The reply codes shall be identical to the reply codes of CWD. See Appendix II for further details.

STRUCTURE MOUNT (SMNT) This command allows the user to mount a different file system data structure without altering his login or accounting information. Transfer parameters are similarly unchanged. The argument is a pathname specifying a directory or other system-dependent file group designator.

REINITIALIZE (REIN) This command terminates a USER, flushing all I/O and account information, except to allow any transfer in progress to be completed. All parameters are reset to the default settings and the control connection is left open. This is identical to the state in which a user finds himself immediately after the control connection is opened. A USER command may be expected to follow.

LOGOUT (QUIT) This command terminates a USER and if file transfer is not in progress, the server closes the control connection. If file transfer is in progress, the connection will remain open for result response and the server will then close it. If the user-process is transferring files for several USERs but does not wish to close and then reopen connections for each, then the REIN command should be used instead of QUIT. An unexpected close on the control connection will cause the server to take the effective action of an abort (ABOR) and a logout (QUIT).

4.1.2 Transfer Parameter Commands. All data transfer parameters have default values, and the commands specifying data transfer parameters are required only if the default parameter values are to be changed. The default value is the last specified value, or if no value has been specified, the standard default value is as stated here. This implies that the server must "remember" the applicable default values. The commands may be in any order except that they must precede the FTP service request. The following commands specify data transfer parameters:

DATA PORT (PORT) The argument is a HOST-PORT specification for the data port to be used in data connection. There are defaults for both the user and server data ports, and under normal circumstances this command and its reply are not needed. If this command is used, the argument is the concatenation of a 32-bit internet host address and a 16-bit TCP port address. This address information is broken into 8-bit fields and the value of each field is transmitted as a decimal number (in character string representation). The fields are separated by commas. A port command would be

PORT h1, h2, h3, h4, p1, p2,

where h1 is the high order 8 bits of the internet host address.

PASSIVE (PASV) This command requests the server-DTP to "listen" on a data port (which is not its default data port) and to wait for a connection rather than initiate one upon receipt of a transfer command. The response to this command includes the host and port address this server is listening on.

Representation type (TYPE) The argument specifies the representation type as described in the section on Data Representation and Storage. Several types take a second parameter. The first parameter is denoted by a single Telnet character, as is the second Format parameter for ASCII and EBCDIC; the second parameter for local byte is a decimal integer to indicate Byte size. The parameters are separated by a <SP> (space, ASCII code 32).

The following codes are assigned for type:

```
                                \    /
A - ASCII |        | N - Nonprint
|-><-| T - Telnet format effectors
                    E - EBCDIC|      | C - Carriage Control (ASA)
                                /    \
I - Image
L <byte size> - Local byte Byte size
```

The default representation type is ASCII Nonprint. If the Format parameter is changed, and later just the first argument is changed, Format then returns to the Nonprint default.

File structure (STRU) The argument is a single Telnet character code specifying file-structure described in the section on Data Representation and Storage.

The following codes are assigned for structure:

F—File (no record-structure)

R—Record-structure

P—Page-structure

The default structure is File.

TRANSFER MODE (MODE) The argument is a single Telnet character code specifying the data transfer modes described in the section on Transmission Modes.

The following codes are assigned for transfer modes:

S—Stream

B—Block

C—Compressed

The default transfer mode is Stream.

4.1.3 FTP service commands. The FTP service commands define the file transfer or the file system function requested by the user. The argument of an FTP service command will normally be a pathname. The syntax of pathnames must conform to server site conventions (with standard defaults applicable), and the language conventions of the control connection. The suggested default handling is to use the last specified device, directory or file name, or the standard default defined for local users. The commands may be in any order except that a "rename from" command must be followed by a "rename to" command and the restart command must be followed by the interrupted service command (e.g., STOR or RETR). The data, when transferred in response to FTP service commands, shall always be sent over the data connection, except for certain informative replies. The following commands specify FTP service requests:

RETRIEVE (RETR) This command causes the server-DTP to transfer a copy of the file, specified in the pathname, to the server- or user-DTP at the other end of the data connection. The status and contents of the file at the server site shall be unaffected.

STORE (STOR) This command causes the server-DTP to accept the data transferred via the data connection and to store the data as a file at the server site. If the file specified in the pathname exists at the server site, then its contents shall be replaced by the data being transferred. A new file is created at the server site if the file specified in the pathname does not already exist.

STORE UNIQUE (STOU) This command behaves like STOR except that the resultant file is to be created in the current directory under a name unique to that directory. The 250 Transfer Started response must include the name generated.

APPEND (with create) (APPE) This command causes the server-DTP to accept the data transferred via the data connection and to store the data in a file at the server site. If the file specified in the pathname exists at the server site, then the data shall be appended to that file; otherwise the file specified in the pathname shall be created at the server site.

ALLOCATE (ALLO) This command may be required by some servers to reserve sufficient storage to accommodate the new file to be transferred. The argument shall be a decimal integer representing the number of bytes (using the logical byte size) of storage to be reserved for the file. For files sent with record- or page-structure a maximum record or page size (in logical bytes) might also be necessary; this is indicated by a decimal integer in a second argument field of the command. This second argument is optional, but when present should be separated from the first by the three Telnet characters <SP> R <SP>. This command shall be followed by a STORe or APPEnd command. The ALLO command should be treated as a NOOP (no operation) by those servers which do not require that the maximum size of the file be declared beforehand, and those servers interested in only the maximum record or page size should accept a dummy value in the first argument and ignore it.

RESTART (REST) The argument field represents the server marker at which file transfer is to be restarted. This command does not cause file transfer but skips over the file to the specified data checkpoint. This command shall be immediately followed by the appropriate FTP service command which shall cause file transfer to resume.

RENAME FROM (RNFR) This command specifies the old pathname of the file which is to be renamed. This command must be immediately followed by a "rename to" command specifying the new file pathname.

RENAME TO (RNTO) This command specifies the new pathname of the file specified in the immediately preceding "rename from" command. Together the two commands cause a file to be renamed.

ABORT (ABOR) This command tells the server to abort the previous FTP service command and any associated transfer of data. The abort command may require "special action," as discussed in the section on FTP Commands, to force recognition by the server. No action is to be taken if the previous command has been completed (including data transfer). The control connection is not to be closed by the server, but the data connection must be closed.

There are two cases for the server upon receipt of this command: (1) the FTP service command was already completed, or (2) the FTP service command is still in progress. In the first case, the server closes the data connection (if it is open) and responds with a 226 reply, indicating that the abort command was successfully processed. In the second case, the server aborts the FTP service in progress and closes the data connection, returning a 426 reply to indicate that the service request terminated abnormally. The server then sends a 226 reply, indicating that the abort command was successfully processed.

DELETE (DELE) This command causes the file specified in the pathname to be deleted at the server site. If an extra level of protection is desired (such as the query, "Do you really wish to delete?"), it should be provided by the user-FTP process.

REMOVE DIRECTORY (RMD) This command causes the directory specified in the pathname to be removed as a directory (if the pathname is absolute) or as a subdirectory of the current working directory (if the pathname is relative). See Appendix II.

MAKE DIRECTORY (MKD) This command causes the directory specified in the pathname to be created as a directory (if the pathname is absolute) or as a subdirectory of the current working directory (if the pathname is relative). See Appendix II.

PRINT WORKING DIRECTORY (PWD) This command causes the name of the current working directory to be returned in the reply. See Appendix II.

LIST (LIST) This command causes a list to be sent from the server to the passive DTP. If the pathname specifies a directory or other group of files, the server should transfer a list of files in the specified directory. If the pathname specifies a file then the server should send current information on the file. A null argument implies the user's current working or default directory. The data transfer is over the data connection in type ASCII or type EBCDIC. (The user must ensure that the TYPE is appropriately ASCII or EBCDIC).

Since the information on a file may vary widely from system to system, this information may be hard to use automatically in a program, but may be quite useful to a human user.

NAME LIST (NLST) This command causes a directory listing to be sent from server to user site. The pathname should specify a directory or other system-specific file group descriptor; a null argument implies the current directory. The server will return a stream of names of files and no other information. The data will be transferred in ASCII or EBCDIC type over the data connection as valid pathname strings separated by <CRLF> or <NL>. (Again the user must ensure that the TYPE is correct.) This command is intended to return information that can be used by a program to further process the files automatically. For example, in the implementation of a "multiple get" function.

SITE PARAMETERS (SITE) This command is used by the server to provide services specific to his system that are essential to file transfer but not sufficiently universal to be included as commands in the protocol. The nature of these services and the specification of their syntax can be stated in a reply to the HELP SITE command.

SYSTEM (SYST) This command is used to find out the type of operating system at the server. The reply shall have as its first word one of the system names listed in the current version of the Assigned Numbers document [4].

STATUS (STAT) This command shall cause a status response to be sent over the control connection in the form of a reply. The command may be sent during a file transfer (along with the Telnet IP and Synch signals—see the section on FTP Commands) in which case the server will respond with the status of the operation in progress, or it may be sent between file transfers. In the latter case, the command may have an argument field. If the argument is a pathname, the command is analogous to the "list" command except that data shall be transferred over the control connection. If a partial pathname is given, the server may respond with a list of file names or attributes associated with that specification. If no argument is given, the server should return general status information about the server FTP process. This should include current values of all transfer parameters and the status of connections.

HELP (HELP) This command shall cause the server to send helpful information regarding its implementation status over the control connection to the user. The command may take an argument (e.g., any command name) and return more specific information as a response. The reply is type 211 or 214. It is suggested that HELP be allowed before entering a USER command. The server may use this reply to specify site-dependent parameters, e.g., in response to HELP SITE.

NOOP (NOOP) This command does not affect any parameters or previously entered commands. It specifies no action other than that the server send an OK reply.

The File Transfer Protocol follows the specifications of the Telnet protocol for all communications over the control connection. Since the language used for Telnet communication may be a negotiated option, all references in the next two sections will be to the "Telnet language" and the corresponding "Telnet end-of-line code." Currently, one may take these to mean NVT-ASCII and <CRLF>. No other specifications of the Telnet protocol will be cited.

FTP commands are "Telnet strings" terminated by the "Telnet end-of-line code." The command codes themselves are alphabetic characters terminated by the character <SP> (Space) if parameters follow, and Telnet-EOL otherwise. The command codes and the semantics of commands are described in this section; the detailed syntax of commands is specified in the section on Commands, the reply sequences are discussed in the section on Sequencing of Commands and Replies, and scenarios illustrating the use of commands are provided in the section on Typical FTP Scenarios.

FTP commands may be partitioned as those specifying access-control identifiers, data transfer parameters, or FTP service requests. Certain commands (such as ABOR, STAT, QUIT) may be sent over the control connection while a data transfer is in progress. Some servers may not be able to monitor the control and data connections simultaneously, in which case some special action will be necessary to get the server's attention. The following ordered format is tentatively recommended:

1. User system inserts the Telnet "Interrupt Process" (IP) signal in the Telnet stream.
2. User system sends the Telnet "Synch" signal.
3. User system inserts the command (e.g., ABOR) in the Telnet stream.
4. Server PI, after receiving "IP," scans the Telnet stream for *exactly one* FTP command.

(For other servers this may not be necessary but the actions listed above should have no unusual effect.)

4.2 FTP replies

Replies to File Transfer Protocol commands are devised to ensure the synchronization of requests and actions in the process of file transfer, and to guarantee that the user process always knows the state of the Server. Every command must generate at least one reply, although there may be more than one; in the latter case, the multiple replies must be easily distinguished. In addition, some commands occur in sequential groups, such as USER, PASS and ACCT, or RNFR and RNTO. The replies show the existence of an intermediate state if all preceding commands have been successful. A failure at any point in the sequence necessitates the repetition of the entire sequence from the beginning.

The details of the command-reply sequence are made explicit in a set of state diagrams (see Section 6).

An FTP reply consists of a three-digit number (transmitted as three alphanumeric characters) followed by some text. The number is intended for use by automata to determine what state to enter next; the text is intended for the human user. It is intended that the three digits contain enough encoded information that the user-process (the User-PI) will not need to examine the text and may either discard it or pass it on to the user, as appropriate. In particular, the text may be server-dependent, so there are likely to be varying texts for each reply code.

A reply is defined to contain the 3-digit code, followed by Space <SP>, followed by one line of text (where some maximum line length has been specified), and terminated by the Telnet end-of-line code. There will be cases however, where the text is longer than a single line. In these cases the complete text must be bracketed so the User-process knows when it may stop reading the reply (i.e., stop processing input on the control connection) and start to do other things. This requires a special format on the first line to indicate that more than one line is coming, and another on the last line to designate it as the last. At least one of these must contain the appropriate reply code to indicate the state of the transaction. To satisfy all factions, it was decided that both the first and last line codes should be the same.

Thus the format for multiline replies is that the first line will begin with the exact required reply code, followed immediately by a Hyphen, "-" (also known as Minus), followed by text. The last line will begin with the same code, followed immediately by Space <SP>, optionally some text, and the Telnet end-of-line code. For example;

123-First line

Second line

234 A line beginning with numbers

123 The last line

The user-process then simply needs to search for the second occurrence of the same reply code, followed by <SP> (Space), at the beginning of a line, and ignore all intermediary lines. If an intermediary line begins with a 3-digit number, the Server must pad the front to avoid confusion.

This scheme allows standard system routines to be used for reply information (such as for the STAT reply), with "artificial" first and last lines tacked on. In rare cases where these routines are able to generate three digits and a Space at the beginning of any line, the beginning of each text line should be offset by some neutral text, like Space. This scheme assumes that multiline replies may not be nested.

The three digits of the reply each have a special significance. This is intended to allow a range of very simple to very sophisticated responses by the user-process. The first digit denotes whether the response is good, bad, or incomplete. An unsophisticated user-process will be able to determine its next action (proceed as planned, redo, retrench, etc.) by simply examining this first digit (referring to the state diagram). A user-process that wants to know approximately what kind of error occurred (e.g., file system error, command syntax error) may examine the second digit, reserving the third digit for the finest gradation of information (e.g., RNTO command without a preceding RNFR).

There are five values for the first digit of the reply code:

1yz: *Positive Preliminary reply*

The requested action is being initiated; expect another reply before proceeding with a new command. (The user-process sending another command before the completion reply would be in violation of protocol; but server-FTP processes should queue any commands that arrive while a preceding command is in progress.) This type of reply can be used to indicate that the command was accepted and the user-process may now pay attention to the data connections for implementations where simultaneous monitoring is difficult. The server-FTP process may send at most one 1yz reply per command.

2yz: *Positive Completion reply*

The requested action has been successfully completed. A new request may be initiated.

3yz: *Positive Intermediate reply*

The command has been accepted, but the requested action is being held in abeyance, pending receipt of further information. The user should send another command specifying this information. This reply is used in command sequence groups.

4yz: *Transient Negative Completion reply*

The command was not accepted and the requested action did not take place, but the error condition is temporary and the action may be requested again. The user should return to the beginning of the command sequence, if any. It is difficult to assign a meaning to "transient," particularly when two distinct sites (Server- and User-processes) have to agree on the interpretation.

Each reply in the 4yz category might have a slightly different time value, but the intent is that the user-process is encouraged to try again. A rule of thumb in determining if a reply fits into the 4yz or the 5yz (Permanent Negative) category is that replies are 4yz if the commands can be repeated without any change in command form or in properties of the User or Server (e.g., the command is spelled the same with the same arguments used; the user does not change his file access or user name; the server does not put up a new implementation.)

5yz: *Permanent Negative Completion reply*

The command was not accepted and the requested action did not take place. The User-process is discouraged from repeating the exact request (in the same sequence). Even some "permanent" error conditions can be corrected, so the human user may want to direct his User-process to reinitiate the command sequence by direct action at some point in the future (e.g., after the spelling has been changed, or the user has altered his directory status.)

The following function groupings are encoded in the second digit:

x0z Syntax—These replies refer to syntax errors, syntactically correct commands that do not fit any functional category, unimplemented or superfluous commands.

x1z Information—These are replies to requests for information, such as status or help.

x2z Connections—Replies referring to the control and data connections.

x3z Authentication and accounting—Replies for the login process and accounting procedures.

x4z Unspecified as yet.

x5z File system—These replies indicate the status of the Server file system vis-a-vis the requested transfer or other file system action.

The third digit gives a finer gradation of meaning in each of the function categories, specified by the second digit. The list of replies below will illustrate this. Note that the text associated with each reply is recommended, rather than mandatory, and may even change according to the command with which it is associated. The reply codes, on the other hand, must strictly follow the specifications in the last section; that is, Server implementations should not invent new codes for situations that are only slightly different from the ones described here, but rather should adapt codes already defined. A command such as TYPE or ALLO whose successful execution does not offer the user-process any new information will cause a 200 reply to be returned. If the command is not implemented by a particular Server-FTP process because it has no relevance to that computer system, for example ALLO at a TOPS20 site, a Positive Completion reply is still desired so that the simple User-process knows it can proceed with its course of action. A 202 reply is used in this case with, for example, the reply text: “No storage allocation necessary.” If, on the other hand, the command requests a non-site-specific action and is unimplemented, the response is 502. A refinement of that is the 504 reply for a command that is implemented, but that requests an unimplemented parameter.

4.2.1 Reply Codes by Function Groups.

200 Command okay.

500 Syntax error, command unrecognized.

This may include errors such as command line too long.

501 Syntax error in parameters or arguments.

202 Command not implemented, superfluous at this site.

502 Command not implemented.

503 Bad sequence of commands.

504 Command not implemented for that parameter.

110 Restart marker reply.

In this case, the text is exact and not left to the particular implementation; it must read:

MARK yyyy = mmmm

where yyyy is User-process data stream marker, and mmmm server's equivalent marker (note the spaces between markers and “=”).

211 System status, or system help reply.

212 Directory status.

213 File status.

214 Help message.

On how to use the server or the meaning of a particular nonstandard command. This reply is useful only to the human user.

215 NAME system type.

Where NAME is an official system name from the list in the Assigned Numbers document.

120 Service ready in nnn minutes.

220 Service ready for new user.

221 Service closing control connection.

Logged out if appropriate.

421 Service not available, closing control connection.

This may be a reply to any command if the service knows it must shut down.

125 Data connection already open; transfer starting.

225 Data connection open; no transfer in progress.

425 Can not open data connection.

226 Closing data connection.

Requested file action successful (for example, file transfer or file abort).

426 Connection closed; transfer aborted.

227 Entering Passive Mode (h1, h2, h3, h4, p1, p2).

230 User logged in, proceed.

530 Not logged in.

331 User name okay, need password.

332 Need account for login.

532 Need account for storing files.

150 File status okay; about to open data connection.

250 Requested file action okay, completed.

257 "PATHNAME" created.

350 Requested file action pending further information.

450 Requested file action not taken.

File unavailable (e.g., file busy).

550 Requested action not taken.

File unavailable (e.g., file not found, no access).

451 Requested action aborted. Local error in processing.

551 Requested action aborted. Page type unknown.

452 Requested action not taken.

Insufficient storage space in system.

552 Requested file action aborted.

Exceeded storage allocation (for current directory or dataset).

553 Requested action not taken.

File name not allowed.

4.2.2 Numeric Order List of Reply Codes.

110 Restart marker reply.

In this case, the text is exact and not left to the particular implementation; it must read:

MARK yyyy = mmmm

where yyyy is User-process data stream marker, and mmmm server's equivalent marker (note the spaces between markers and "=").

120 Service ready in nnn minutes.

125 Data connection already open; transfer starting.

150 File status okay; about to open data connection.

200 Command okay.

202 Command not implemented, superfluous at this site.

211 System status, or system help reply.

212 Directory status.

213 File status.

214 Help message.

On how to use the server or the meaning of a particular nonstandard command. This reply is useful only to the human user.

215 NAME system type.

Where NAME is an official system name from the list in the Assigned Numbers document.

220 Service ready for new user.

221 Service closing control connection.

Logged out if appropriate.

225 Data connection open; no transfer in progress.

226 Closing data connection.

Requested file action successful (for example, file transfer or file abort).

227 Entering Passive Mode (h1, h2, h3, h4, p1, p2).

230 User logged in, proceed.

250 Requested file action okay, completed.

257 "PATHNAME" created.

331 User name okay, need password.

332 Need account for login.

350 Requested file action pending further information.

421 Service not available, closing control connection.

This may be a reply to any command if the service knows it must shut down.

425 Cannot open data connection.

426 Connection closed; transfer aborted.

450 Requested file action not taken.

File unavailable (e.g., file busy).

451 Requested action aborted: local error in processing.

452 Requested action not taken.

Insufficient storage space in system.

500 Syntax error, command unrecognized.

This may include errors such as command line too long.

501 Syntax error in parameters or arguments.

502 Command not implemented.

503 Bad sequence of commands.

504 Command not implemented for that parameter.

530 Not logged in.

532 Need account for storing files.

550 Requested action not taken.

File unavailable (e.g., file not found, no access).

551 Requested action aborted: page type unknown.

552 Requested file action aborted.

Exceeded storage allocation (for current directory or dataset).

553 Requested action not taken.

File name not allowed.

5 Declarative Specifications

5.1 Minimum implementation

In order to make FTP workable without needless error messages, the following minimum implementation is required for all servers:

TYPE — ASCII Non-print

MODE — Stream

STRUCTURE— File, Record

COMMANDS— USER, QUIT, PORT, TYPE, MODE, STRU, for the default values RETR, STOR, NOOP.

The default values for transfer parameters are:

TYPE —ASCII Nonprint

MODE—Stream

STRU —File

All hosts must accept the above as the standard defaults.

5.2 Connections

The server protocol interpreter shall “listen” on Port L. The user or user protocol interpreter shall initiate the full-duplex control connection. Server- and user-processes should follow the conventions of the Telnet protocol as specified in the ARPA-Internet Protocol Workbook [1]. Servers are under no obligation to provide for editing of command lines and may require that it be done in the user host. The control connection shall be closed by the server at the user’s request after all transfers and replies are completed.

The user-DTP must “listen” on the specified data port; this may be the default user port (U) or a port specified in the PORT command. The server shall initiate the data connection from his own default data port (L-1) using the specified user data port. The direction of the transfer and the port used will be determined by the FTP service command.

Note that all FTP implementation must support data transfer using the default port, and that only the USER-PI may initiate the use of nondefault ports.

When data is to be transferred between two servers, A and B (refer to Figure 2), the user-PI, C, sets up control connections with both server-PI’s. One of the servers, say A, is then sent a PASV command telling him to “listen” on his data port rather than initiate a connection when he receives a transfer service command. When the user-PI receives an acknowledgment to the PASV command, which includes the identity of the host and port being listened on, the user-PI then sends A’s port, a, to B in a PORT command; a reply is returned. The user-PI may then send the corresponding service commands to A and B. Server B initiates the connection and the transfer proceeds. The command-reply sequence is listed below where the messages are vertically synchronous but horizontally asynchronous:

```
        User-PI - Server A                  User-PI - Server B
        ------------------                  ------------------

        C->A : Connect                      C->B : Connect
C->A : PASV
A->C : 227 Entering Passive Mode. A1,A2,A3,A4,a1,a2
C->B : PORT A1,A2,A3,A4,a1,a2
B->C : 200 Okay
C->A : STOR                      C->B : RETR
B->A : Connect to HOST-A, PORT-a
```

Figure 3

The data connection shall be closed by the server under the conditions described in the section on Establishing Data Connections. If the data connection is to be closed following a data transfer, where closing the connection is not required to indicate the end-of-file, the server must do so immediately. Waiting until after a new transfer command is not permitted because the user-process will have already tested the data connection to see if it needs to do a "listen"; (remember that the user must "listen" on a closed data port BEFORE sending the transfer request). To prevent a race condition here, the server sends a reply (226) after closing the data connection (or if the connection is left open, a "file transfer completed" reply [250] and the user-PI should wait for one of these replies before issuing a new transfer command).

Any time either the user or the server sees that the connection is being closed by the other side, it should promptly read any remaining data queued on the connection and issue the close on its own side.

5.3 Commands

The commands are Telnet character strings transmitted over the control connections as described in the section on FTP Commands. The command functions and semantics are described in the sections on Access Control Commands, Transfer Parameter Commands, FTP Service Commands, and Miscellaneous Commands. The command syntax is specified here.

The commands begin with a command code followed by an argument field. The command codes are four or fewer alphabetic characters. Upper and lower case alphabetic characters are to be treated identically. Thus, any of the following may represent the retrieve command:

RETR   Retr   retr   ReTr   rETr

This also applies to any symbols representing parameter values, such as A or a for ASCII TYPE. The command codes and the argument fields are separated by one or more spaces.

The argument field consists of a variable length character string ending with the character sequence <CRLF> (Carriage Return, Line Feed) for NVT-ASCII representation; for other negotiated languages a different end-of-line character might be used. It should be noted that the server is to take no action until the end-of-line code is received.

The syntax is specified below in NVT-ASCII. All characters in the argument field are ASCII characters including any ASCII represented decimal integers. Square brackets denote an optional argument field. If the option is not taken, the appropriate default is implied.

5.3.1 FTP commands.

The following are the FTP commands:

USER <SP> <username> <CRLF>

PASS <SP> <password> <CRLF>

ACCT <SP> <account-information> <CRLF>

CWD <SP> <pathname> <CRLF>

CDUP <CRLF>

SMNT <SP> <pathname> <CRLF>

QUIT <CRLF>

REIN <CRLF>

PORT <SP> <host-port> <CRLF>

PASV <CRLF>

TYPE <SP> <type-code> <CRLF>

STRU <SP> <structure-code> <CRLF>

MODE <SP> <mode-code> <CRLF>

RETR <SP> <pathname> <CRLF>

STOR <SP> <pathname> <CRLF>

STOU <CRLF>

APPE <SP> <pathname> <CRLF>

ALLO <SP> <decimal-integer>

[<SP> R <SP> <decimal-integer>] <CRLF>

REST <SP> <marker> <CRLF>

RNFR <SP> <pathname> <CRLF>

RNTO <SP> <pathname> <CRLF>

ABOR <CRLF>

DELE <SP> <pathname> <CRLF>

RMD <SP> <pathname> <CRLF>

MKD <SP> <pathname> <CRLF>
PWD <CRLF>
LIST [<SP> <pathname>] <CRLF>
NLST [<SP> <pathname>] <CRLF>
SITE <SP> <string> <CRLF>
SYST <CRLF>
STAT [<SP> <pathname>] <CRLF>
HELP [<SP> <string>] <CRLF>
NOOP <CRLF>

5.3.2 FTP command arguments. The syntax of the above argument fields (using BNF notation where applicable) is:

<username> ::= <string>
<password> ::= <string>
<account-information> ::= <string>
<string> ::= <char> | <char><string>
<char> ::= any of the 128 ASCII characters except <CR> and <LF>
<marker> ::= <pr-string>
<pr-string> ::= <pr-char> | <pr-char><pr-string>
<pr-char> ::= printable characters, any ASCII code 33 through 126
<byte-size> ::= <number>
<host-port> ::= <host-number>,<port-number>
<host-number> ::= <number>,<number>,<number>,<number>
<port-number> ::= <number>,<number>
<number> ::= any decimal integer 1 through 255
<form-code> ::= N | T | C
<type-code> ::= A [<sp> <form-code>]
| E [<sp> <form-code>]
| I
| L <sp> <byte-size>
<structure-code> ::= F | R | P
<mode-code> ::= S | B | C
<pathname> ::= <string>
<decimal-integer> ::= any decimal integer

5.4 Sequencing of commands and replies

The communication between the user and server is intended to be an alternating dialogue. As such, the user issues an FTP command and the server responds with a prompt primary reply. The user should wait for this initial primary success or failure response before sending further commands.

Certain commands require a second reply for which the user should also wait. These replies may, for example, report on the progress or completion of file transfer or the closing of the data connection. They are secondary replies to file transfer commands.

One important group of informational replies is the connection greetings. Under normal circumstances, a server will send a 220 reply, "awaiting input," when the connection is completed. The user should wait for this greeting message before sending any commands. If the server is unable to accept input right away, a 120 "expected delay" reply should be sent immediately and a 220 reply when ready. The user will then know not to hang up if there is a delay.

Spontaneous Replies

Sometimes "the system" spontaneously has a message to be sent to a user (usually all users). For example, "System going down in 15 minutes." There is no provision in FTP for such spontaneous information to be sent from the server to the user. It is recommended that such information be queued in the server-PI and delivered to the user-PI in the next reply (possibly making it a multiline reply).

The following table lists alternative success and failure replies for each command. These must be strictly adhered to; a server may substitute text in the replies, but the meaning and action implied by the code numbers and by the specific command reply sequence cannot be altered.

Command-Reply Sequences

In this section, the command-reply sequence is presented. Each command is listed with its possible replies; command groups are listed together. Preliminary replies are listed first (with their succeeding replies indented and under them), then positive and negative completion, and finally intermediary replies with the remaining commands from the sequence following. This listing forms the basis for the state diagrams, which will be presented separately.

Connection Establishment

120

220

220

421

Login

USER

230

530

500, 501, 421

331, 332

PASS

230

202

530

500, 501, 503, 421

332

ACCT

230

202

530

500, 501, 503, 421

CWD

250

500, 501, 502, 421, 530, 550

CDUP

200

500, 501, 502, 421, 530, 550

SMNT

202, 250

500, 501, 502, 421, 530, 550

Logout

REIN

120

220

220

421

500, 502

QUIT

221

500

Transfer parameters

PORT

200

500, 501, 421, 530

PASV

227

500, 501, 502, 421, 530

MODE

200

500, 501, 504, 421, 530

TYPE

200

500, 501, 504, 421, 530

STRU

200

500, 501, 504, 421, 530

File action commands

ALLO

200

202

500, 501, 504, 421, 530

REST

500, 501, 502, 421, 530

350

STOR

125, 150

(110)

226, 250

425, 426, 451, 551, 552

532, 450, 452, 553

500, 501, 421, 530

STOU

125, 150

(110)

226, 250

425, 426, 451, 551, 552

532, 450, 452, 553

500, 501, 421, 530

RETR

125, 150

(110)

226, 250

425, 426, 451

450, 550

500, 501, 421, 530

LIST

125, 150

226, 250

425, 426, 451

450

500, 501, 502, 421, 530

NLST

125, 150

226, 250

425, 426, 451

450

500, 501, 502, 421, 530

APPE

125, 150

(110)

226, 250

425, 426, 451, 551, 552

532, 450, 550, 452, 553

500, 501, 502, 421, 530

RNFR

450, 550

500, 501, 502, 421, 530

350

RNTO

250

532, 553

500, 501, 502, 503, 421, 530

DELE

250

450, 550

500, 501, 502, 421, 530

RMD

250

500, 501, 502, 421, 530, 550

MKD

257

500, 501, 502, 421, 530, 550

PWD

257

500, 501, 502, 421, 550

ABOR

225, 226

500, 501, 502, 421

Informational commands

SYST

215

500, 501, 502, 421

STAT

211, 212, 213

450

500, 501, 502, 421, 530

HELP

211, 214

500, 501, 502, 421

Miscellaneous commands

SITE

200

202

500, 501, 530

NOOP

200

500 421

6 State Diagrams

Here we present state diagrams for a very simple-minded FTP implementation. Only the first digit of the reply codes is used. There is one state diagram for each group of FTP commands or command sequences.

The command groupings were determined by constructing a model for each command then collecting together the commands with structurally identical models.

For each command or command sequence there are three possible outcomes: success (S), failure (F), and error (E). In the state diagrams below we use the symbol B for "begin," and the symbol W for "wait for reply."

We first present the diagram that represents the largest group of FTP commands:

```
1,3    +---+
----------->| E |
                                 |              +---+
+---+     cmd     +---+     2          +---+
| B |---------->| W |---------->| S |
      +---+              +---+              +---+
                                 |
|     4,5     +---+
----------->| F |
                                                +---+
```

This diagram models the commands:
ABOR, ALLO, DELE, CWD, CDUP, SMNT, HELP, MODE, NOOP, PASV, QUIT, SITE, PORT, SYST, STAT, RMD, MKD, PWD, STRU, and TYPE.

The other large group of commands is represented by a very similar diagram:

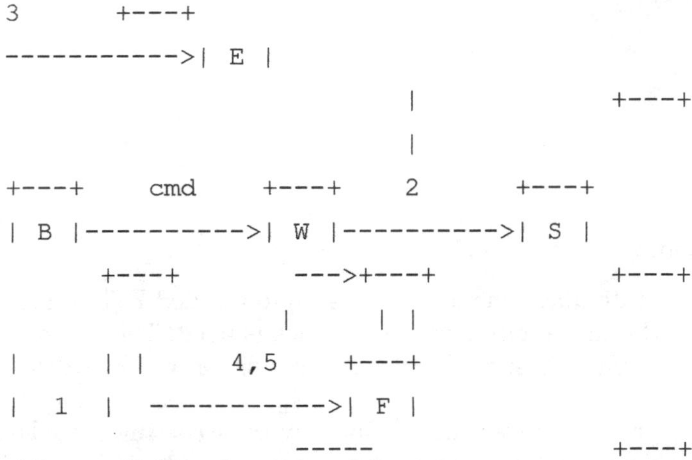

This diagram models the commands:
APPE, LIST, NLST, REIN, RETR, STOR, and STOU.

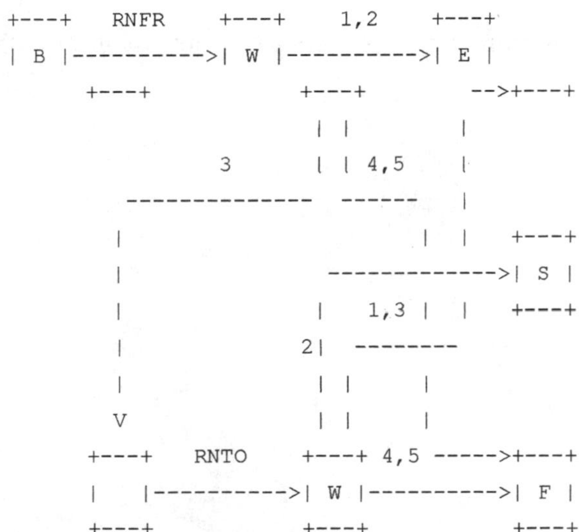

Note that this second model could also be used to represent the first group of commands, the only difference being that in the first group the 100 series replies are unexpected and therefore treated as error, while the second group expects (some may require) 100 series replies. Remember that at most, one 100 series reply is allowed per command. The remaining diagrams model command sequences, perhaps the simplest of these is the rename sequence on page 343.

The next diagram is a simple model of the Restart command:

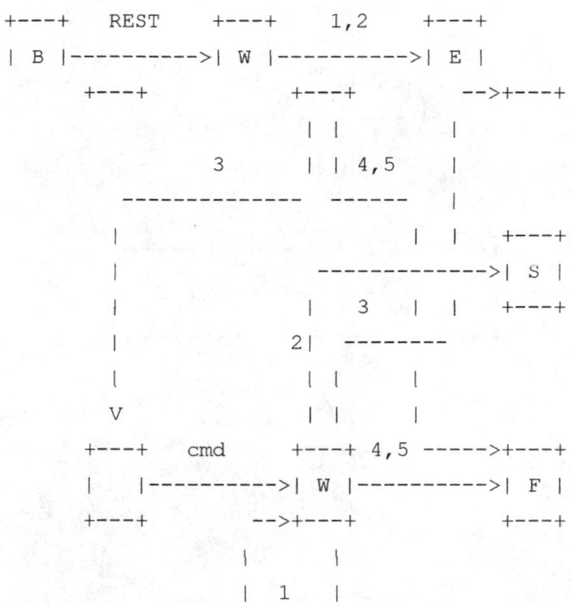

where "cmd" is APPE, STOR, or RETR.

We note that the above three models are similar. The Restart differs from the Rename two only in the treatment of 100 series replies at the second stage, while the second group expects (some may require) 100 series replies. Remember that, at most, one 100 series reply is allowed per command.

The most complicated diagram is for the Login sequence:

```
+---+   USER    +---+------------->+---+
| B |---------->| W | 2       ---->| E |
+---+           +---+------  |  -->+---+
                 | |       | | |
               3 | | 4,5   | | |
   --------------   -----  | | |
  |                      | | | |
  |                      | | | |
  |                 ---------  |
  |               1|     | |   |
  V                |     | |   |
+---+   PASS    +---+ 2  |  ------>+---+
|   |---------->| W |------------->| S |
+---+           +---+   ---------->+---+
                 | |   | |     |
               3 | |4,5| |     |
   --------------   --------   |
  |                    | |  |  |
  |                    | |  |  |
  |                 -----------
  |             1,3|   | |  |
  V                |  2| |  |
+---+   ACCT    +---+--  |   ----->+---+
|   |---------->| W | 4,5 -------->| F |
+---+           +---+------------->+---+
```

Finally, we present a generalized diagram that could be used to model the command and reply interchange.

```
            |                                 |
    Begin   |                                 |
        |   V                                 |
        |  +---+  cmd   +---+ 2        +---+    |
•     >|   |------->|   |---------->|   |     |
|   |      | W |            | S |-----|
•     >|   |     -->| |----        |   |     |
        |  +---+    |   +---+ 4,5 |     +---+     |
        |   |       |    | |      |                |
        |   |       |   1| |3     |     +---+     |
        |   |       |    | |      |     |   |     |
        |   |        ----  |       ---->| F |-----
        |   |              |            |   |
        |   |              |            +---+
            |
            |
V
End
```

7 Typical FTP Scenario

User at host U wanting to transfer files to/from host S:
In general, the user will communicate to the server via a mediating user-FTP process. The following may be a typical scenario. The user-FTP prompts are shown in parentheses, '---->' represents commands from host U to host S, and '<----' represents replies from host S to host U.

LOCAL COMMANDS BY USER	ACTION INVOLVED
ftp (host) multics<CR>	Connect to host S, port L, establishing control connections.
	<---- 220 Service ready <CRLF>.
username Doe <CR>	USER Doe<CRLF>---->
	<---- 331 User name ok, need password<CRLF>.
password mumble <CR>	PASS mumble<CRLF>---->
	<---- 230 User logged in<CRLF>.
retrieve (local type) ASCII<CR>	
(local pathname) test 1 <CR>	User-FTP opens local file in ASCII.
(for. pathname) test.p11<CR>	RETR test.p11<CRLF>---->
	<---- 150 File status okay;

about to open data
connection<CRLF>.
Server makes data connection
to port U.
<---- 226 Closing data connection,
file transfer successful<CRLF>.

type Image<CR>	TYPE I<CRLF>---->
	<---- 200 Command OK<CRLF>
store (local type) image<CR>	
(local pathname) file dump<CR>	User-FTP opens local file in Image.
(for.pathname) >udd>cn>fd<CR>	STOR >udd>cn>fd<CRLF>---->
<---- 550 Access denied<CRLF>	
terminate	QUIT <CRLF>---->
	Server closes all connections.

8 Connection Establishment

The FTP control connection is established via TCP between the user process port U and the server process port L. This protocol is assigned the service port 21 (25 octal), that is L = 21.

Appendix I: Page-Structure

The need for FTP to support page-structure derives principally from the need to support efficient transmission of files between TOPS-20 systems, particularly the files used by NLS.

The file system of TOPS-20 is based on the concept of pages. The operating system is most efficient at manipulating files as pages. The operating system provides an interface to the file system so that many applications view files as sequential streams of characters. However, a few applications use the underlying page structure directly, and some of these create holey files.

A TOPS-20 disk file consists of four things: a pathname, a page table, a (possibly empty) set of pages, and a set of attributes. The pathname is specified in the RETR or STOR command. It includes the directory name, file name, file name extension, and generation number.

The page table contains up to 2**18 entries. Each entry may be EMPTY, or may point to a page. If it is not empty, there are also some page-specific access bits; not all pages of a file need have the same access protection.

A page is a contiguous set of 512 words of 36 bits each. The attributes of the file, in the File Descriptor Block (FDB), contain such things as creation time, write time, read time, writer's byte-size, end-of-file pointer, count of reads and writes, backup system tape numbers, etc.

Note that there is NO requirement that entries in the page table be contiguous. There may be empty page table slots between occupied ones. Also, the end-of-file pointer is simply a number. There is no requirement that it in fact point at the "last" datum in the file. Ordinary sequential I/O calls in TOPS-20 will cause the end-of-file pointer to be left after the last datum written, but other operations may cause it not to be so, if a particular programming system so requires.

In fact, in both of these special cases, "holey" files and end-of-file pointers NOT at the end of the file, occur with NLS data files.

The TOPS-20 paged files can be sent with the FTP transfer parameters: TYPE L 36, STRU P, and MODE S (in fact, any mode could be used). Each page of information has a header. Each header field, which is a logical byte, is a TOPS-20 word, since the TYPE is L 36.

The header fields are:

Word 0: Header Length.

The header length is 5.

Word 1: Page Index.

If the data is a disk file page, this is the number of that page in the file's page map. Empty pages (holes) in the file are simply not sent. Note that a hole is NOT the same as a page of zeros.

Word 2: Data Length.

The number of data words in this page, following the header. Thus, the total length of the transmission unit is the Header Length plus the Data Length.

Word 3: Page Type.

A code for what type of chunk this is. A data page is type 3, the FDB page is type 2.

Word 4: Page Access Control.

The access bits associated with the page in the file's page map. (This full word quantity is put into AC2 of a SPACS by the program reading from net to disk.)

After the header are Data Length data words. Data Length is currently either 512 for a data page or 31 for an FDB. Trailing zeros in a disk file page may be discarded, making Data Length less than 512 in that case.

Appendix II: Directory Commands

Since UNIX has a tree-like directory structure in which directories are as easy to manipulate as ordinary files, it is useful to expand the FTP servers on these machines to include commands which deal with the creation of directories. Since there are other hosts on the ARPA-Internet which have tree-like directories (including TOPS-20 and Multics), these commands are as general as possible.

Four directory commands have been added to FTP:

MKD pathname

Make a directory with the name "pathname."

RMD pathname

Remove the directory with the name "pathname."

PWD

Print the current working directory name.

CDUP

Change to the parent of the current working directory.

The "pathname" argument should be created (removed) as a subdirectory of the current working directory, unless the "pathname" string contains sufficient information to specify otherwise to the server, e.g., "pathname" is an absolute pathname (in UNIX and Multics), or pathname is something like "<abso.lute.path>" to TOPS-20.

Reply codes

The CDUP command is a special case of CWD, and is included to simplify the implementation of programs for transferring directory trees between operating systems having different syntaxes for naming the parent directory. The reply codes for CDUP are identical to the reply codes of CWD.

The reply codes for RMD are identical to the reply codes for its file analogue, DELE.

The reply codes for MKD, however, are a bit more complicated. A freshly created directory will probably be the object of a future CWD command. Unfortunately, the argument to MKD may not always be a suitable argument for CWD. This is the case, for example, when a TOPS-20 subdirectory is created by giving just the subdirectory name. That is, with a TOPS-20 server FTP, the command sequence

MKD MYDIR

CWD MYDIR

will fail. The new directory may only be referred to by its "absolute" name; e.g., if the MKD command above were issued while connected to the directory <DFRANKLIN>, the new subdirectory could only be referred to by the name <DFRANKLIN.MYDIR>.

Even on UNIX and Multics, however, the argument given to MKD may not be suitable. If it is a "relative" pathname (i.e., a pathname which is interpreted relative to the current directory), the user would need to be in the same current directory in order to reach the subdirectory. Depending on the application, this may be inconvenient. It is not very robust in any case.

To solve these problems, upon successful completion of an MKD command, the server should return a line of the form:

257<space>"<directory-name>"<space><commentary>

That is, the server will tell the user what string to use when referring to the created directory. The directory name can contain any character; embedded double-quotes should be escaped by double-quotes (the "quote-doubling" convention).

For example, a user connects to the directory /usr/dm, and creates a subdirectory, named pathname:

CWD /usr/dm

200 directory changed to /usr/dm

MKD pathname

257 "/usr/dm/pathname" directory created

An example with an embedded double quote:

MKD foo"bar

257 "/usr/dm/foo" "bar" directory created

CWD /usr/dm/foo"bar

200 directory changed to /usr/dm/foo"bar

The prior existence of a subdirectory with the same name is an error, and the server must return an "access denied" error reply in that case.

CWD /usr/dm

200 directory changed to /usr/dm

MKD pathname

521- "/usr/dm/pathname" directory already exists;

521 taking no action.

The failure replies for MKD are analogous to its file creating cousin, STOR. Also, an “access denied” return is given if a file name with the same name as the subdirectory will conflict with the creation of the subdirectory (this is a problem on UNIX, but should not be one on TOPS-20).

Essentially because the PWD command returns the same type of information as the successful MKD command, the successful PWD command uses the 257 reply code as well.

Subtleties

Because these commands will be most useful in transferring subtrees from one machine to another, carefully observe that the argument to MKD is to be interpreted as a subdirectory of the current working directory, unless it contains enough information for the destination host to tell otherwise. A hypothetical example of its use in the TOPS-20 world:

CWD <some.where>

200 Working directory changed

MKD overrainbow

257 “<some.where.overrainbow>” directory created

CWD overrainbow

431 No such directory

CWD <some.where.overrainbow>

200 Working directory changed

CWD <some.where>

200 Working directory changed to <some.where>

MKD <unambiguous>

257 “<unambiguous>” directory created

CWD <unambiguous>

Note that the first example results in a subdirectory of the connected directory. In contrast, the argument in the second example contains enough information for TOPS-20 to tell that the <unambiguous> directory is a top-level directory. Note also that in the first example the user “violated” the protocol by attempting to access the freshly created directory with a name other than the one returned by TOPS-20. Problems could have resulted in this case had there been an <overrainbow> directory; this is an ambiguity inherent in some TOPS-20 implementations.

Similar considerations apply to the RMD command. The point is this: except where to do so would violate a host's conventions for denoting relative versus absolute pathnames, the host should treat the operands of the MKD and RMD commands as subdirectories. The 257 reply to the MKD command must always contain the absolute pathname of the created directory.

Appendix III: RFCs on FTP

Bhushan, Abhay, "A File Transfer Protocol," RFC ***114*** (NIC 5823), MIT-Project MAC, 16 April 1971.

Harslem, Eric, and John Heafner, "Comments on RFC ***114*** (A File Transfer Protocol)," RFC *141* (NIC 6726), RAND, 29 April 1971.

Bhushan, Abhay, et al, "The File Transfer Protocol," RFC ***172*** (NIC 6794), MIT-Project MAC, 23 June 1971.

Braden, Bob, "Comments on DTP and FTP Proposals," RFC ***238*** (NIC 7663), UCLA/CCN, 29 September 1971.

Bhushan, Abhay, et al, "The File Transfer Protocol," RFC ***265*** (NIC 7813), MIT-Project MAC, 17 November 1971.

McKenzie, Alex, "A Suggested Addition to File Transfer Protocol," RFC ***281*** (NIC 8163), BBN, 8 December 1971.

Bhushan, Abhay, "The Use of 'Set Data Type' Transaction in File Transfer Protocol," RFC ***294*** (NIC 8304), MIT-Project MAC, 25 January 1972.

Bhushan, Abhay, "The File Transfer Protocol," RFC ***354*** (NIC 10596), MIT-Project MAC, 8 July 1972.

Bhushan, Abhay, "Comments on the File Transfer Protocol (RFC ***354***)," RFC ***385*** (NIC 11357), MIT-Project MAC, 18 August 1972.

Hicks, Greg, "User FTP Documentation," RFC ***412*** (NIC 12404), Utah, 27 November 1972.

Bhushan, Abhay, "File Transfer Protocol (FTP) Status and Further Comments," RFC ***414*** (NIC 12406), MIT-Project MAC, 20 November 1972.

Braden, Bob, "Comments on File Transfer Protocol," RFC ***430*** (NIC 13299), UCLA/CCN, 7 February 1973.

Thomas, Bob, and Bob Clements, "FTP Server-Server Interaction," RFC ***438*** (NIC 13770), BBN, 15 January 1973.

Braden, Bob, "Print Files in FTP," RFC ***448*** (NIC 13299), UCLA/CCN, 27 February 1973.

McKenzie, Alex, "File Transfer Protocol," RFC ***454*** (NIC 14333), BBN, 16 February 1973.

Bressler, Bob, and Bob Thomas, "Mail Retrieval via FTP," RFC *458* (NIC 14378), BBN-NET and BBN-TENEX, 20 February 1973.

Neigus, Nancy, "File Transfer Protocol," RFC ***542*** (NIC 17759), BBN, 12 July 1973.

Krilanovich, Mark, and George Gregg, "Comments on the File Transfer Protocol," RFC ***607*** (NIC 21255), UCSB, 7 January 1974.

Pogran, Ken, and Nancy Neigus, "Response to RFC ***607***—Comments on the File Transfer Protocol," RFC ***614*** (NIC 21530), BBN, 28 January 1974.

Krilanovich, Mark, George Gregg, Wayne Hathaway, and Jim White, "Comments on the File Transfer Protocol," RFC ***624*** (NIC 22054), UCSB, Ames Research Center, SRI-ARC, 28 February 1974.

Bhushan, Abhay, "FTP Comments and Response to RFC ***430,***" RFC ***463*** (NIC 14573), MIT-DMCG, 21 February 1973.

Braden, Bob, "FTP Data Compression," RFC ***468*** (NIC 14742), UCLA/CCN, 8 March 1973.

Bhushan, Abhay, "FTP and Network Mail System," RFC ***475*** (NIC 14919), MIT-DMCG, 6 March 1973.

Bressler, Bob, and Bob Thomas "FTP Server-Server Interaction—II," RFC ***478*** (NIC 14947), BBN-NET and BBN-TENEX, 26 March 1973.
White, Jim, "Use of FTP by the NIC Journal," RFC ***479*** (NIC 14948), SRI-ARC, 8 March 1973.
White, Jim, "Host-Dependent FTP Parameters," RFC ***480*** (NIC 14949), SRI-ARC, 8 March 1973.
Padlipsky, Mike, "An FTP Command-Naming Problem," RFC ***506*** (NIC 16157), MIT-Multics, 26 June 1973.
Day, John, "Memo to FTP Group (Proposal for File Access Protocol)," RFC ***520*** (NIC 16819), Illinois, 25 June 1973.
Merryman, Robert, "The UCSD-CC Server-FTP Facility," RFC ***532*** (NIC 17451), UCSD-CC, 22 June 1973.
Braden, Bob, "TENEX FTP Problem," RFC ***571*** (NIC 18974), UCLA/CCN, 15 November 1973.
McKenzie, Alex, and Jon Postel, "Telnet and FTP Implementation—Schedule Change," RFC ***593*** (NIC 20615), BBN and MITRE, 29 November 1973.
Sussman, Julie, "FTP Error Code Usage for More Reliable Mail Service," RFC ***630*** (NIC 30237), BBN, 10 April 1974.
Postel, Jon, "Revised FTP Reply Codes," RFC ***640*** (NIC 30843), UCLA/NMC, 5 June 1974.
Harvey, Brian, "Leaving Well Enough Alone," RFC ***686*** (NIC 32481), SU-AI, 10 May 1975.
Harvey, Brian, "One More Try on the FTP," RFC ***691*** (NIC 32700), SU-AI, 28 May 1975.
Lieb, J., "CWD Command of FTP," RFC ***697*** (NIC 32963), 14 July 1975.
Harrenstien, Ken, "FTP Extension: XSEN," RFC ***737*** (NIC 42217), SRI-KL, 31 October 1977.
Harrenstien, Ken, "FTP Extension: XRSQ/XRCP," RFC ***743*** (NIC 42758), SRI-KL, 30 December 1977.
Lebling, P. David, "Survey of FTP Mail and MLFL," RFC ***751,*** MIT, 10 December 1978.
Postel, Jon, "File Transfer Protocol Specification," RFC ***765,*** ISI, June 1980.
Mankins, David, Dan Franklin, and Buzz Owen, "Directory Oriented FTP Commands," RFC ***776,*** BBN, December 1980.
Padlipsky, Michael, "FTP Unique-Named Store Command," RFC ***949,*** MITRE, July 1985.

References

1. Feinler, Elizabeth, "Internet Protocol Transition Workbook," Network Information Center, SRI International, March 1982.
2. Postel, Jon, "Transmission Control Protocol—DARPA Internet Program Protocol Specification," RFC ***793,*** DARPA, September 1981.
3. Postel, Jon, and Joyce Reynolds, "Telnet Protocol Specification," RFC ***854,*** ISI, May 1983.
4. Reynolds, Joyce, and Jon Postel, "Assigned Numbers," RFC ***943,*** ISI, April 1985.

Appendix

D

RFC 1157: A Simple Network Management Protocol (SNMP)

SNMP is another Application layer protocol and is included here for reference. This protocol was developed in order to create a basic network structure which was uniform and avoided any system incompatibilities across network connections. Very few systems use this protocol; it has provided a basis for the development of more sophisticated systems. It should be reviewed for the basic network concepts it describes. SNMP is explained in Chapter 2.

M. Schoffstall
J. Davin
J. Case
M. Fedor

1 Status of this Memo

This RFC is a rerelease of RFC 1098, with a changed "Status of this Memo" section plus a few minor typographical corrections. This memo defines a simple protocol by which management information for a network element may be inspected or altered by logically remote users. In particular, together with its companion memos which describe the structure of management information along with the management information base, these documents provide a simple, workable architecture and system for managing TCP/IP-based internets and in particular the Internet. (May 1990)

The Internet Activities Board recommends that all IP and TCP implementations be network manageable. This implies implementation of the Internet MIB (RFC-1156) and at least one of the two recommended management protocols SNMP (RFC-1157) or CMOT (RFC-1095). It should be noted that, at this time, SNMP is a full Internet standard and CMOT is a draft standard. See also the Host and Gateway Requirements RFCs for more specific information on the applicability of this standard.

Please refer to the latest edition of the "IAB Official Protocol Standards" RFC for current information on the state and status of standard Internet protocols. Distribution of this memo is unlimited.

2 Introduction

As reported in RFC 1052, IAB Recommendations for the Development of Internet Network Management Standards [1], a two-prong strategy for network management of TCP/IP-based internets was undertaken. In the short-term, the Simple Network Management Protocol (SNMP) was to be used to manage nodes in the Internet community. In the long-term, the use of the OSI network management framework was to be examined. Two documents were produced to define the management information: RFC 1065, which defined the Structure of Management Information (SMI) [2], and RFC 1066, which defined the Management Information Base (MIB) [3]. Both of these documents were designed so as to be compatible with both the SNMP and the OSI network management framework.

This strategy was quite successful in the short-term: Internet-based network management technology was fielded, by both the research and commercial communities, within a few months. As a result of this, portions of the Internet community became network manageable in a timely fashion.

As reported in RFC 1109, Report of the Second Ad Hoc Network Management Review Group [4], the requirements of the SNMP and the OSI network management frameworks were more different than anticipated. As such, the requirement for compatibility between the SMI/MIB and both frameworks was suspended. This action permitted the operational network management framework, the SNMP, to respond to new operational needs in the Internet community by producing documents defining new MIB items.

The IAB has designated the SNMP, SMI, and the initial Internet MIB to be full "Standard Protocols" with "Recommended" status. By this action, the IAB recommends that all IP and TCP implementations be network manageable and that the implementations that are network manageable are expected to adopt and implement the SMI, MIB, and SNMP.

As such, the current network management framework for TCP/IP-based internets consists of: Structure and Identification of Management Information for TCP/IP-based Internets, which describes how managed objects contained in the MIB are defined as set forth in RFC 1155 [5]; Management Information Base for Network Management of TCP/IP-based Internets, which describes the managed objects contained in the MIB as set forth in RFC 1156 [6]; and, the Simple Network Management Protocol, which defines the protocol used to manage these objects, as set forth in this memo.

As reported in RFC 1052, IAB Recommendations for the Development of Internet Network Management Standards [1], the Internet Activities Board has directed the Internet Engineering Task Force

(IETF) to create two new working groups in the area of network management. One group was charged with the further specification and definition of elements to be included in the Management Information Base (MIB). The other was charged with defining the modifications to the Simple Network Management Protocol (SNMP) to accommodate the short-term needs of the network vendor and operations communities, and to align with the output of the MIB working group.

The MIB working group produced two memos, one which defines a Structure for Management Information (SMI) [2] for use by the managed objects contained in the MIB. A second memo [3] defines the list of managed objects.

The output of the SNMP Extensions working group is this memo, which incorporates changes to the initial SNMP definition [7] required to attain alignment with the output of the MIB working group. The changes should be minimal in order to be consistent with the IAB's directive that the working groups be "extremely sensitive to the need to keep the SNMP simple." Although considerable care and debate has gone into the changes to the SNMP which are reflected in this memo, the resulting protocol is not backwardly compatible with its predecessor, the Simple Gateway Monitoring Protocol (SGMP) [8].

Although the syntax of the protocol has been altered, the original philosophy, design decisions, and architecture remain intact. In order to avoid confusion, new UDP ports have been allocated for use by the protocol described in this memo.

3 The SNMP Architecture

Implicit in the SNMP architectural model is a collection of network management stations and network elements. Network management stations execute management applications which monitor and control network elements. Network elements are devices such as hosts, gateways, terminal servers, and the like, which have management agents responsible for performing the network management functions requested by the network management stations. The Simple Network Management Protocol (SNMP) is used to communicate management information between the network management stations and the agents in the network elements.

3.1 Goals of the architecture

The SNMP explicitly minimizes the number and complexity of management functions realized by the management agent itself. This goal is attractive in at least four respects:

1. The development cost for management agent software necessary to support the protocol is accordingly reduced.

2. The degree of management function that is remotely supported is accordingly increased, thereby admitting fullest use of internet resources in the management task.
3. The degree of management function that is remotely supported is accordingly increased, thereby imposing the fewest possible restrictions on the form and sophistication of management tools.
4. Simplified sets of management functions are easily understood and used by developers of network management tools.

A second goal of the protocol is that the functional paradigm for monitoring and control be sufficiently extensible to accommodate additional, possibly unanticipated aspects of network operation and management. A third goal is that the architecture be, as much as possible, independent of the architecture and mechanisms of particular hosts or particular gateways.

3.2 Elements of the architecture

The SNMP architecture articulates a solution to the network management problem in terms of:

1. the scope of the management information communicated by the protocol,
2. the representation of the management information communicated by the protocol,
3. operations on management information supported by the protocol,
4. the form and meaning of exchanges among management entities,
5. the definition of administrative relationships among management entities, and
6. the form and meaning of references to management information.

3.2.1 Scope of management information. The scope of the management information communicated by operation of the SNMP is exactly that represented by instances of all nonaggregate object types either defined in Internet-standard MIB or defined elsewhere according to the conventions set forth in Internet-standard SMI [5].

Support for aggregate object types in the MIB is neither required for conformance with the SMI nor realized by the SNMP.

3.2.2 Representation of management information. Management information communicated by operation of the SNMP is represented according to the subset of the ASN.1 language [9] that is specified for the definition of nonaggregate types in the SMI.

The SGMP adopted the convention of using a well-defined subset of the ASN.1 language [9]. The SNMP continues and extends this tradition by utilizing a moderately more complex subset of ASN.1 for describing managed objects and for describing the protocol data units used for managing those objects. In addition, the desire to ease eventual transition to OSI-based network management protocols led to the definition in the ASN.1 language of an Internet-standard Structure of Management Information (SMI) [5] and Management Information Base (MIB) [6]. The use of the ASN.1 language was, in part, encouraged by the successful use of ASN.1 in earlier efforts, in particular, the SGMP. The restrictions on the use of ASN.1 that are part of the SMI contribute to the simplicity espoused and validated by experience with the SGMP.

Also for the sake of simplicity, the SNMP uses only a subset of the basic encoding rules of ASN.1 [10]. Namely, all encodings use the definite-length form. Further, whenever permissible, nonconstructor encodings are used rather than constructor encodings. This restriction applies to all aspects of ASN.1 encoding, both for the top-level protocol data units and the data objects they contain.

3.2.3 Operations supported on management information. The SNMP models all management agent functions as alterations or inspections of variables. Thus, a protocol entity on a logically remote host (possibly the network element itself) interacts with the management agent resident on the network element in order to retrieve (get) or alter (set) variables. This strategy has at least two positive consequences:

1. It has the effect of limiting the number of essential management functions realized by the management agent to two: one operation to assign a value to a specified configuration or other parameter and another to retrieve such a value.
2. A second effect of this decision is to avoid introducing into the protocol definition support for imperative management commands: the number of such commands is in practice ever-increasing, and the semantics of such commands are in general arbitrarily complex.

The strategy implicit in the SNMP is that the monitoring of network state at any significant level of detail is accomplished primarily by polling for appropriate information on the part of the monitoring center(s). A limited number of unsolicited messages (traps) guide the timing and focus of the polling. Limiting the number of unsolicited messages is consistent with the goal of simplicity and minimizing the amount of traffic generated by the network management function.

The exclusion of imperative commands from the set of explicitly supported management functions is unlikely to preclude any desirable management agent operation. Currently, most commands are requests either to set the value of some parameter or to retrieve such a value, and the function of the few imperative commands currently supported is easily accommodated in an asynchronous mode by this management model. In this scheme, an imperative command might be realized as the setting of a parameter value that subsequently triggers the desired action. For example, rather than implementing a "reboot command," this action might be invoked by simply setting a parameter indicating the number of seconds until system reboot.

3.2.4 Form and meaning of protocol exchanges. The communication of management information among management entities is realized in the SNMP through the exchange of protocol messages. The form and meaning of those messages is defined below in Section 4.

Consistent with the goal of minimizing complexity of the management agent, the exchange of SNMP messages requires only an unreliable datagram service, and every message is entirely and independently represented by a single transport datagram. While this document specifies the exchange of messages via the UDP protocol [11], the mechanisms of the SNMP are generally suitable for use with a wide variety of transport services.

3.2.5 Definition of administrative relationships. The SNMP architecture admits a variety of administrative relationships among entities that participate in the protocol. The entities residing at management stations and network elements which communicate with one another using the SNMP are termed SNMP application entities. The peer processes which implement the SNMP, and thus support the SNMP application entities, are termed protocol entities.

A pairing of an SNMP agent with some arbitrary set of SNMP application entities is called an SNMP community. Each SNMP community is named by a string of octets, that is called the community name for said community.

An SNMP message originated by an SNMP application entity that in fact belongs to the SNMP community named by the community component of said message is called an authentic SNMP message. The set of rules by which an SNMP message is identified as an authentic SNMP message for a particular SNMP community is called an authentication scheme. An implementation of a function that identifies authentic SNMP messages according to one or more authentication schemes is called an authentication service.

Clearly, effective management of administrative relationships among SNMP application entities requires authentication services that (by the use of encryption or other techniques) are able to identify authentic SNMP messages with a high degree of certainty. Some SNMP implementations may wish to support only a trivial authentication service that identifies all SNMP messages as authentic SNMP messages. For any network element, a subset of objects in the MIB that pertain to that element is called a SNMP MIB view. Note that the names of the object types represented in a SNMP MIB view need not belong to a single subtree of the object type name space.

An element of the set {READ-ONLY, READ-WRITE} is called an SNMP access mode. A pairing of a SNMP access mode with a SNMP MIB view is called an SNMP community profile. A SNMP community profile represents specified access privileges to variables in a specified MIB view. For every variable in the MIB view in a given SNMP community profile, access to that variable is represented by the profile according to the following conventions:

1. if said variable is defined in the MIB with "Access:" of "none," it is unavailable as an operand for any operator;
2. if said variable is defined in the MIB with "Access:" of "read-write" or "write-only" and the access mode of the given profile is READ-WRITE, that variable is available as an operand for the get, set, and trap operations;
3. otherwise, the variable is available as an operand for the get and trap operations.
4. In those cases where a "write-only" variable is an operand used for the get or trap operations, the value given for the variable is implementation-specific. A pairing of a SNMP community with a SNMP community profile is called a SNMP access policy. An access policy represents a specified community profile afforded by the SNMP agent of a specified SNMP community to other members of that community. All administrative relationships among SNMP application entities are architecturally defined in terms of SNMP access policies.

For every SNMP access policy, if the network element on which the SNMP agent for the specified SNMP community resides is not that to which the MIB view for the specified profile pertains, then that policy is called a SNMP proxy access policy. The SNMP agent associated with a proxy access policy is called a SNMP proxy agent. While careless definition of proxy access policies can result in management loops, prudent definition of proxy policies is useful in at least two ways:

1. It permits the monitoring and control of network elements which are otherwise not addressable using the management protocol and the transport protocol. That is, a proxy agent may provide a proto-

col conversion function allowing a management station to apply a consistent management framework to all network elements, including devices such as modems, multiplexors, and other devices which support different management frameworks.

2. It potentially shields network elements from elaborate access control policies. For example, a proxy agent may implement sophisticated access control whereby diverse subsets of variables within the MIB are made accessible to different management stations without increasing the complexity of the network element.

By way of example, Figure 1 illustrates the relationship between management stations, proxy agents, and management agents. In this example, the proxy agent is envisioned to be a Normal Internet Network Operations Center (INOC) of some administrative domain which has a standard managerial relationship with a set of management agents.

3.2.6 Form and meaning of references to managed objects. The SMI requires that the definition of a conformant management protocol address:

1. the resolution of ambiguous MIB references,
2. the resolution of MIB references in the presence of multiple MIB versions, and
3. the identification of particular instances of object types defined in the MIB.

3.2.6.1 Resolution of ambiguous MIB references. Because the scope of any SNMP operation is conceptually confined to objects relevant to a single network element, and because all SNMP references to MIB objects are (implicitly or explicitly) by unique variable names, there is no possibility that any SNMP reference to any object type defined in the MIB could resolve to multiple instances of that type.

3.2.6.2 Resolution of references across MIB versions. The object instance referred to by any SNMP operation is exactly that specified as part of the operation request or (in the case of a get-next operation) its immediate successor in the MIB as a whole. In particular, a reference to an object as part of some version of the Internet-standard MIB does not resolve to any object that is not part of said version of the Internet-standard MIB, except in the case that the requested operation is get-next and the specified object name is lexicographically last among the names of all objects presented as part of said version of the Internet-Standard MIB.

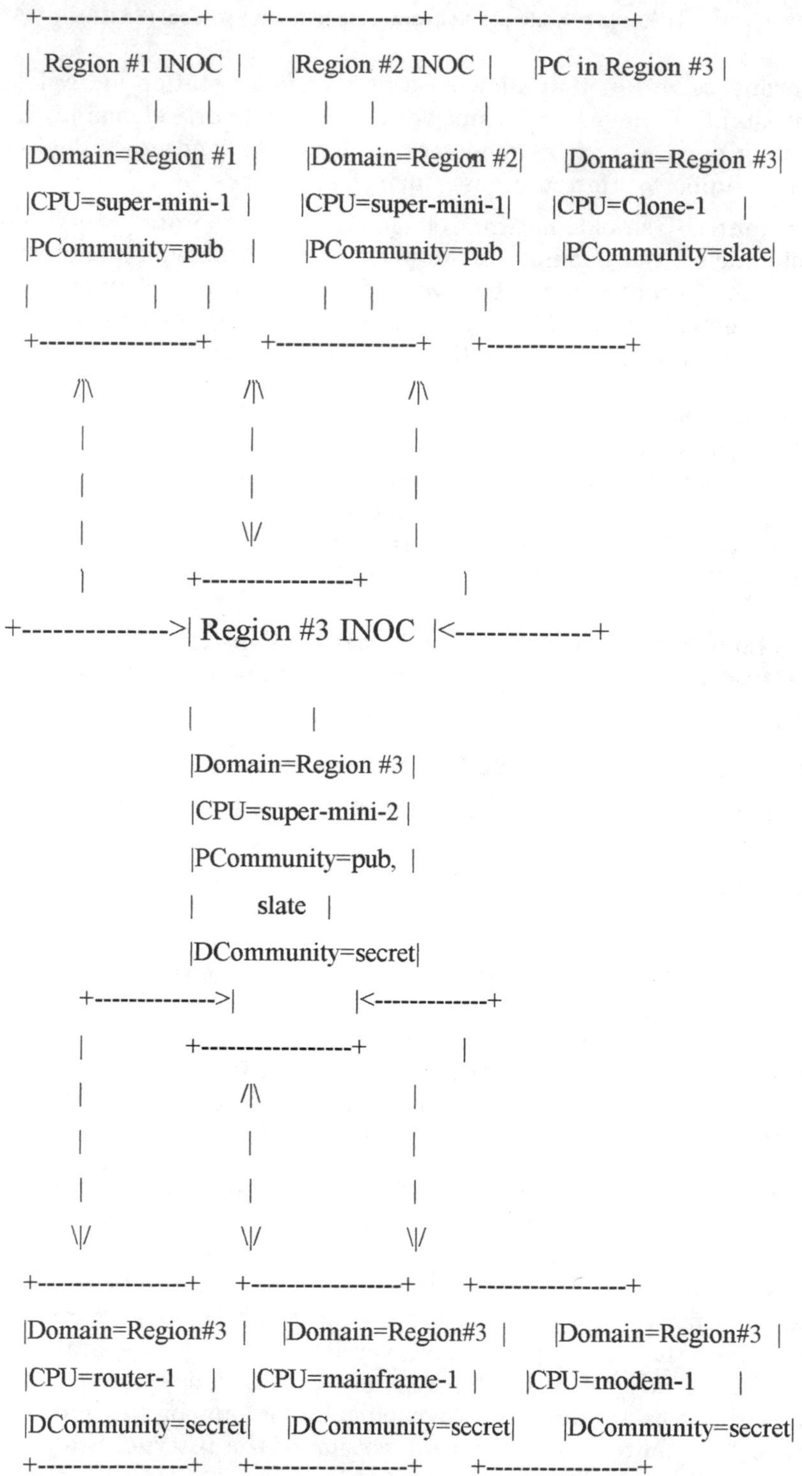

Figure 1 Example Network Management Configuration. Domain: the administrative domain of the element; PCommunity: the name of a community utilizing a proxy agent; DCommunity: the name of a direct community.

3.2.6.3 Identification of object instances. The names for all object types in the MIB are defined explicitly either in the Internet-standard MIB or in other documents which conform to the naming conventions of the SMI. The SMI requires that conformant management protocols define mechanisms for identifying individual instances of those object types for a particular network element.

Each instance of any object type defined in the MIB is identified in SNMP operations by a unique name called its "variable name." In general, the name of an SNMP variable is an OBJECT IDENTIFIER of the form x.y, where x is the name of a nonaggregate object type defined in the MIB and y is an OBJECT IDENTIFIER fragment that, in a way specific to the named object type, identifies the desired instance.

This naming strategy admits the fullest exploitation of the semantics of the GetNextRequest-PDU (see Section 4), because it assigns names for related variables so as to be contiguous in the lexicographical ordering of all variable names known in the MIB.

The type-specific naming of object instances is defined below for a number of classes of object types. Instances of an object type to which none of the following naming conventions are applicable are named by OBJECT IDENTIFIERs of the form x.0, where x is the name of said object type in the MIB definition.

For example, suppose one wanted to identify an instance of the variable sysDescr. The object class for sysDescr is

iso	org	dod	internet	mgmt	mib	system	sysDescr
1	3	6	1	2	1	1	1

Hence, the object type, x, would be 1.3.6.1.2.1.1.1 to which is appended an instance subidentifier of 0. That is, 1.3.6.1.2.1.1.1.0 identifies the one and only instance of sysDescr.

3.2.6.3.1 ifTable object type names. The name of a subnet interface, s, is the OBJECT IDENTIFIER value of the form i, where i has the value of that instance of the ifIndex object type associated with s.

For each object type, t, for which the defined name, n, has a prefix of ifEntry, an instance, i, of t is named by an OBJECT IDENTIFIER of the form n.s, where s is the name of the subnet interface about which i represents information.

For example, suppose one wanted to identify the instance of the variable ifType associated with interface 2. Accordingly, ifType.2 would identify the desired instance.

3.2.6.3.2 atTable object type names. The name of an AT-cached network address, x, is an OBJECT IDENTIFIER of the form 1.a.b.c.d, where a.b.c.d is the value (in the familiar "dot" notation) of the atNet-

Address object type associated with x. The name of an address translation equivalence e is an OBJECT IDENTIFIER value of the form s.w, such that s is the value of that instance of the atIndex object type associated with e and such that w is the name of the AT-cached network address associated with e.

For each object type, t, for which the defined name, n, has a prefix of atEntry, an instance, i, of t is named by an OBJECT IDENTIFIER of the form n.y, where y is the name of the address translation equivalence about which i represents information.

For example, suppose one wanted to find the physical address of an entry in the address translation table (ARP cache) associated with an IP address of 89.1.1.42 and interface 3. Accordingly, atPhysAddress.3.1.89.1.1.42 would identify the desired instance.

3.2.6.3.3 ipAddrTable object type names. The name of an IP-addressable network element, x, is the OBJECT IDENTIFIER of the form a.b.c.d such that a.b.c.d is the value (in the familiar "dot" notation) of that instance of the ipAdEntAddr object type associated with x.

For each object type, t, for which the defined name, n, has a prefix of ipAddrEntry, an instance, i, of t is named by an OBJECT IDENTIFIER of the form n.y, where y is the name of the IP-addressable network element about which i represents information.

For example, suppose one wanted to find the network mask of an entry in the IP interface table associated with an IP address of 89.1.1.42.

Accordingly, ipAdEntNetMask.89.1.1.42 would identify the desired instance.

3.2.6.3.4 ipRoutingTable object type names. The name of an IP route, x, is the OBJECT IDENTIFIER of the form a.b.c.d such that a.b.c.d is the value (in the familiar "dot" notation) of that instance of the ipRouteDest object type associated with x.

For each object type, t, for which the defined name, n, has a prefix of ipRoutingEntry, an instance, i, of t is named by an OBJECT IDENTIFIER of the form n.y, where y is the name of the IP route about which i represents information.

For example, suppose one wanted to find the next hop of an entry in the IP routing table associated with the destination of 89.1.1.42. Accordingly, ipRouteNextHop.89.1.1.42 would identify the desired instance.

3.2.6.3.5 tcpConnTable object type names. The name of a TCP connection, x, is the OBJECT IDENTIFIER of the form a.b.c.d.e.f.g.h.i.j such that a.b.c.d is the value (in the familiar "dot" notation) of that instance of the tcpConnLocalAddress object type associated with x and such that f.g.h.i is the value (in the familiar "dot" notation) of that instance of the tcpConnRemoteAddress object type associated with x and such that e is the value of that instance of the tcpConnLocalPort object type associated with x and such that j is the value of that instance of the tcpConnRemotePort object type associated with x.

For each object type, t, for which the defined name, n, has a prefix of tcpConnEntry, an instance, i, of t is named by an OBJECT IDENTIFIER of the form n.y, where y is the name of the TCP connection about which i represents information.

For example, suppose one wanted to find the state of a TCP connection between the local address of 89.1.1.42 on TCP port 21 and the remote address of 10.0.0.51 on TCP port 2059. Accordingly, tcpConnState.89.1.1.42.21.10.0.0.51.2059 would identify the desired instance.

3.2.6.3.6 egpNeighTable object type names. The name of an EGP neighbor, x, is the OBJECT IDENTIFIER of the form a.b.c.d such that a.b.c.d is the value (in the familiar "dot" notation) of that instance of the egpNeighAddr object type associated with x.

For each object type, t, for which the defined name, n, has a prefix of egpNeighEntry, an instance, i, of t is named by an OBJECT IDENTIFIER of the form n.y, where y is the name of the EGP neighbor about which i represents information.

For example, suppose one wanted to find the neighbor state for the IP address of 89.1.1.42. Accordingly, egpNeighState.89.1.1.42 would identify the desired instance.

4 Protocol Specification

The network management protocol is an application protocol by which the variables of an agent's MIB may be inspected or altered.

Communication among protocol entities is accomplished by the exchange of messages, each of which is entirely and independently represented within a single UDP datagram using the basic encoding rules of ASN.1 (as discussed in Section 3.2.2). A message consists of a version identifier, an SNMP community name, and a protocol data unit (PDU). A protocol entity receives messages at UDP port 161 on the host with which it is associated for all messages except for those which report traps (i.e., all messages except those which contain the Trap-PDU).

Messages which report traps should be received on UDP port 162 for further processing. An implementation of this protocol need not accept messages whose length exceeds 484 octets. However, it is recommended that implementations support larger datagrams whenever feasible. It is mandatory that all implementations of the SNMP support the five PDUs: GetRequest-PDU, GetNextRequest-PDU, GetResponse-PDU, SetRequest-PDU, and Trap-PDU.

RFC1157-SNMP DEFINITIONS ::= BEGIN

IMPORTS

ObjectName, ObjectSyntax, NetworkAddress, IpAddress, TimeTicks FROM RFC1155-SMI;

- top-level message

```
Message ::=
SEQUENCE {
version        --version-1 for this RFC
INTEGER {
version-1(0)
},

community   --community name
OCTET STRING,
data   --e.g., PDUs if trivial
ANY   --authentication is being used
}
```

- protocol data units

```
PDUs ::=
CHOICE {
get-request
GetRequest-PDU,
get-next-request
GetNextRequest-PDU,
get-response
GetResponse-PDU,
set-request
SetRequest-PDU,
trap
Trap-PDU.
}
```

- the individual PDUs and commonly used
- data types will be defined later

END

4.1 Elements of procedure

This section describes the actions of a protocol entity implementing the SNMP. Note, however, that it is not intended to constrain the internal architecture of any conformant implementation. In the text that follows, the term transport address is used. In the case of the UDP, a transport address consists of an IP address along with a UDP port. Other transport services may be used to support the SNMP. In these cases, the definition of a transport address should be made accordingly.

The top-level actions of a protocol entity which generates a message are as follows:

1. It first constructs the appropriate PDU, e.g., the GetRequest-PDU, as an ASN.1 object.
2. It then passes this ASN.1 object along with a community name, its source transport address, and the destination transport address, to the service which implements the desired authentication scheme. This authentication service returns another ASN.1 object.
3. The protocol entity then constructs an ASN.1 Message object, using the community name and the resulting ASN.1 object.
4. This new ASN.1 object is then serialized, using the basic encoding rules of ASN.1, and then sent using a transport service to the peer protocol entity.

Similarly, the top-level actions of a protocol entity which receives a message are as follows:

1. It performs a rudimentary parse of the incoming datagram to build an ASN.1 object corresponding to an ASN.1 Message object. If the parse fails, it discards the datagram and performs no further actions.
2. It then verifies the version number of the SNMP message. If there is a mismatch, it discards the datagram and performs no further actions.
3. The protocol entity then passes the community name and user data found in the ASN.1 Message object, along with the datagram's source and destination transport addresses, to the service which implements the desired authentication scheme. This entity returns another ASN.1 object, or signals an authentication failure. In the latter case, the protocol entity notes this failure, (possibly) generates a trap, and discards the datagram and performs no further actions.
4. The protocol entity then performs a rudimentary parse on the ASN.1 object returned from the authentication service to build an ASN.1 object corresponding to an ASN.1 PDUs object. If the parse fails, it discards the datagram and performs no further actions. Otherwise, using the named SNMP community, the appropriate

profile is selected, and the PDU is processed accordingly. If, as a result of this processing, a message is returned, then the source transport address that the response message is sent from shall be identical to the destination transport address that the original request message was sent to.

4.1.1 Common constructs. Before introducing the six PDU types of the protocol, it is appropriate to consider some of the ASN.1 constructs used frequently:

- request/response information

```
RequestID ::=
INTEGER
ErrorStatus ::=
INTEGER {
noError(0),
tooBig(1),
noSuchName(2),
badValue(3),
readOnly(4)
genErr(5)
}
ErrorIndex ::=
INTEGER
```

- variable bindings

```
VarBind ::=
SEQUENCE {
name
ObjectName,
value
ObjectSyntax
}
VarBindList ::=
SEQUENCE OF
VarBind
```

RequestIDs are used to distinguish among outstanding requests. By use of the RequestID, an SNMP application entity can correlate incoming responses with outstanding requests. In cases where an unreliable datagram service is being used, the RequestID also provides a simple means of identifying messages duplicated by the network.

A nonzero instance of ErrorStatus is used to indicate that an exception occurred while processing a request. In these cases, ErrorIndex may provide additional information by indicating which variable in a list caused the exception.

The term variable refers to an instance of a managed object. A variable binding, or VarBind, refers to the pairing of the name of a variable to the variable's value. A VarBindList is a simple list of variable names and corresponding values. Some PDUs are concerned only with the name of a variable and not its value (e.g., the GetRequest-PDU). In this case, the value portion of the binding is ignored by the protocol entity. However, the value portion must still have valid ASN.1 syntax and encoding. It is recommended that the ASN.1 value NULL be used for the value portion of such bindings.

4.1.2 The GetRequest-PDU. The form of the GetRequest-PDU is

```
GetRequest-PDU ::=
[0]
IMPLICIT SEQUENCE {
    request-id
    RequestID,
    error-status   --always 0
    ErrorStatus,
    error-index        --always 0
    ErrorIndex,
    variable-bindings
    VarBindList
    }
```

The GetRequest-PDU is generated by a protocol entity only at the request of its SNMP application entity. Upon receipt of the GetRequest-PDU, the receiving protocol entity responds according to any applicable rule in the list below:

1. If, for any object named in the variable-bindings field, the object's name does not exactly match the name of some object available for get operations in the relevant MIB view, then the receiving entity sends to the originator of the received message the GetResponse-PDU of identical form, except that the value of the error-status field is noSuchName, and the value of the error-index field is the index of said object name component in the received message.
2. If, for any object named in the variable-bindings field, the object is an aggregate type (as defined in the SMI), then the receiving entity sends to the originator of the received message the GetResponse-PDU of identical form, except that the value of the error-status field is noSuchName, and the value of the error-index field is the index of said object name component in the received message.
3. If the size of the GetResponse-PDU generated as described below would exceed a local limitation, then the receiving entity sends to

the originator of the received message the GetResponse-PDU of identical form, except that the value of the error-status field is tooBig, and the value of the error-index field is zero.

4. If, for any object named in the variable-bindings field, the value of the object cannot be retrieved for reasons not covered by any of the foregoing rules, then the receiving entity sends to the originator of the received message the GetResponse-PDU of identical form, except that the value of the error-status field is genErr and the value of the error-index field is the index of said object name component in the received message. If none of the foregoing rules apply, then the receiving protocol entity sends to the originator of the received message the GetResponse-PDU such that, for each object named in the variable-bindings field of the received message, the corresponding component of the GetResponse-PDU represents the name and value of that variable. The value of the error-status field of the GetResponse-PDU is noError and the value of the error-index field is zero. The value of the request-id field of the GetResponse-PDU is that of the received message.

4.1.3 The GetNextRequest-PDU. The form of the GetNextRequest-PDU is identical to that of the GetRequest-PDU except for the indication of the PDU type. In the ASN.1 language:

```
GetNextRequest-PDU::=
[1]
IMPLICIT SEQUENCE {
    request-id
    RequestID,
  error-status    --always 0
    ErrorStatus,
    error-index   --always 0
    ErrorIndex,
    variable-bindings
    VarBindList
    }
```

The GetNextRequest-PDU is generated by a protocol entity only at the request of its SNMP application entity.

Upon receipt of the GetNextRequest-PDU, the receiving protocol entity responds according to any applicable rule in the list below:

1. If, for any object name in the variable-bindings field, that name does not lexicographically precede the name of some object available for get operations in the relevant MIB view, then the receiving

entity sends to the originator of the received message the GetResponse-PDU of identical form, except that the value of the error-status field is noSuchName, and the value of the error-index field is the index of said object name component in the received message.

2. If the size of the GetResponse-PDU generated as described below would exceed a local limitation, then the receiving entity sends to the originator of the received message the GetResponse-PDU of identical form, except that the value of the error-status field is tooBig, and the value of the error-index field is zero.
3. If, for any object named in the variable-bindings field, the value of the lexicographical successor to the named object cannot be retrieved for reasons not covered by any of the foregoing rules, then the receiving entity sends to the originator of the received message the GetResponse-PDU of identical form, except that the value of the error-status field is genErr and the value of the error-index field is the index of said object name component in the received message.

If none of the foregoing rules apply, then the receiving protocol entity sends to the originator of the received message the GetResponse-PDU such that, for each name in the variable-bindings field of the received message, the corresponding component of the GetResponse-PDU represents the name and value of that object whose name is, in the lexicographical ordering of the names of all objects available for get operations in the relevant MIB view, together with the value of the name field of the given component, the immediate successor to that value. The value of the error-status field of the GetResponse-PDU is noError and the value of the errorindex field is zero. The value of the request-id field of the GetResponse-PDU is that of the received message.

4.1.3.1 Example of table traversal. One important use of the GetNext-Request-PDU is the traversal of conceptual tables of information within the MIB. The semantics of this type of SNMP message, together with the protocol-specific mechanisms for identifying individual instances of object types in the MIB, affords access to related objects in the MIB as if they enjoyed a tabular organization.

By the SNMP exchange sketched below, an SNMP application entity might extract the destination address and next hop gateway for each entry in the routing table of a particular network element. Suppose that this routing table has three entries:

Destination	NextHop	Metric
10.0.099	89.1.1.42	5
9.1.2.3	99.0.0.3	3
10.0.0.51	89.1.1.42	5

The management station sends to the SNMP agent a GetNextRequest-PDU containing the indicated OBJECT IDENTIFIER values as the requested variable names:

GetNextRequest (ipRouteDest, ipRouteNextHop, ipRouteMetric1)
The SNMP agent responds with a GetResponse-PDU:

GetResponse ((ipRouteDest.9.1.2.3 = "9.1.2.3"),

(ipRouteNextHop.9.1.2.3 = "99.0.0.3"),
(ipRouteMetric1.9.1.2.3 = 3)).

The management station continues with

GetNextRequest (ipRouteDest.9.1.2.3,

ipRouteNextHop.9.1.2.3,
ipRouteMetric1.9.1.2.3).

The SNMP agent responds with

GetResponse ((ipRouteDest.10.0.0.51 = "10.0.0.51"),

(ipRouteNextHop.10.0.0.51 = "89.1.1.42"),
(ipRouteMetric1.10.0.0.51 = 5)).

The management station continues with

GetNextRequest (ipRouteDest.10.0.0.51,

ipRouteNextHop.10.0.0.51,
ipRouteMetric1.10.0.0.51).

The SNMP agent responds with

GetResponse ((ipRouteDest.10.0.0.99 = "10.0.0.99"),

(ipRouteNextHop.10.0.0.99 = "89.1.1.42"),
(ipRouteMetric1.10.0.0.99 = 5)).

The management station continues with

GetNextRequest (ipRouteDest.10.0.0.99,
ipRouteNextHop.10.0.0.99,
ipRouteMetric1.10.0.0.99).

As there are no further entries in the table, the SNMP agent returns those objects that are next in the lexicographical ordering of the known object names. This response signals the end of the routing table to the management station.

4.1.4 The GetResponse-PDU. The form of the GetResponse-PDU is identical to that of the GetRequest-PDU except for the indication of the PDU type. In the ASN.1 language,

```
GetResponse-PDU ::=
[2]
IMPLICIT SEQUENCE {
    request-id
    RequestID,
    error-status
    ErrorStatus,
    error-index
    ErrorIndex,
    variable-bindings
    VarBindList
    }
```

The GetResponse-PDU is generated by a protocol entity only upon receipt of the GetRequest-PDU, GetNextRequest-PDU, or SetRequest-PDU, as described elsewhere in this document.

Upon receipt of the GetResponse-PDU, the receiving protocol entity presents its contents to its SNMP application entity.

4.1.5 The SetRequest-PDU. The form of the SetRequest-PDU is identical to that of the GetRequest-PDU except for the indication of the PDU type. In the ASN.1 language,

```
SetRequest-PDU ::=
[3]
IMPLICIT SEQUENCE {
    request-id
    RequestID,
    error-status   --always 0
    ErrorStatus,
    error-index       --always 0
    ErrorIndex,
    variable-bindings
    VarBindList
    }
```

The SetRequest-PDU is generated by a protocol entity only at the request of its SNMP application entity.

Upon receipt of the SetRequest-PDU, the receiving entity responds according to any applicable rule in the list below:

1. If, for any object named in the variable-bindings field, the object is not available for set operations in the relevant MIB view, then the receiving entity sends to the originator of the received message the GetResponse-PDU of identical form, except that the value of the error-status field is noSuchName, and the value of the error-index field is the index of said object name component in the received message.

2. If, for any object named in the variable-bindings field, the contents of the value field does not, according to the ASN.1 language, manifest a type, length, and value that is consistent with that required for the variable, then the receiving entity sends to the originator of the received message the GetResponse-PDU of identical form, except that the value of the error-status field is badValue, and the value of the error-index field is the index of said object name in the received message.

3. If the size of the GetResponse type message generated as described below would exceed a local limitation, then the receiving entity sends to the originator of the received message the GetResponse-PDU of identical form, except that the value of the error-status field is tooBig, and the value of the error-index field is zero.

4. If, for any object named in the variable-bindings field, the value of the named object cannot be altered for reasons not covered by any of the foregoing rules, then the receiving entity sends to the originator of the received message the GetResponse-PDU of identical form, except that the value of the error-status field is genErr and the value of the error-index field is the index of said object name component in the received message. If none of the foregoing rules apply, then for each object named in the variable-bindings field of the received message, the corresponding value is assigned to the variable. Each variable assignment specified by the SetRequest-PDU should be effected as if simultaneously set with respect to all other assignments specified in the same message.

The receiving entity then sends to the originator of the received message the GetResponse-PDU of identical form except that the value of the error-status field of the generated message is noError and the value of the error-index field is zero.

4.1.6 The Trap-PDU. The form of the Trap-PDU is

Trap-PDU ::=
[4]

```
IMPLICIT SEQUENCE {
enterprise    --type of object generating
        • trap, see sysObjectID in [5]
OBJECT IDENTIFIER,
            agent-addr            --address of object generating
            NetworkAddress,       --trap
            generic-trap          --generic trap type
            INTEGER {
            coldStart(0),
            warmStart(1),
            linkDown(2),
            linkUp(3),
            authenticationFailure(4),
            egpNeighborLoss(5),
            enterpriseSpecific(6)
            },
      specific-trap               --specific code, present even
            INTEGER,              --if generic-trap is not
              • enterpriseSpecific
      time-stamp                  --time elapsed between the last
        TimeTicks,                --(re)initialization of the network
              • entity and the generation of the trap
        variable-bindings --"interesting" information
        VarBindList
      }
```

The Trap-PDU is generated by a protocol entity only at the request of the SNMP application entity. The means by which an SNMP application entity selects the destination addresses of the SNMP application entities is implementation-specific.

Upon receipt of the Trap-PDU, the receiving protocol entity presents its contents to its SNMP application entity. The significance of the variable-bindings component of the Trap-PDU is implementation-specific.

Interpretations of the value of the generic-trap field are described in Sections 4.1.6.1–4.1.6.7.

4.1.6.1 The coldStart trap. A coldStart(0) trap signifies that the sending protocol entity is reinitializing itself such that the agent's configuration or the protocol entity implementation may be altered.

4.1.6.2 The warmStart trap. A warmStart(1) trap signifies that the sending protocol entity is reinitializing itself such that neither the agent configuration nor the protocol entity implementation is altered.

4.1.6.3 The linkDown trap. A linkDown(2) trap signifies that the sending protocol entity recognizes a failure in one of the communication links represented in the agent's configuration.

The Trap-PDU of type linkDown contains as the first element of its variable-bindings, the name and value of the ifIndex instance for the affected interface.

4.1.6.4 The linkUp trap. A linkUp(3) trap signifies that the sending protocol entity recognizes that one of the communication links represented in the agent's configuration has come up.

The Trap-PDU of type linkUp contains as the first element of its variable-bindings, the name and value of the ifIndex instance for the affected interface.

4.1.6.5 The authenticationFailure trap. An authenticationFailure(4) trap signifies that the sending protocol entity is the addressee of a protocol message that is not properly authenticated. While implementations of the SNMP must be capable of generating this trap, they must also be capable of suppressing the emission of such traps via an implementation-specific mechanism.

4.1.6.6 The egpNeighborLoss trap. An egpNeighborLoss(5) trap signifies that an EGP neighbor for whom the sending protocol entity was an EGP peer has been marked down and the peer relationship no longer obtains.

The Trap-PDU of type egpNeighborLoss contains as the first element of its variable-bindings, the name and value of the egpNeighAddr instance for the affected neighbor.

4.1.6.7 The enterpriseSpecific trap. An enterpriseSpecific(6) trap signifies that the sending protocol entity recognizes that some enterprise-specific event has occurred. The specific-trap field identifies the particular trap which occurred.

5 Definitions

RFC1157-SNMP DEFINITIONS ::= BEGIN

IMPORTS

ObjectName, ObjectSyntax, NetworkAddress, IpAddress, TimeTicks FROM RFC1155-SMI;

- top-level message

  Message ::=

  SEQUENCE {

  version --version-1 for this RFC

  INTEGER {

```
version-1(0)
              },
community    --community name
OCTET STRING,
        data       --e.g., PDUs if trivial
          ANY    --authentication is being used
      }
```

- protocol data units

```
PDUs ::=
CHOICE {
get-request
GetRequest-PDU,
get-next-request
GetNextRequest-PDU,
get-response
GetResponse-PDU,
set-request
SetRequest-PDU,
trap
Trap-PDU
                 }
```

- PDUs

```
GetRequest-PDU ::=
[0]
```

IMPLICIT PDU

GetNextRequest-PDU ::=
[1]

IMPLICIT PDU

GetResponse-PDU ::=
[2]

IMPLICIT PDU

SetRequest-PDU ::=
[3]

IMPLICIT PDU

```
PDU ::=
SEQUENCE {
request-id
INTEGER,
error-status   – sometimes ignored
INTEGER {
noError(0),
tooBig(1),
noSuchName(2),
badValue(3),
readOnly(4),
genErr(5)
                          },

error-index   – sometimes ignored
INTEGER,
variable-bindings   – values are sometimes ignored
VarBindList
                          }

Trap-PDU ::=
[4]
```

IMPLICIT SEQUENCE {

```
enterprise      -- type of object generating
```

- trap, see sysObjectID in [5]

```
                      OBJECT IDENTIFIER
```

Appendix

E

RFC 1180: A TCP/IP Tutorial

Theodore John Socolofsky
Claudia Jeanne Kale

Following is a useful tutorial on the TCP/IP protocol suite. The reader is urged to review this tutorial before moving on to other material. The goal in providing this material is to ensure that a complete foundation is provided for further study of this subject.

Status of this Memo

This RFC is a tutorial on the TCP/IP protocol suite, focusing particularly on the steps in forwarding an IP datagram from source host to destination host through a router. It does not specify an Internet standard. Distribution of this memo is unlimited. (January 1991)

1 Introduction

This tutorial contains only one view of the salient points of TCP/IP, and therefore it is the "bare bones" of TCP/IP technology. It omits the history of development and funding, the business case for its use, and its future as compared to ISO OSI. Indeed, a great deal of technical information is also omitted. What remains is a minimum of information that must be understood by the professional working in a TCP/IP environment. These professionals include the systems administrator, the systems programmer, and the network manager.

This tutorial uses examples from the UNIX TCP/IP environment, however the main points apply across all implementations of TCP/IP. Note that the purpose of this memo is explanation, not definition. If any question arises about the correct specification of a protocol, please refer to the actual standards defining RFC.

The next section is an overview of TCP/IP, followed by detailed descriptions of individual components.

2 TCP/IP Overview

The generic term "TCP/IP" usually means anything and everything related to the specific protocols of TCP and IP. It can include other protocols, applications, and even the network medium. A sample of these protocols are: UDP, ARP, and ICMP. A sample of these applications are: TELNET, FTP, and rcp. A more accurate term is "internet technology." A network that uses internet technology is called an "internet."

2.1 Basic structure

To understand this technology you must first understand the following logical structure:

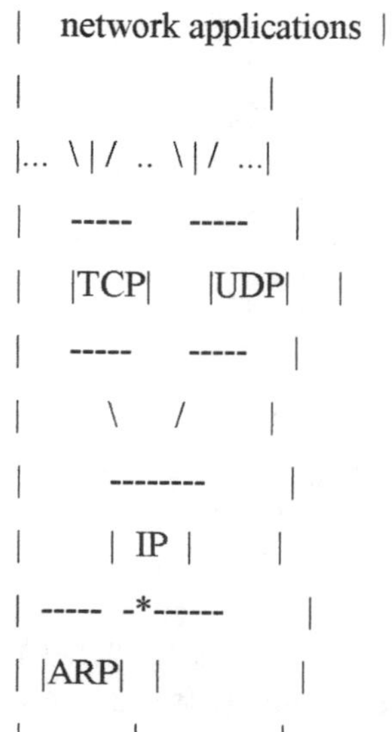

Figure 1 Basic TCP/IP Network Node.

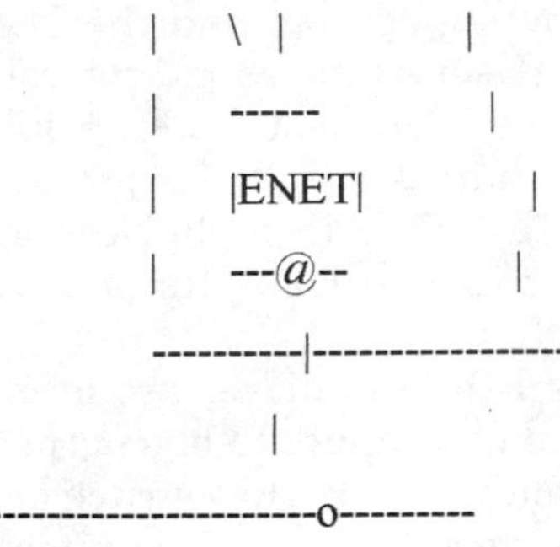

Ethernet Cable

Figure 1 *(Continued)*

This is the logical structure of the layered protocols inside a computer on an internet. Each computer that can communicate using internet technology has such a logical structure. It is this logical structure that determines the behavior of the computer on the internet. The boxes represent processing of the data as it passes through the computer, and the lines connecting boxes show the path of data. The horizontal line at the bottom represents the Ethernet cable; the "o" is the transceiver. The "*" is the IP address and the "@" is the Ethernet address. Understanding this logical structure is essential to understanding internet technology; it is referred to throughout this tutorial.

2.2 Terminology

The name of a unit of data that flows through an internet is dependent upon where it exists in the protocol stack. In summary: if it is on an Ethernet it is called an Ethernet frame; if it is between the Ethernet driver and the IP module it is called an IP packet; if it is between the IP module and the UDP module it is called a UDP datagram; if it is between the IP module and the TCP module it is called a TCP segment (more generally, a transport message); and if it is in a network application it is called an application message. These definitions are imperfect. Actual definitions vary from one publication to the next. More specific definitions can be found in RFC 1122, section 1.3.3.

A driver is software that communicates directly with the network interface hardware. A module is software that communicates with a driver, with network applications, or with another module. The terms driver, module, Ethernet frame, IP packet, UDP datagram, TCP message, and application message are used where appropriate throughout this tutorial.

2.3 Flow of data

Let us follow the data as it flows down through the protocol stack shown in Figure 1. For an application that uses TCP (Transmission

Control Protocol), data passes between the application and the TCP module. For applications that use UDP (User Datagram Protocol), data passes between the application and the UDP module. FTP (File Transfer Protocol) is a typical application that uses TCP. Its protocol stack in this example is FTP/TCP/IP/ENET. SNMP (Simple Network Management Protocol) is an application that uses UDP. Its protocol stack in this example is SNMP/UDP/IP/ENET.

The TCP module, UDP module, and the Ethernet driver are n-to-1 multiplexers. As multiplexers they switch many inputs to one output. They are also 1-to-n demultiplexers. As demultiplexers they switch one input to many outputs according to the type field in the protocol header.

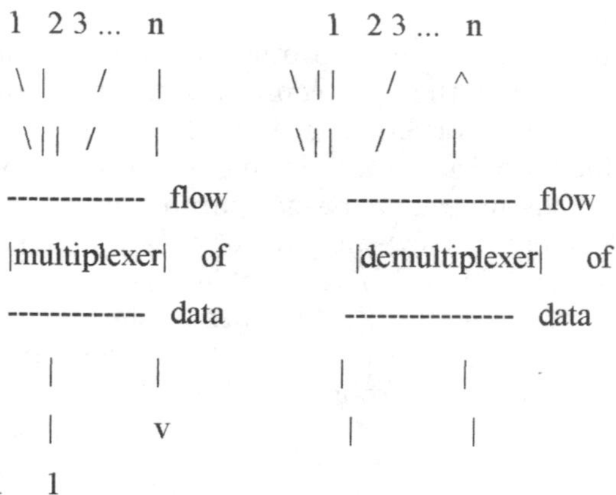

Figure 2 n-to-1 multiplexer and 1-to-n demultiplexer.

If an Ethernet frame comes up into the Ethernet driver off the network, the packet can be passed upward to either the ARP (Address Resolution Protocol) module or to the IP (Internet Protocol) module. The value of the type field in the Ethernet frame determines whether the Ethernet frame is passed to the ARP or the IP module.

If an IP packet comes up into IP, the unit of data is passed upward to either TCP or UDP, as determined by the value of the protocol field in the IP header.

If the UDP datagram comes up into UDP, the application message is passed upward to the network application based on the value of the port field in the UDP header. If the TCP message comes up into TCP, the application message is passed upward to the network application based on the value for the port field in the TCP header.

The downward multiplexing is simple to perform because from each starting point there is only the one downward path; each protocol module adds its header information so the packet can be demultiplexed at the destination computer.

Data passing out from the applications through either TCP or UDP converges on the IP module and is sent downward through the lower network interface driver.

Although internet technology supports many different network media, Ethernet is used for all examples in this tutorial because it is the most common physical network used under IP. The computer in Figure 1 has a single Ethernet connection. The 6-byte Ethernet address is unique for each interface on an Ethernet and is located at the lower interface of the Ethernet driver.

The computer also has a 4-byte IP address. This address is located at the lower interface to the IP module. The IP address must be unique for an internet. A running computer always knows its own IP address and Ethernet address.

2.4 Two network interfaces

If a computer is connected to two separate Ethernets it is as depicted in Figure 3.

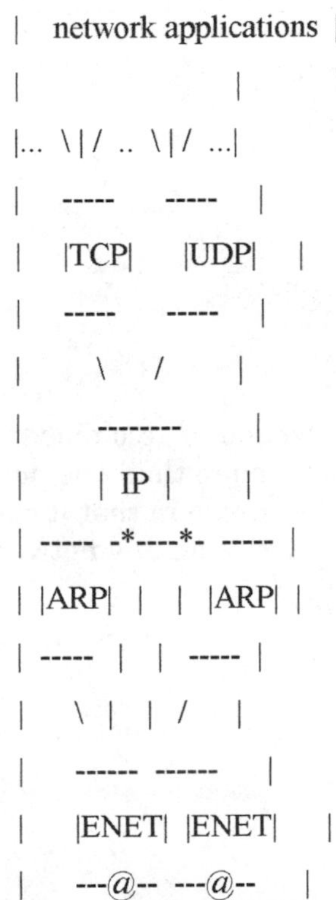

Figure 3 TCP/IP Network Node on two Ethernets. Note that this computer has two Ethernet addresses and two IP addresses.

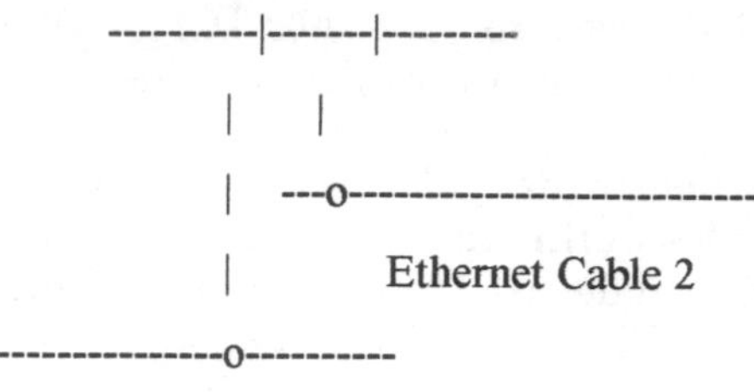

Ethernet Cable 1

Figure 3 *(Continued)*

It is seen from this structure that for computers with more than one physical network interface, the IP module is both an n-to-m multiplexer and an m-to-n demultiplexer.

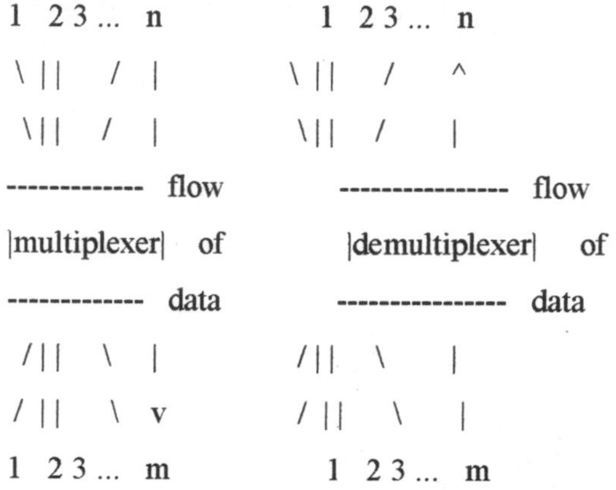

Figure 4 n-to-m multiplexer and m-to-n demultiplexer.

It performs this multiplexing in either direction to accommodate incoming and outgoing data. An IP module with more than one network interface is more complex than our original example in that it can forward data onto the next network. Data can arrive on any network interface and be sent out on any other.

```
  TCP   UDP
   \   /
    \ /
 |  IP  |
 |      |
 |  --- |
```

Figure 5 Example of IP Forwarding an IP Packet.

```
 | / \ |
 |/   v |
  /     \
 /       \
data      data
comes in    goes out
here        here
```

Figure 5 Example of IP Forwarding a IP Packet.

The process of sending an IP packet out onto another network is called "forwarding" an IP packet. A computer that has been dedicated to the task of forwarding IP packets is called an "IP-router."

As you can see from the figure, the forwarded IP packet never touches the TCP and UDP modules on the IP-router. Some IP-router implementations do not have a TCP or UDP module.

2.5 IP creates a single logical network

The IP module is central to the success of internet technology. Each module or driver adds its header to the message as the message passes down through the protocol stack. Each module or driver strips the corresponding header from the message as the message climbs the protocol stack up toward the application. The IP header contains the IP address, which builds a single logical network from multiple physical networks. This interconnection of physical networks is the source of the name: internet. A set of interconnected physical networks that limit the range of an IP packet is called an "internet."

2.6 Physical network independence

IP hides the underlying network hardware from the network applications. If you invent a new physical network, you can put it into service by implementing a new driver that connects to the internet underneath IP. Thus, the network applications remain intact and are not vulnerable to changes in hardware technology.

2.7 Interoperability

If two computers on an internet can communicate, they are said to "interoperate"; if an implementation of internet technology is good, it is said to have "interoperability." Users of general-purpose computers benefit from the installation of an internet because of the interoperability in computers on the market. Generally, when you buy a computer, it will interoperate. If the computer does not have interoperability, and interoperability cannot be added, it occupies a rare and special niche in the market.

2.8 After the overview

With the background set, we will answer the following questions:

When sending out an IP packet, how is the destination Ethernet address determined?

How does IP know which of multiple lower network interfaces to use when sending out an IP packet?

How does a client on one computer reach the server on another?

Why do both TCP and UDP exist, instead of just one or the other?

What network applications are available?

These will be explained, in turn, after an Ethernet refresher.

3 Ethernet

This section is a short review of Ethernet technology. An Ethernet frame contains the destination address, source address, type field, and data.

An Ethernet address is 6 bytes. Every device has its own Ethernet address and listens for Ethernet frames with that destination address. All devices also listen for Ethernet frames with a wild-card destination address of "FF-FF-FF-FF-FF-FF" (in hexadecimal), called a "broadcast" address.

Ethernet used CSMA/CD (Carrier Sense and Multiple Access with Collision Detection). CSMA/CD means that all devices communicate on a single medium, that only one can transmit at a time, and that they can all receive simultaneously. If two devices try to transmit at the same instant, the transmit collision is detected, and both devices wait a random (but short) period before trying to transmit again.

3.1 A human analogy

A good analogy of Ethernet technology is a group of people talking in a small, completely darkened room. In this analogy, the physical network medium is sound waves on air in the room instead of electrical signals on a coaxial cable.

Each person can hear the words when another is talking (Carrier Sense). Everyone in the room has equal capability to talk (Multiple Access), but none of them gives lengthy speeches because they are polite. If a person is impolite, he is asked to leave the room (i.e., thrown off the net).

No one talks while another is speaking. But if two people start speaking at the same instant, each of them knows this because each hears something they have not said (Collision Detection). When these two people notice this condition, they wait for a moment, then one begins talking. The other hears the talking and waits for the first to finish before beginning his own speech.

Each person has an unique name (unique Ethernet address) to avoid confusion. Every time one of them talks, he prefaces the message with the name of the person he is talking to and with his own name (Ethernet destination and source address, respectively), i.e., "Hello Jane, this is Jack, . . . blah blah blah . . . ." If the sender wants to talk to everyone he might say "everyone" (broadcast address), i.e., "Hello Everyone, this is Jack, . . . blah blah blah. . . ."

4 ARP

When sending out an IP packet, how is the destination Ethernet address determined? ARP (Address Resolution Protocol) is used to translate IP addresses to Ethernet addresses. The translation is done only for outgoing IP packets, because this is when the IP header and the Ethernet header are created.

4.1 ARP table for address translation

The translation is performed with a table look-up. The table, called the ARP table, is stored in memory and contains a row for each computer. There is a column for IP address and a column for Ethernet address. When translating an IP address to an Ethernet address, the table is searched for a matching IP address. The following is a simplified ARP table:

TABLE E.1 Example ARP Table

IP address	Ethernet address
223.1.2.1	08-00-39-00-2F-C3
223.1.2.3	08-00-5A-21-A7-22
223.1.2.4	08-00-10-99-AC-54

The human convention when writing out the 4-byte IP address is each byte in decimal and separating bytes with a period. When writing out the 6-byte Ethernet address, the conventions are each byte in hexadecimal and separating bytes with either a minus sign or a colon.

The ARP table is necessary because the IP address and Ethernet address are selected independently; you cannot use an algorithm to translate IP address to Ethernet address. The IP address is selected by the network manager based on the location of the computer on the internet. When the computer is moved to a different part of an internet, its IP address must be changed. The Ethernet address is selected by the manufacturer based on the Ethernet address space licensed by the manufactuer. When the Ethernet hardware interface board changes, the Ethernet address changes.

4.2 Typical translation scenario

During normal operation, a network application, such as TELNET, sends an application message to TCP, then TCP sends the corresponding TCP message to the IP module. The destination IP address is known by the application, the TCP module, and the IP module. At this point the IP packet has been constructed and is ready to be given to the Ethernet driver, but first the destination Ethernet address must be determined. The ARP table is used to look-up the destination Ethernet address.

4.3 ARP request/response pair

But how does the ARP table get filled in the first place? The answer is that it is filled automatically by ARP on an "as-needed" basis.

Two things happen when the ARP table cannot be used to translate an address:

1. An ARP request packet with a broadcast Ethernet address is sent out on the network to every computer.
2. The outgoing IP packet is queued.

Every computer's Ethernet interface receives the broadcast Ethernet frame. Each Ethernet driver examines the Type field in the Ethernet frame and passes the ARP packet to the ARP module. The ARP request packet says "If your IP address matches this target IP address, then please tell me your Ethernet address." An ARP request packet looks something like this:

TABLE E.2 Example ARP Request

| Sender IP Address 223.1.2.1 |
| Sender Enet Address 08-00-39-00-2F-C3 |
| Target IP Address 223.1.2.2 |
| Target Enet Address <blank> |

Each ARP module examines the IP address and if the Target IP address matches its own IP address, it sends a response directly to the source Ethernet address. The ARP response packet says "Yes, that target IP address is mine, let me give you my Ethernet address." An ARP response packet has the sender/targer field contents swapped as compared to the request. It looks something like this:

TABLE E.3 Example ARP Response

I Sender IP Address 223.1.2.2 I
I Sender Enet Address 08-00-28-00-38-A9 I
I Target IP Address 223.1.2.1 I
I Target Enet Address 08-00-39-00-2F-C3 I

The response is received by the original sender computer. The Ethernet driver looks at the Type field in the Ethernet frame then passes the ARP packet to the ARP module. The ARP module examines the ARP packet and adds the sender's IP and Ethernet addresses to its ARP table.

The updated table now looks like this:

TABLE E.4 ARP Table after Response

I IP address address I	Ethernet
I 223.1.2.1	08-00-39-00-2F-C3 I
I 223.1.2.2	08-00-28-00-38-A9 I
I 223.1.2.3	08-00-5A-21-A7-22 I
I 223.1.2.4	08-00-10-99-AC-54 I

4.4 Scenario continued

The new translation has now been installed automatically in the table, just milli-seconds after it was needed. As you remember from step 2 above, the outgoing IP packet was queued. Next, the IP address to Ethernet address translation is performed by look-up in the ARP table then the Ethernet frame is transmitted on the Ethernet. Therefore, with the new steps 3, 4, and 5, the scenario for the sender computer is:

1. An ARP request packet with a broadcast Ethernet address is sent out on the network to every computer.
2. The outgoing IP packet is queued.
3. The ARP response arrives with the IP-to-Ethernet address translation for the ARP table.

4. For the queued IP packet, the ARP table is used to translate the IP address to the Ethernet address.
5. The Ethernet frame is transmitted on the Ethernet.

In summary, when the translation is missing from the ARP table, one IP packet is queued. The translation data is quickly filled in with ARP request/response and the queued IP packet is transmitted. Each computer has a separate ARP table for each of its Ethernet interfaces. If the target computer does not exist, there will be no ARP response and no entry in the ARP table. IP will discard outgoing IP packets sent to that address. The upper layer protocols cannot tell the difference between a broken Ethernet and the absence of a computer with the target IP address.

Some implementations of IP and ARP do not queue the IP packet while waiting for the ARP response. Instead the IP packet is discarded and the recovery from the IP packet loss is left to the TCP module or the UDP network application. This recovery is performed by time-out and retransmission. The retransmitted message is successfully sent out onto the network because the first copy of the message has already caused the ARP table to be filled.

5 Internet Protocol

The IP module is central to internet technology and the essence of IP is its route table. IP uses this in-memory table to make all decisions about routing an IP packet. The content of the route table is defined by the network administrator. Mistakes block communication.

To understand how a route table is used is to understand internetworking. This understanding is necessary for the successful administration and maintenance of an IP network.

The route table is best understood by first having an overview of routing, then learning about IP network addresses, and then looking at the details.

5.1 Direct routing

Figure 6 is of a tiny internet with 3 computers: A, B, and C. Each computer has the same TCP/IP protocol stack as in Figure 1. Each computer's Ethernet interface has its own Ethernet address. Each computer has an IP address assigned to the IP interface by the network manager, who also has assigned an IP network number to the Ethernet.

When A sends an IP packet to B, the IP header contains A's IP address as the source IP address, and the Ethernet header contains A's Ethernet

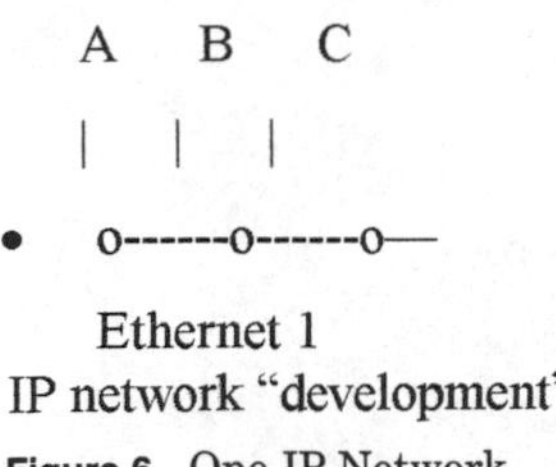

Figure 6 One IP Network.

address as the source Ethernet address. Also, the IP header contains B's IP address as the destination IP address and the Ethernet header contains B's Ethernet address as the destination Ethernet address.

TABLE E.5 Addresses in an Ethernet Frame for an IP Packet from A to B

address	source	destination
IP header	A	B
Ethernet header	A	B

For this simple case, IP is overhead because the IP adds little to the service offered by Ethernet. However, IP does add cost: the extra CPU processing and network bandwidth to generate, transmit, and parse the IP header.

When B's IP module receives the IP packet from A, it checks the destination IP address against its own, looking for a match, then it passes the datagram to the upper-level protocol. This communication between A and B uses direct routing.

5.2 Indirect routing

The figure below is a more realistic view of an internet. It is composed of three Ethernets and three IP networks connected by an IP-router called computer D. Each IP network has four computers; each computer has its own IP address and Ethernet address.

Except for computer D, each computer has a TCP/IP protocol stack like that in Figure 1. Computer D is the IP-router; it is connected to all three networks and therefore has three IP addresses and three Ethernet addresses. Computer D has a TCP/IP protocol stack similar to that in Figure 3, except that it has three ARP modules and three Ethernet drivers instead of two. Please note that computer D has only one IP module. The network manager has assigned a unique number, called an IP network number, to each of the Ethernets. The IP network numbers are not shown in this diagram, just the network names.

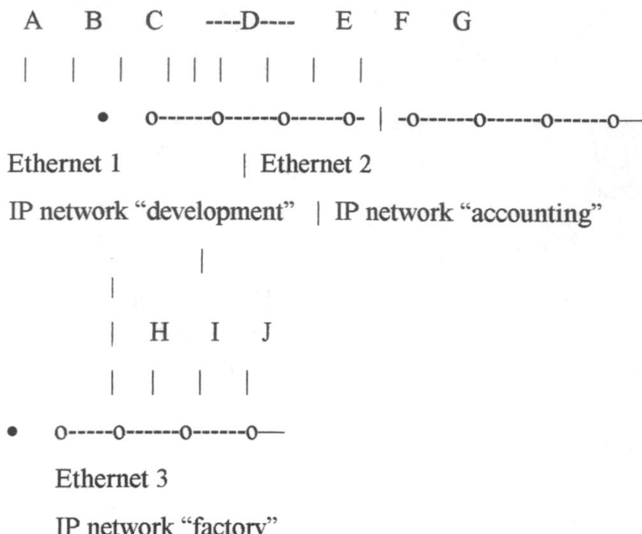

Figure 7 Three IP Networks; One internet.

When computer A sends an IP packet to computer B, the process is identical to the single network example above. Any communication between computers located on a single IP network matches the direct routing example discussed previously.

When computers D and A communicate, it is direct communication. When computers D and E communicate, it is direct communication. When computers D and H communicate, it is direct communication. This is because each of these pairs of computers is on the same IP network. However, when computer A communicates with a computer on the far side of the IP-router, communication is no longer direct. A must use D to forward the IP packet to the next IP network. This communication is called "indirect."

This routing of IP packets is done by IP modules and happens transparently to TCP, UDP, and the network applications. If A sends an IP packet to E, the source IP address and the source Ethernet address are A's. The destination IP address is E's, but because A's IP module sends the IP packet to D for forwarding, the destination Ethernet address is D's.

TABLE E.6 Addresses in an Ethernet Frame for an IP Packet from A to E (before D)

address	source	destination
IP header	A	E
Ethernet header	A	D

D's IP module receives the IP packet and upon examining the destination IP address, says "This is not my IP address," and sends the IP packet directly to E.

TABLE E.7 Addresses in an Ethernet Frame for an IP Packet from A to E (after D)

address	source	destination
IP header	A	E
Ethernet header	D	E

In summary, for direct communication, both the source IP address and the source Ethernet address is the sender's, and the destination IP address and the destination Ethernet address is the recipient's. For indirect communication, the IP address and Ethernet addresses do not pair up in this way.

This example internet is a very simple one. Real networks are often complicated by many factors, resulting in multiple IP-routers and several types of physical networks. This example internet might have come about because the network manager wanted to split a large Ethernet in order to localize Ethernet broadcast traffic.

5.3 IP module routing rules

This overview of routing has shown what happens, but not how it happens. Now let us examine the rules, or algorithm, used by the IP module.

For an outgoing IP packet, entering IP from an upper layer, IP must decide whether to send the IP packet directly or indirectly, and IP must choose a lower network interface. These choices are made by consulting the route table.

For an incoming IP packet, entering IP from a lower interface, IP must decide whether to forward the IP packet or pass it to an upper layer. If the IP packet is being forwarded, it is treated as an outgoing IP packet.

When an incoming IP packet arrives it is never forwarded back out through the same network interface.

These decisions are made before the IP packet is handed to the lower interface and before the ARP table is consulted.

5.4 IP address

The network manager assigns IP addresses to computers according to the IP network to which the computer is attached. One part of a 4-byte

IP address is the IP network number, the other part is the IP computer number (or host number). For the computer in Table E1, with an IP address of 223.1.2.1, the network number is 223.1.2 and the host number is number 1.

The portion of the address that is used for network number and for host number is defined by the upper bits in the 4-byte address. All example IP addresses in this tutorial are of type class C, meaning that the upper 3 bits indicate that 21 bits are the network number and 8 bits are the host number. This allows 2,097,152 class C networks of up to 254 hosts on each network.

The IP address space is administered by the NIC (Network Information Center). All internets that are connected to the single worldwide Internet must use network numbers assigned by the NIC. If you are setting up your own internet and you are not intending to connect it to the Internet, you should still obtain your network numbers from the NIC. If you pick your own number, you run the risk of confusion and chaos in the eventuality that your internet is connected to another internet.

5.5 Names

People refer to computers by names, not numbers. A computer called alpha might have the IP address of 223.1.2.1. For small networks, this name-to-address translation data is often kept on each computer in the "hosts" file. For larger networks, this translation data file is stored on a server and accessed across the network when needed. A few lines from that file might look like this:

```
223.1.2.1   alpha
223.1.2.2   beta
223.1.2.3   gamma
223.1.2.4   delta
223.1.3.2   epsilon
223.1.4.2   iota
```

The IP address is the first column and the computer name is the second column.

In most cases, you can install identical "hosts" files on all computers. You may notice that "delta" has only one entry in this file even though it has 3 IP addresses. Delta can be reached with any of its IP addresses; it does not matter which one is used. When delta receives an IP packet and looks at the destination address, it will recognize any of its own IP addresses. IP networks are also given names. If you have three IP networks, your "networks" file for documenting these names might look something like this:

223.1.2	development
223.1.3	accounting
223.1.4	factory

The IP network number is in the first column and its name is in the second column. From this example you can see that alpha is computer number 1 on the development network, beta is computer number 2 on the development network and so on. You might also say that alpha is development.1, beta is development.2, and so on.

The above hosts file is adequate for the users, but the network manager will probably replace the line for delta with

223.1.2.4	devnetrouter	delta
223.1.3.1	facnetrouter	
223.1.4.1	accnetrouter	

These three new lines for the hosts file give each of delta's IP addresses a meaningful name. In fact, the first IP address listed has two names; "delta" and "devnetrouter" are synonyms. In practice "delta" is the general-purpose name of the computer and the other three names are only used when administering the IP route table.

These files are used by network administration commands and network applications to provide meaningful names. They are not required for operation of an internet, but they do make it easier for us.

5.6 IP route table

How does IP know which lower network interface to use when sending out an IP packet? IP looks it up in the route table using a search key of the IP network number extracted from the IP destination address.

The route table contains one row for each route. The primary columns in the route table are IP network number, direct/indirect flag, router IP address, and interface number. This table is referred to by IP for each outgoing IP packet.

On most computers the route table can be modified with the "route" command. The content of the route table is defined by the network manager, because the network manager assigns the IP addresses to the computers.

5.7 Direct routing details

To explain how it is used, let us visit in detail the routing situations we have reviewed previously.

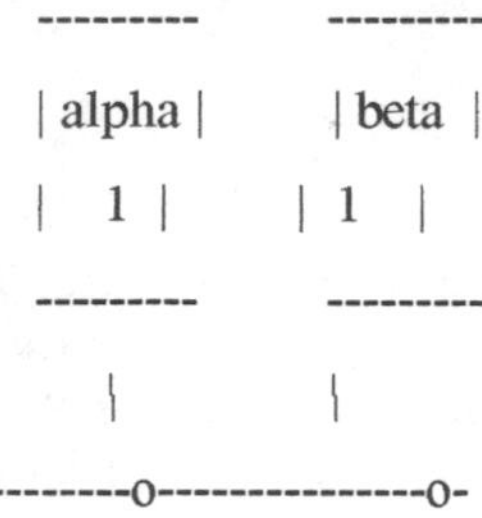

Ethernet 1

IP network "development"

Figure 8 Close-up View of One IP Network.

The route table inside alpha looks like this:

TABLE E.8 Example Simple Route Table

network	direct/indirect flag	router	interface number
development	direct	<blank>	1

This view can be seen on some UNIX systems with the "netstat-r" command. With this simple network, all computers have identical routing tables.

For discussion, the table is printed again without the network number translated to its network name.

TABLE E.9 Example Simple Route Table with Numbers

network	direct/indirect flag	router	interface number
223.1.2	direct	<blank>	1

5.8 Direct scenario

Alpha is sending an IP packet to beta. The IP packet is in alpha's IP module and the destination IP address is beta or 223.1.2.2. IP extracts the network portion of this IP address and scans the first column of the table looking for a match. With this network a match is found on the first entry.

The other information in this entry indicates that computers on this network can be reached directly through interface number 1. An ARP table translation is done on beta's IP address, then the Ethernet frame is sent directly to beta via interface number 1. If an application tries to send data to an IP address that is not on the development network, IP will be unable to find a match in the route table. IP then discards the IP packet. Some computers provide a "Network not reachable" error message.

5.9 Indirect routing details

Now, let us take a closer look at the more complicated routing scenario that we examined previously.

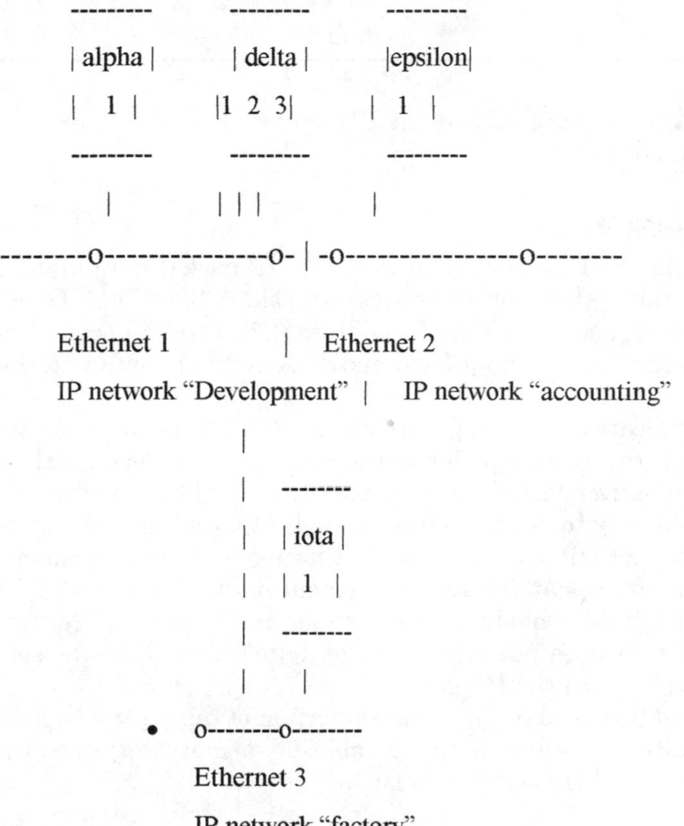

Figure 9 Close-up View of Three IP Networks.

The route table inside alpha looks like this:

TABLE E.10 Alpha Route Table

network	direct/indirect flag	router	interface number
development	direct	<blank>	1
accounting	indirect	devnetrouter	1
factory	indirect	devnetrouter	1

For discussion the table is printed again using numbers instead of names.

TABLE E.11 Alpha Route Table with Numbers

network	direct/indirect flag	router	interface number
223.1.2	direct	<blank>	1
223.1.3	indirect	223.1.2.4	1
223.1.4	indirect	223.1.2.4	1

The router in Alpha's route table is the IP address of delta's connection to the development network.

5.10 Indirect scenario

Alpha is sending an IP packet to epsilon. The IP packet is in alpha's IP module and the destination IP address is epsilon (223.1.3.2). IP extracts the network portion of this IP address (223.1.3) and scans the first column of the table looking for a match. A match is found on the second entry.

This entry indicates that computers on the 223.1.3 network can be reached through the IP-router devnetrouter. Alpha's IP module then does an ARP table translation for devnetrouter's IP address and sends the IP packet directly to devnetrouter through Alpha's interface number 1. The IP packet still contains the destination address of epsilon.

The IP packet arrives at delta's development network interface and is passed up to delta's IP module. The destination IP address is examined and because it does not match any of delta's own IP addresses, delta decides to forward the IP packet.

Delta's IP module extracts the network portion of the destination IP address (223.1.3) and scans its route table for a matching network field. Delta's route table looks like this:

TABLE E.12 Delta's Route Table

network	direct/indirect flag	router	interface number
development	direct	<blank>	1
factory	direct	<blank>	3
accounting	direct	<blank>	2

Below is delta's table printed again, without the translation to names.

The match is found on the second entry. IP then sends the IP packet directly to epsilon through interface number 3. The IP packet contains

TABLE E.13 Delta's Route Table with Numbers

network	direct/indirect flag	router	interface number
223.1.2	direct	<blank>	1
223.1.3	direct	<blank>	3
223.1.4	direct	<blank>	2

the IP destination address of epsilon and the Ethernet destination address of epsilon.

The IP packet arrives at epsilon and is passed up to epsilon's IP module. The destination IP address is examined and found to match with epsilon's IP address, so the IP packet is passed to the upper protocol layer.

5.11 Routing summary

When an IP packet travels through a large internet it may go through many IP-routers before it reaches its destination. The path it takes is not determined by a central source but is a result of consulting each of the routing tables used in the journey. Each computer defines only the next hop in the journey and relies on that computer to send the IP packet on its way.

5.12 Managing the routes

Maintaining correct routing tables on all computers in a large internet is a difficult task; network configuration is being modified constantly by the network managers to meet changing needs. Mistakes in routing tables can block communication in ways that are excruciatingly tedious to diagnose.

Keeping a simple network configuration goes a long way toward making a reliable internet. For instance, the most straightforward method of assigning IP networks to Ethernet is to assign a single IP network number to each Ethernet.

Help is also available from certain protocols and network applications. ICMP (Internet Control Message Protocol) can report some routing problems. For small networks the route table is filled manually on each computer by the network administrator. For larger networks the network administrator automates this manual operation with a routing protocol to distribute routes throughout a network.

When a computer is moved from one IP network to another, its IP address must change. When a computer is removed from an IP network its old address becomes invalid. These changes require frequent updates to the "hosts" filc. This flat file can become difficult to maintain for even medium-size networks. The Domain Name System helps solve these problems.

6 User Datagram Protocol

UDP is one of the two main protocols to reside on top of IP. It offers service to the user's network applications. Example network applications that use UDP are Network File System (NFS) and Simple Network Management Protocol (SNMP). The service is little more than an interface to IP.

UDP is a connectionless datagram delivery service that does not guarantee delivery. UDP does not maintain an end-to-end connection with the remote UDP module; it merely pushes the datagram out on the net and accepts incoming datagrams off the net.

UDP adds two values to what is provided by IP. One is the multiplexing of information between applications based on port number. The other is a checksum to check the integrity of the data.

6.1 Ports

How does a client on one computer reach the server on another? The path of communication between an application and UDP is through UDP ports. These ports are numbered, beginning with zero. An application that is offering service (the server) waits for messages to come in on a specific port dedicated to that service. The server waits patiently for any client to request service.

For instance, the SNMP server, called an SNMP agent, always waits on port 161. There can be only one SNMP agent per computer because there is only one UDP port number 161. This port number is well known; it is a fixed number, an internet assigned number. If an SNMP client wants service, it sends its request to port number 161 of UDP on the destination computer.

When an application sends data out through UDP it arrives at the far end as a single unit. For example, if an application does 5 writes to the UDP port, the application at the far end will do 5 reads from the UDP port. Also, the size of each write matches the size of each read.

UDP preserves the message boundary defined by the application. It never joins two application messages together or divides a single application message into parts.

6.2 Checksum

An incoming IP packet with an IP header type field indicating "UDP" is passed up to the UDP module by IP. When the UDP module receives the UDP datagram from IP it examines the UDP checksum. If the checksum is zero, it means that checksum was not calculated by the sender and can be ignored. Thus the sending computer's UDP module may or may not generate checksums. If Ethernet is the only network between the 2 UDP modules communicating, then you may not need checksumming. However, it is recommended that checksum generation always be enabled because at some point in the future a route table change may send the data across less reliable media.

If the checksum is valid (or zero), the destination port number is examined and if an application is bound to that port, an application message is queued for the application to read. Otherwise the UDP datagram is discarded. If the incoming UDP datagrams arrive faster than the application can read them and if the queue fills to a maximum value, UDP datagrams are discarded by UDP. UDP will continue to discard UDP datagrams until there is space in the queue.

7 Transmission Control Protocol

TCP provides a different service than UDP. TCP offers a connection-oriented byte stream instead of a connectionless datagram delivery service. TCP guarantees delivery, whereas UDP does not.

TCP is used by network applications that require guaranteed delivery and cannot be bothered with doing time-outs and retransmissions. The two most typical network applications that use TCP are File Transfer Protocol (FTP) and the TELNET. Other popular TCP network applications include X-Window System, rcp (remote copy), and the r-series commands. TCP's greater capability is not without cost: it requires more CPU and network bandwidth. The internals of the TCP module are much more complicated than those in a UDP module.

Similar to UDP, network applications connect to TCP ports. Well-defined port numbers are dedicated to specific applications. For instance, the TELNET server uses port number 23. The TELNET client can find the server simply by connecting to port 23 of TCP on the specified computer.

When the application first starts using TCP, the TCP module on the client's computer and the TCP module on the server's computer start communicating with each other. These two end-point TCP modules contain state information that defines a virtual circuit. This virtual circuit consumes resources in both TCP end-points. The virtual circuit is full duplex; data can go in both directions simultaneously. The application writes data to the TCP port; the data traverses the network and is read by the application at the far end.

TCP packetizes the byte stream at will; it does not retain the boundaries between writes. For example, if an application does 5 writes to the TCP port, the application at the far end might do 10 reads to get all the data. Or it might get all the data with a single read. There is no correlation between the number and size of writes at one end to the number and size of reads at the other end.

TCP is a sliding window protocol with time-out and retransmits. Outgoing data must be acknowledged by the far-end TCP. Acknowledgments can be piggybacked on data. Both receiving ends can flow control the far end, thus preventing a buffer overrun. As with all sliding window protocols, the protocol has a window size. The window size determines the amount of data that can be transmitted before an acknowledgment is required. For TCP, this amount is not a number of TCP segments but a number of bytes.

8 Network Applications

Why do both TCP and UDP exist, instead of just one or the other? They supply different services. Most applications are implemented to use only one or the other. You, the programmer, choose the protocol that best meets your needs. If you need a reliable stream delivery service, TCP might be best. If you need a datagram service, UDP might be best. If you need efficiency over long-haul circuits, TCP might be best. If you need efficiency over fast networks with short latency, UDP might be best. If your needs do not fall nicely into these categories, then the "best" choice is unclear. However, applications can make up for deficiencies in the choice. For instance, if you choose UDP and you need reliability, then the application must provide reliability. If you choose TCP and you need a record-oriented service, then the application must insert markers in the byte stream to delimit records.

What network applications are available? There are far too many to list. The number is growing continually. Some of the applications have existed since the beginning of internet technology: TELNET and FTP. Others are relatively new: X-Windows and SNMP. The following is a brief description of the applications mentioned in this tutorial.

8.1 TELNET

TELNET provides a remote login capability on TCP. The operation and appearance is similar to keyboard dialing through a telephone switch.

On the command line the user types "telnet delta" and receives a login prompt from the computer called "delta."

TELNET works well; it is an old application and has widespread interoperability. Implementations of TELNET usually work between different operating systems. For instance, a TELNET client may be on VAX/VMS and the server on UNIX System V.

8.2 FTP

File Transfer Protocol (FTP), as old as TELNET, also uses TCP and has widespread interoperability. The operation and appearance is as if you TELNETed to the remote computer. But instead of typing your usual commands, you have to make do with a short list of commands for directory listings and the like. FTP commands allow you to copy files between computers.

8.3 rsh

Remote shell (rsh or remsh) is one of an entire family of remote UNIX style commands. The UNIX copy command, cp, becomes rcp. The UNIX "who is logged in" command, who, becomes rwho. The list continues and is referred to collectively as the "r" series commands or the "r*" (r star) commands.

The r* commands mainly work between UNIX systems and are designed for interaction between trusted hosts. Little consideration is given to security, but they provide a convenient user environment.

To execute the "cc file.c" command on a remote computer called delta, type "rsh delta cc file.c". To copy the "file.c" file to delta, type "rcp file.c delta:". To login to delta, type "rlogin delta", and if you administered the computers in a certain way, you will not be challenged with a password prompt.

8.4 NFS

Network File System, first developed by Sun Microsystems Inc., uses UDP and is excellent for mounting UNIX file systems on multiple computers. A diskless workstation can access its server's hard disk as if the disk were local to the workstation. A single disk copy of a database on mainframe "alpha" can also be used by mainframe "beta" if the database's file system is NFS mounted on "beta."

NFS adds significant load to a network and has poor utility across slow links, but the benefits are strong. The NFS client is implemented in the kernel, allowing all applications and commands to use the NFS mounted disk as if it were a local disk.

8.5 SNMP

Simple Network Management Protocol (SNMP) uses UDP and is designed for use by central network management stations. It is a well known fact that if given enough data, a network manager can detect and diagnose network problems. The central station uses SNMP to collect this data from other computers on the network. SNMP defines the format for the data; it is left to the central station or network manager to interpret the data.

8.6 X-Window

The X-Window System uses the X-Window protocol on TCP to draw windows on a workstation's bitmap display. X-Window is much more than a utility for drawing windows; it is an entire philosophy for designing a user interface.

9 Other Information

Much information about internet technology was not included in this tutorial. This section lists information that is considered the next level of detail for the reader who wishes to learn more.

- administration commands: arp, route, and netstat
- ARP: permanent entry, publish entry, time-out entry, spoofing
- IP route table: host entry, default gateway, subnets
- IP: time-to-live counter, fragmentation, ICMP
- RIP, routing loops
- Domain Name System

10 References

1 Comer, D., "Internetworking with TCP/IP Principles, Protocols, and Architecture," Prentice Hall, Englewood Cliffs, New Jersey, U.S.A., 1988.

2 Feinler, E., et al, "DDN Protocol Handbook," Volume 2 and 3, DDN Network Information Center, SRI International, 333 Ravenswood Avenue, Room EJ291, Menlo Park, California, U.S.A., 1985.

3 Spider Systems, Ltd., "Packets and Protocols," Spider Systems Ltd., Stanwell Street, Edinburgh, U.K. EH6 5NG, 1990.

11 Relation to Other RFCs

This RFC is a tutorial and it does not UPDATE or OBSOLETE any other RFC.

12 Security Considerations

There are security considerations within the TCP/IP protocol suite. To some people these considerations are serious problems, to others they are not; it depends on the user requirements.

This tutorial does not discuss these issues, but if you want to learn more you should start with the topic of ARP-spoofing, then use the "Security Considerations" section of RFC 1122 to lead you to more information.

13 Authors' Addresses

Theodore John Socolofsky
Spider Systems Limited
Spider Park, Stanwell Street
Edinburgh EH6 5NG
United Kingdom
Phone:
from UK 031-554-9424
from USA 011-44-31-554-9424
Fax:
from UK 031-554-0649
from USA 011-44-31-554-0649
E-mail: TEDS@SPIDER.CO.UK

Claudia Jeanne Kale
12 Gosford Place
Edinburgh EH6 4BJ
United Kingdom
Phone:
from UK 031-554-7432
from USA 011-44-31-554-7432
E-mail: CLAUDIAK@SPIDER.CO.UK

Appendix

RFC 2151: A Primer on Internet and TCP/IP Tools and Utilities

Rounding out this appendix section is a primer on the Internet and some useful utilities such as Ping, Finger, Telnet, and FTP. This appendix is intended to provide a useful reference for those who use the Internet regularly, or for those interested in becoming more familiar with the nuts and bolts of the Internet.

Status of this Memo

This memo provides information for the Internet community. This memo does not specify an Internet standard of any kind. Distribution of this memo is unlimited. (June 1997)

Abstract

This memo is an introductory guide to many of the most commonly available TCP/IP and Internet tools and utilities. It also describes discussion lists accessible from the Internet, ways to obtain Internet and TCP/IP documents, and some resources that help users weave their way through the Internet.

1 Introduction

This memo is an introductory guide to some of the most commonly available TCP/IP and Internet tools and utilities that allow users to access the wide variety of information on the network, from determining if a particular host is up to viewing a multimedia thesis on foreign policy. It also describes discussion lists accessible from the Internet, ways to obtain Internet and TCP/IP documents, and some resources that help users weave their way through the Internet. This memo may be used as a tutorial for individual self-learning, a step-by-step laboratory manual for a course, or as the basis for a site's users manual. It is intended as a basic guide only and will refer to other sources for more detailed information.

2 Nomenclature

The following sections provide descriptions and detailed examples of several TCP/IP utilities and applications, including the reproduction of actual sessions using these utilities (with some extraneous information removed). Each section describes a single TCP/IP-based tool, its application, and, in some cases, how it works. The text description is usually followed by an actual sample session. The sample dialogues shown below were obtained from a variety of software and hardware systems, including AIX running on an IBM RS/6000, Linux on an Intel 486, Multinet TCP/IP over VMS on a VAX, and FTP Software's OnNet (formerly PC/TCP) running on a DOS/Windows PC. While the examples below can be used as a guide to using and learning about the capabilities of TCP/IP tools, the reader should understand that not all of these utilities may be found at all TCP/IP hosts nor in all commercial software packages. Furthermore, the user interface for different packages will be different and the actual command line may appear differently than shown here; this will be particularly true for graphical user interfaces running over Windows, X-Windows, OS/2, or Macintosh systems. Windows-based sessions are not shown in this RFC because of the desire to have a text version of this document; in addition, most GUI-based TCP/IP packages obscure some of the detail that is essential for understanding what is really happening when you click on a button or drag a file. The Internet has many exciting things to offer but standardized interfaces to the protocols is not yet one of them! This guide will not provide any detail or motivation about the Internet Protocol Suite; more information about the TCP/IP protocols and related issues may be found in RFC 1180 [29], Comer [6], Feit [7], Kessler [14], and Stevens [30].

In the descriptions below, commands are shown in a `Courier font` (Postscript and HTML versions); items appearing in square brackets (`[ ]`) are optional, the vertical-bar (`|`) means "or," parameters appearing with no brackets or within curly brackets (`{ }`) are mandatory, and parameter names that need to be replaced with a specific value will be shown in italics (Postscript and HTML versions) or within angle brackets (< >, text version). In the sample dialogues, user input is in bold (Postscript and HTML versions) or denoted with asterisks (`**`) in the margin (text version).

3 Finding Information about Internet Hosts and Domains

There are several tools that teach you information about Internet hosts and domains. These tools provide the ability for an application or a user to perform host name/address reconciliation (NSLOOKUP), determine whether another host is up and available (PING), learn about another host's users (Finger), and learn the route that packets will take to another host (Traceroute).

3.1 NSLOOKUP

NSLOOKUP is the name server lookup program that comes with many TCP/IP software packages. A user can use NSLOOKUP to examine entries in the Domain Name System (DNS) database that pertain to a particular host or domain; one common use is to determine a host system's IP address from its name or the host's name from its IP address. The general form of the command to make a single query is

```
nslookup [IP_address/host_name]
```

If the program is started without any parameters, the user will be prompted for input; the user can enter either an IP address or host name at that time, and the program will respond with the name and address of the default name server, the name server actually used to resolve each request, and the IP address and host name that was queried. Exit is used to quit the NSLOOKUP application.

Three simple queries are shown in the list below:

1. Requests the address of the host named www.hill.com, the World Wide Web server at Hill Associates. As it turns out, this is not the true name of the host, but an alias. The full name of the host and the IP address are listed by NSLOOKUP.
2. Requests the address of host syrup.hill.com, which is the same host as in the first query. Note that NSLOOKUP provides a "nonauthoritative" answer. Since NSLOOKUP just queried this same address, the information is still in its cache memory. Rather than send additional messages to the name server, the answer is one that it remembers from before; the server did not look up the information again, however, so it is not guaranteed to still be accurate (because the information might have changed within the last few milliseconds!).
3. Requests the name of the host with the given IP address. The result points to the Internet gateway to Australia, munnari.oz.au. One additional query is shown in the dialogue below. NSLOOKUP examines information that is stored by the DNS. The default NSLOOKUP queries examine basic address records (called "A records") to reconcile the host name and IP address, although other information is also available. In the final query below, for example, the user wants to know where electronic mail addressed to the hill.com domain actually gets delivered, since hill.com is not the true

name of an actual host. This is accomplished by changing the query type to look for mail exchange (MX) records by issuing a set type command (which must be in lower case). The query shows that mail addressed to hill.com is actually sent to a mail server called mail.hill.com. If that system is not available, mail delivery will be attempted to first mailme.hill.com and then to netcomsv.netcom.com; the order of these attempts is controlled by the "preference" value. This query also returns the name of the domain's name servers and all associated IP addresses.

The DNS is beyond the scope of this introduction, although more information about the concepts and structure of the DNS can be found in STD 13/RFC 1034 [19], RFC 1591 [21], and Kessler [16]. The help command can be issued at the program prompt for information about NSLOOKUP's more advanced commands.

TECHNICAL NOTE: There are other tools that might be available on your system or with your software for examining the DNS. Alternatives to NSLOOKUP include HOST and DIG.

```
**SMCVAX$ nslookup
Default Server: ns1.ner.bbnplanet.net
Address:192.52.71.5

** > www.hill.com
Name:    syrup.hill.com
Address:199.182.20.3
Aliases:www.hill.com

** > syrup.hill.com
Nonauthoritative answer:
Name:    syrup.hill.com
Address:199.182.20.3

** > 128.250.1.21
Name:    munnari.oz.AU
Address:128.250.1.21

** > set type = MX
** > hill.com
hill.com preference = 20, mail exchanger = mail.hill.com
```

```
hill.com preference = 40, mail exchanger = mailme.hill.com
hill.com preference = 60, mail exchanger =
netcomsv.netcom.com
hill.com nameserver = nameme.hill.com
hill.com nameserver = ns1.noc.netcom.net
hill.com nameserver = ns.netcom.com
mail.hill.com internet address = 199.182.20.4
mailme.hill.com     internet address = 199.182.20.3
netcomsv.netcom.com internet address = 192.100.81.101
ns1.noc.netcom.net internet address = 204.31.1.1
ns.netcom.com internet address = 192.100.81.105

** > exit
SMCVAX$
```

3.2 Ping

Ping, reportedly an acronym for the Packet Internetwork Groper, is one of the most widely available tools bundled with TCP/IP software packages. Ping uses a series of Internet Control Message Protocol (ICMP) [22] Echo messages to determine if a remote host is active or inactive, and to determine the round-trip delay in communicating with it.

A common form of the Ping command, showing some of the more commonly available options that are of use to general users, is

```
ping [-q] [-v] [-R] [-c Count] [-i Wait]
          [-s PacketSize] Host
```

where:

- q   Quiet output; nothing is displayed except summary lines at startup and completion.
- v   Verbose output, which lists ICMP packets that are received in addition to Echo Responses.
- R   Record route option; includes the RECORD_ROUTE option in the Echo Request packet and displays the route buffer on returned packets.
- c   Count Specifies the number of Echo Requests to be sent before concluding test (default is to run until interrupted with a control−C).
- i   Wait   Indicates the number of seconds to wait between sending each packet (default = 1).

• s PacketSize Specifies the number of data bytes to be sent; the total ICMP packet size will be PacketSize+8 bytes due to the ICMP header (default = 56, or a 64 byte packet).

Host IP address or host name of target system.

In the first example below, the user pings the host thumper.bellcore.com, requesting that 6 (−c) messages be sent, each containing 64 bytes (−s) of user data. The display shows the round-trip delay of each Echo message returned to the sending host; at the end of the test, summary statistics are displayed.

In the second example, the user pings the host smcvax.smcvt.edu, requesting that 10 messages be sent in quiet mode (−q). In this case, a summary is printed at the conclusion of the test and individual responses are not listed.

TECHNICAL NOTE: Older versions of the Ping command, which are still available on some systems, had the following general format:

```
ping [-s] {IP_address|host_name} [PacketSize] [Count]
```

In this form, the optional "−s" string tells the system to continually send an ICMP Echo message every second; the optional PacketSize parameter specifies the number of bytes in the Echo message (the message will contain PacketSize-8 bytes of data; the default is 56 bytes of data and a 64 byte message); and the optional Count parameter indicates the number of Echo messages to send before concluding the test (the default is to run the test continuously until interrupted).

```
**syrup:/home$ ping -c 6 -s 64 thumper.bellcore.com
PING thumper.bellcore.com (128.96.41.1): 64 data bytes
72 bytes from 128.96.41.1: icmp_seq=0 tt1=240 time=641.8 ms
72 bytes from 128.96.41.1: icmp_seq=2 tt1=240 time=1072.7 ms
72 bytes from 128.96.41.1: icmp_seq=3 tt1=240 time=1447.4 ms
72 bytes from 128.96.41.1: icmp_seq=4 tt1=240 time=758.5 ms
72 bytes from 128.96.41.1: icmp_seq=5 tt1=240 time=482.1 ms
---thumper.bellcore.com ping statistics---
6 packets transmitted, 5 packets received, 16% packet loss
round-trip min/avg/max = 482.1/880.5/1447.4 ms
```

```
**syrup:/home$ ping -q -c 10 smcvax.smcvt.edu
PING smcvax.smcvt.edu (192.80.64.1): 56 data bytes
---smcvax.smcvt.edu ping statistics---
10 packets transmitted, 8 packets received, 20% packet
loss
round-trip min/avg/max = 217.8/246.4/301.5 ms
```

3.3 Finger

The Finger program may be used to find out who is logged in on another system or to find out detailed information about a specific user. This command has also introduced a brand new verb; fingering someone on the Internet is not necessarily a rude thing to do! The Finger User Information Protocol is described in RFC 1288 [32]. The most general format of the Finger command is

```
finger [username]@host_name
```

The first example below shows the result of fingering an individual user at a remote system. The first line of the response shows the username, the user's real name, their process identifier, application, and terminal port number. Additional information may be supplied at the option of the user in "plan" and/or "project" files that they supply; these files are often named PLAN.TXT or PROJECT.TXT, respectively, and reside in a user's root directory (or somewhere in an appropriate search path).

The second example shows the result of fingering a remote system. This lists all of the processes currently running at the fingered system or other information, depending upon how the remote system's administrator set up the system to respond to the Finger command.

```
**C:> finger kumquat@smcvax.smcvt.edu
[smcvax.smcvt.edu]
KUMQUAT Gary Kessler            KUMQUAT not logged in
Last login Fri 16-Sep-1996 3:47PM-EDT Plan:
Gary C. Kessler
Adjunct Faculty Member, Graduate College

INTERNET: kumquat@smcvt.edu
**C:> finger @smcvax.smcvt.edu
{smcvax.smcvt.edu}
```

```
  Tuesday, September 17, 1996 10:12AM-EDT Up 30 09:40:18
  5+1 Jobs on SMCVAX Load ave 0.16 0.19 0.21
User     Personal Name    Subsys   Terminal      Console Location
GOODWIN  Dave Goodwin     LYNX     6.NTY2        waldo.smcvt.edu
JAT      John Tronoan     TELNET   1.TXA5
HELPDESK  System Manager  EDT      2:08.NTY4     [199.93.35.182]
SMITH    Lorraine Smith   PINE     .NTY3         [199.93.34.139]
SYSTEM   System Manager   MAIL     23.OPA0       The VAX Console
                          DCL      SMCVX1$OPA0   The VAX Console
```

3.4 Traceroute

Traceroute is another common TCP/IP tool, this one allowing users to learn about the route that packets take from their local host to a remote host. Although used often by network and system managers as a simple, yet powerful, debugging tool, traceroute can be used by end users to learn something about the ever-changing structure of the Internet.

The classic Traceroute command has the following general format (where "#" represents a positive integer value associated with the qualifier):

```
traceroute [-m #] [-q #] [-w #] [-p #]
            {IP_address/host_name}
```

where

- m is the maximum allowable TTL value, measured as the number of hops allowed before the program terminates (default 5 30)
- q is the number of UDP packets that will be sent with each time-to-live setting (default 5 3)
- w is the amount of time, in seconds, to wait for an answer from a particular router before giving up (default 5 5)
- p is the invalid port address at the remote host (default 5 33434)

The Traceroute example below shows the route between a host at St. Michael's College (domain smcvt.edu) and a host at Hill Associates (www.hill.com), both located in Colchester, VT, but served by different Internet service providers (ISP).

1. St. Michael's College is connected to the Internet via BBN Planet; since the mid-1980s, BBN operated the NSF's regional ISP, called

the New England Academic and Research Network (NEARNET), which was renamed in 1994. The first hop, then, goes to St. Mike's BBN Planet gateway router (smc.bbnplanet.net). The next hop goes to another BBN Planet router (denoted here only by IP address since a name was not assigned to the device), until the packet reaches the BBN Planet T3 backbone.

2. The packet takes two hops through routers at BBN Planet's Cambridge, MA, facility and is then forwarded to BBN Planet in New York City, where the packet takes four more hops. The packet is then forwarded to BBN Planet in College Park, MD.

3. The packet is sent to BBN Planet's router at MAE-East, MFS Datanet's Network Access Point (NAP) in Washington, D.C. MAE stands for Metropolitan Area Exchange, and is a Fiber Distributed Data Interface (FDDI) ring interconnecting routers from subscribing ISPs. The packet is then forwarded to NETCOM, Hill Associates' ISP.

4. The packet now travels through NETCOM's T3 backbone, following links from Washington, D.C. to Chicago to Santa Clara (CA), to San Jose (CA).

5. The packet is now sent to Hill Associates router (again, a system designated only by an IP address since the NETCOM side of the router was not named) and then passed to the target system. Note that the host's real name is not www.hill.com, but syrup.hill.com.

TECHNICAL NOTE: The original version of Traceroute works by sending a sequence of User Datagram Protocol (UDP) datagrams to an invalid port address at the remote host. Using the default settings, three datagrams are sent, each with a Time-To-Live (TTL) field value set to one. The TTL value of 1 causes the datagram to "timeout" as soon as it hits the first router in the path; this router will then respond with an ICMP Time Exceeded Message (TEM) indicating that the datagram has expired. Another three UDP messages are now sent, each with the TTL value set to 2, which causes the second router to return ICMP TEMs. This process continues until the packets actually reach the other destination. Since these datagrams are trying to access an invalid port at the destination host, ICMP Destination Unreachable Messages are returned indicating an unreachable port; this event signals the Traceroute program that it is finished! The Traceroute program displays the round-trip delay associated with each of the attempts. (Note that some current implementations of Traceroute use the Record-Route option in IP rather than the method described above.)

As an aside, Traceroute did not begin life as a general-purpose utility, but as a quick-and-dirty debugging aid used to find a routing problem. The code (complete with comments!) is available by anonymous FTP in the file traceroute.tar.Z from the host ftp.ee.lbl.gov. (See Section 4.2 for a discussion of anonymous FTP.)

```
**SMCVAX$ traceroute www.hill.com
traceruote to syrup.hill.com (199.182.20.3), 30 hops
max, 38 byte packets
 1 smc.bbnplanet.net (192.80.64.5) 10 ms 0 ms 0 ms
 2 131.192.48.105 (131.192.48.105) 0 ms 10 ms 10 ms
 3 cambridgel-cr4.bbnplanet.net (199.94.204.77) 40 ms
40 ms 50 ms
 4 cambridgel-br1.bbnplanet.net (4.0.1.205) 30 ms 50
ms 50 ms
 5 nyc1-br2.bbnplanet.net (4.0.1.121) 60 ms 60 ms 40
ms
 6 nyc2-br2.bbnplanet.net (4.0.1.154) 60 ms 50 ms 60 ms
 7 nyc2-br2.bbnplanet.net (4.0.1.154) 60 ms 40 ms 50 ms
 8 nyc2-br1.bbnplanet.net (4.0.1.54) 70 ms 60 ms 30 ms
 9 collegepk-br2.bbnplanet.net (4.0.1.21) 50 ms 50 ms
40 ms
10 maeeast.bbnplanet.net (4.0.1.18) 200 ms 170 ms 210
ms
11 fddi.mae-east.netcom.net (192.41.177.210) 60 ms 50
ms 70 ms
12 t3-2.was-dc-gw1.netcom.net (163.179.220.181) 70 ms
60 ms 50 ms
13 t3-2.chw-il-gw1.netcom.net (163.179.220.186) 70 ms
80 ms 80 ms
14 t3-2.scl-ca-gw1.netcom.net (163.179.220.190) 140 ms
110 ms 160 ms
15 t3-1.sjx-ca-gw1.netcom.net (163.179.220.193) 120 ms
130 ms 120 ms
16 198.211.141.8 (198.211.141.8) 220 ms 260 ms 240 ms
17 syrup.hill.com (199.182.20.3) 220 ms 240 ms 219 ms
SMCVAX$
```

4 The Two Fundamental Tools

The two most basic tools for Internet applications are TELNET and the File Transfer Protocol (FTP). TELNET allows a user to login to a remote host over a TCP/IP network, while FTP, as the name implies, allows a user to move files between two TCP/IP hosts. These two utilities date back to the very early days of the ARPANET.

4.1 TELNET

TELNET [27] is TCP/IP's virtual terminal protocol. Using TELNET, a user connected to one host can login to another host, appearing like a directly attached terminal at the remote system; this is TCP/IP's definition of a virtual terminal. The general form of the TELNET command is

```
telnet [IP_address|host_name] [port]
```

As shown, a TELNET connection is initiated when the user enters the telnet command and supplies either a host_name or IP_address; if neither are given, TELNET will ask for one once the application begins.

In the example below, a user of a PC uses TELNET to attach to the remote host smcvax.smcvt.edu. Once logged in via TELNET, the user can do anything on the remote host that would be possible if connected via a directly attached terminal or via modem. The commands that are subsequently used are those available on the remote system to which the user is attached. In the sample dialogue below, the user attached to SMCVAX will use basic VAX/VMS commands:

- The dir command lists the files having a "COM" file extension.
- The mail command enters the VMS MAIL subsystem; the dir command here lists waiting mail.
- Ping checks the status of another host.

When finished, the logout command logs the user off the remote host; TELNET automatically closes the connection to the remote host and returns control to the local system.

It is important to note that TELNET is a very powerful tool, one that may provide users with access to many Internet utilities and services that might not be available otherwise. Many of these features are accessed by specifying a port number with the TELNET command, in addition to a host's address, and knowledge of port numbers provides another mechanism for users to access information with TELNET.

This guide discusses several TCP/IP and Internet utilities that require local client software, such as Finger, Whois, Archie, and Gopher. But what if your software does not include a needed client? In some cases, TELNET may be used to access a remote client and provide the same functionality.

This is done by specifying a port number with the TELNET command. Just as TCP/IP hosts have a unique IP address, applications on the host are associated with an address, called a port. Finger (see Section 3.3), for example, is associated with the well-known port number 79. In the absence of a Finger client, TELNETing to port 79 at a remote host may provide the same information. You can finger another host with TELNET by using a command like: telnet host_name 79.

Other well-known TCP port numbers include 25 (Simple Mail Transfer Protocol), 43 (whois), 80 (Hypertext Transfer Protocol), and 119 (Network News Transfer Protocol).

Some services are available on the Internet using TELNET and special port numbers. A geographical information database, for example, may be accessed by TELNETing to port 3000 at host martini.eecs.umich.edu and current weather information is available at port 3000 at host downwind.sprl.umich.edu.

```
**C:> telnet smcvax.smcvt.edu

FTP Software PC/TCP tn 3.10 01/24/95 02:40

Copyright © 1986-1995 by FTP Software, Inc. All
rights reserved.

• Connected to St. Michael's College -

**Username: kumquat

**Password:


St. Michael's College VAX/VMS System.

Node SMCVAX.


Last interactive login on Monday, 16-SEP-1996 15:47

Last noninteractive login on Wednesday, 6-MAR-1996
08:19

You have 1 new Mail message.

Good Afternoon User KUMQUAT. Logged in on 17-SEP-1996
at 1:10 PM.

User [GUEST,KUMQUAT] has 3225 blocks used, 6775
available, of 10000 authorized and permitted overdraft
of 100 blocks on $1$DIA2

To see a complete list of news items, type: NEWS DIR
```

To read a particular item, type NEWS followed by the name of the item you wish to read.

```
**SMCVAX$ dir *.com

Directory $1$DIA2: [GUEST.KUMQUAT]

  BACKUP.COM;24       24 16-JUL-1990 16:22:46.68 (RWED, RWED, RE,)
  DELTREE.COM;17       3 16-JUL-1990 16:22:47.58 (RWED, RWED, RE,)
  EXPANDZ.COM;7        2 22-FEB-1993 10:00:04.35 (RWED, RWED, RE,)
  FTSLOGBLD.COM;3      1 16-JUL-1990 16:22:48.57 (RWED, RWED, RE,)
  FTSRRR.COM;2         1 16-JUL-1990 16:22:48.73 (RWED, RWED, RE,)
  LOGIN.COM;116        5  1-DEC-1993 09:33:21.61 (RWED, RWED, RE,)
  SNOOPY.COM;6         1 16-JUL-1990 16:22:52.06 (RWED, RWED, RE,)
  SYLOGIN.COM;83       8 16-JUL-1990 16:22:52.88 (RWED, RWED, RE,RE)
  SYSTARTUP.COM;88    15 16-JUL-1990 16:22:53.21 (RWED, RWED, RE,)
  WATCH_MAIL.COM;1   173 10-MAY-1994 09:59:52.65 (RWED, RWED, RE,)

Total of 10 files, 233 blocks.

**SMCVAX$ mail

You have 1 new message.

**MAIL> dir

NEWMAIL

# From      Date     Subject

1 IN%" ibug@plainfield. 15-SEP-1996 ANNOUNCE: Burlington WWW Conference

**MAIL> exit

**SMCVAX$ ping kestrel.hill.com /n=5

PING HILL.COM (199.182.20.24): 56 data bytes

64 bytes from 199.182.20.24: icmp_seq=0 time=290 ms

64 bytes from 199.182.20.24: icmp_seq=1 time=260 ms

64 bytes from 199.182.20.24: icmp_seq=2 time=260 ms

64 bytes from 199.182.20.24: icmp_seq=3 time=260 ms

64 bytes from 199.182.20.24: icmp_seq=4 time=260 ms

--KESTREL.HILL.COM PING Statistics--

5 packets transmitted, 5 packets received, 0% packet loss round-trip (ms) min/avg/max = 260/266/290

**SMCVAX$ logout

KUMQUAT logged out at 17-SEP-1996 13:17:04.29
```

```
Connection #0 closed
C:>
```

4.2 FTP

FTP [26] is one of the most useful and powerful TCP/IP utilities for the general user. FTP allows users to upload and download files between local and remote hosts. Anonymous FTP, in particular, is commonly available at file archive sites to allow users to access files without having to preestablish an account at the remote host. TELNET might, in fact, be used for this purpose but TELNET gives the user complete access to the remote system; FTP limits the user to file transfer activities.

The general form of the FTP command is

```
ftp [IP_address|host_name]
```

An FTP session can be initiated in several ways. In the example shown below, an FTP control connection is initiated to a host (the Defense Data Network's Network Information Center) by supplying a host name with the FTP command; optionally, the host's IP address in dotted decimal (numeric) form could be used. If neither host name nor IP address are supplied in the command line, a connection to a host can be initiated by typing open host_name or open IP_address once the FTP application has been started.

The remote host will ask for a username and password. If a bona fide registered user of this host supplies a valid username and password, then the user will have access to any files and directories to which this username has privilege. For anonymous FTP access, the username anonymous is used. Historically, the password for the anonymous user (not shown in actual use) has been guest, although most systems today ask for the user's Internet e-mail address (and several sites attempt to verify that packets are coming from that address before allowing the user to login).

The "help?" command may be used to obtain a list of FTP commands and help topics available with your software; although not always shown, nearly all TCP/IP applications have a help command. An example of the help for FTP's type command is shown in the sample dialogue. This command is a very important one, by the way; if transferring a binary or executable file, be sure to set the type to image (or binary on some systems).

The dir command provides a directory listing of the files in the current directory at the remote host; the UNIX ls command may also usually be used. Note that an FTP data transfer connection is established for the transfer of the directory information to the local host. The output from the dir command will show a file listing that is consistent with the native operating system of the remote host. Although the TCP/IP suite is often associated with UNIX, it can (and does) run with nearly all common operating systems. The directory information shown in the sample dialogue happens to be in UNIX format and includes the following information:

◦ File attributes. The first character identifies the type of file entry as a directory (d), link or symbolic name (l), or individual file (−). The next nine characters are the file access permissions list; the first three characters are for the owner, the next three for the owner's group, and the last three for all other users. Three access privileges may be assigned to each file for each of these groups: read ®, write (w), and execute (x).

◦ Number of entries, or hard links, in this structure. This value will be a "1" if the entry refers to a file or link, or will be the number of files in the listed directory.

◦ File owner

◦ File owner's group.

◦ File size, in bytes.

◦ Date and time of last modification. If the date is followed by a timestamp, then the date is from the current year.

◦ File name.

After the directory information has been transferred, FTP closes the data transfer connection.

The command cd is used to change to another working directory, in this case the rfc directory (note that file and directory names may be case-sensitive). As in DOS, "cd . ." will change to the parent of the current directory. The CWD command successful is the only indication that the user's cd command was correctly executed; the show-directory (may be truncated to fewer characters, as shown) command, if available, may be used to see which working directory you are in.

Another dir command is used to find all files with the name rfc173*.txt; note the use of the * wildcard character. We can now copy (download) the file of choice (RFC 1739 is the previous version of this primer) by using the get (or receive) command, which has the following general format: get remote_file_name local_file_name

FTP opens another data transfer connection for this file transfer purpose; note that the effective data transfer rate is 93.664 kbps. FTP's put (or send) command allows uploading from the local host to the remote. Put is often not available when using anonymous FTP. Finally, we terminate the FTP connection by using the close command.

The user can initiate another FTP connection using the open command or can leave FTP by issuing a quit command. Quit can also be used to close a connection and terminate a session.

TECHNICAL NOTE: It is important to note that different FTP packages have different commands available and even those with similar names may act differently. In the example shown here (using MultiNet for VMS), the show command will display the current working directory; in FTP Software's OnNet, show will display a file from the remote host at the local host. Some packages have nothing equivalent to either of these commands.

```
  **SMCVAX$ ftp nic.ddn.mil
  SMCVAX.SMCVT.EDU MultiNet FTP user process 3.4(111)
  Connection opened (Assuming 8-bit connections)
  <*****Welcome to the DOD Network Information Cen-
ter*****
  <      *****Login with username "anonymous" and
password "guest"
  **Username: anonymous <Guest login ok, send "guest"
as password.
  **Password: guest                    <--- Not displayed
  <Guest login ok, access restrictions apply.
**NIC.DDN.MIL >help type
  TYPE
  Set the transfer type to type.
  Format
  TYPE type
  Additional information available:
  Parameters Example   Restrictions
  **TYPE Subtopic? parameters
  TYPE
  Parameters
  type
  Specify a value of ASCII, BACKUP, BINARY, IMAGE or
LOGICAL-BYTE.
  Use TYPE ASCII (the default) for transferring text
files.
```

Use TYPE BACKUP to set the transfer type to IMAGE and write the local file with 2048-byte fixed length records. Use this command to transfer VAX/VMS BACKUP save sets.

Use TYPE BINARY to transfer binary files (same as TYPE IMAGE).

Use TYPE IMAGE to transfer binary files (for example, .EXE).

Use TYPE LOGICAL-BYTE to transfer binary files to or from a TOPS-20 machine.

```
**TYPE Subtopic?

**Topic?

**NIC.DDN.MIL. dir
<Opening ASCII mode data connection for /bin/ls.
total 58
  drwxr-xr-x 2 nic  1     512 Sep 16 23:00 bcp
  drwxr-xr-x 2 root 1     512 Mar 19 1996 bin
  drwxr-xr-x 2 nic  1    1536 Jul 15 23:00 ddn-news
  drwxr-xr-x 2 nic  1     512 Mar 19 1996 demo
  drwxr-xr-x 2 nic  1     512 Mar 25 14:25 dev
  drwxr-xr-x 2 nic  10    512 Mar 19 1996 disn_info
  drwxr-xr-x 2 nic  1     512 Sep 17 07:01 domain
  drwxr-xr-x 2 nic  1     512 Mar 19 1996 etc
  lrwxrwxrwx 1 nic  1       3 Mar 19 1996 fyi → rfc
  drwxr-xr-x 2 nic  10   1024 Sep 16 23:00 gosip
  drwxr-xr-x 2 nic  1     512 Mar 19 1996 home
  drwxr-xr-x 2 nic  1     512 Mar 19 1996 lost+found
  lrwxrwxrwx 1 nic  1       8 Mar 19 1996 mgt → ddn-
                                             news
  drwxr-xr-x 2 nic  1    1024 Sep 13 12:11 netinfo
  drwxr-xr-x 4 nic  1     512 May 3 23:00 netprog
  drwxr-xr-x 2 nic  1    1024 Mar 19 1996 protocols
  drwxr-xr-x 2 nic  1     512 Mar 19 1996 pub
  drwxr-xr-x 3 140  10    512 Aug 27 21:03 registrar
  drwxr-xr-x 2 nic  1   29696 Sep 16 23:00 rfc
  drwxr-xr-x 2 nic  1    5632 Sep 9 23:00 scc
```

```
 drwxr-xr-x 2 nic 1      1536 Sep 16 23:00 std
 drwxr-xr-x 2 nic 1      1024 Sep 16 23:00 templates
 drwxr-xr-x 3 nic 1       512 Mar 19 1996  usr
<Transfer complete.
1437 bytes transferred at 33811 bps.
Run time = 20. ms, Elapsed time = 340. ms.

**NIC.DDN.MIL> cd rfc
<CWD command successful.
**NIC.DDN.MIL> show
< "/rfc" is current directory.
**NIC.DDN.MIL> dir rfc173*.txt
<Opening ASCII mode data connection for /bin/ls.
• rw-r-r- 1 nic 10 156660 Dec 20 1994 rfc1730. txt
• rw-r-r- 1 nic 10  11433 Dec 20 1994 rfc1731. txt
• rw-r-r- 1 nic 10   9276 Dec 20 1994 rfc1732. txt
• rw-r-r- 1 nic 10   6205 Dec 20 1994 rfc1733. txt
• rw-r-r- 1 nic 10   8499 Dec 20 1994 rfc1734. txt
• rw-r-r- 1 nic 10  24485 Sep 15 1995 rfc1735. txt
• rw-r-r- 1 nic 10  22415  Feb 8 1995 rfc1736. txt
• rw-r-r- 1 nic 10  16337 Dec 15 1994 rfc1737. txt
• rw-r-r- 1 nic 10  51348 Dec 15 1994 rfc1738. txt
• rw-r-r- 1 nic 10 102676 Dec 21 1994 rfc1739. txt
<Transfer complete.
670 bytes transferred at 26800 bps.
Run time = 10. ms, Elapsed time = 200. ms.

**NIC.DDN.MIL> get rfc1739.txt primer.txt
<Opening ASCII mode data connection for rfc1739.txt
(102676 bytes).
<Transfer complete.
105255 bytes transferred at 93664 bps.
Run time = 130. ms, Elapsed time = 8990. ms.
**NIC.DDN.MIL> quit
<Goodbye.
SMCVAX$
```

5 User Database Lookup Tools

Finding other users on the Internet is an art, not a science. Although there is a distributed database listing all of the 16+ million hosts on the Internet, no similar database yet exists for the tens of millions of users. While many commercial ISPs provide directories of the users of their network, these databases are not yet linked. The paragraphs below will discuss some of the tools available for finding users on the Internet.

5.1 WHOIS/NICNAME

WHOIS and NICNAME are TCP/IP applications that search databases to find the name of network and system administrators, RFC authors, system and network points-of-contact, and other individuals who are registered in appropriate databases. The original NICNAME/WHOIS protocol is described in RFC 954 [10].

WHOIS may be accessed by TELNETing to an appropriate WHOIS server and logging in as whois (no password is required); the most common Internet name server is located at the Internet Network Information Center (InterNIC) at rs.internic.net. This specific database only contains INTERNET domains, IP network numbers, and domain points of contact; policies governing the InterNIC database are described in RFC 1400 [31]. The MILNET database resides at nic.ddn.mil and PSI's White Pages pilot service is located at psi.com.

Many software packages contain a WHOIS/NICNAME client that automatically establishes the TELNET connection to a default name server database, although users can usually specify any name server database that they want.

The accompanying dialogue shows several types of WHOIS/NICNAME information queries. In the session below, we request information about an individual (Denis Stratford) by using WHOIS locally, a specific domain (hill.com) by using NICNAME locally, and a network address (199.182.20.0) and high-level domain (com) using TELNET to a WHOIS server.

```
**SMCVAX$ whois stratford, denis
Stratford, Denis (DS378)     denis@@SMCVAX.SMCVT.EDU
St. Michael's College
Jemery Hall, Room 274
Winooski Park
Colchester, VT 05439
(802) 654-2384
```

```
Record last updated on 02-Nov-92.
SMCVAX$
**C:> nicname hill.com
[198.41.0.5]
Hill Associates (HILL-DOM)
17 Roosevelt Hwy.
Colchester, Vermont 05446
US

Domain Name: HILL.COM.Administrative Contact:
Kessler, Gary C. (GK34) g.kessler@HILL.COM
802-655-0940
Technical Contact, Zone Contact:
Monaghan, Carol A. (CAM4) c.monaghan@HILL.COM
802-655-0940

Billing Contact:
Parry, Amy (AP1257) a.parry@HILL.COM
802-655-0940

Record last updated on 11-Jun-96.
Record created on 11-Jan-93.

Domain servers in listed order:
     SYRUP.HILL.COM      199.182.20.3
     NS1.NOC.NETCOM.NET  204.31.1.1

**C:> telnet rs.internic.net
SunOS UNIX 4.1 (rs1) (ttypb)
-------------------------------------------------------
 * -- InterNIC Registration Services Center --
  *
 • For wais, type:            WAIS <search string> <return>
 • For the original whois type: WHOIS [search string] <return>
 • For referral whois type:   RWHOIS [search string] <return>
  *
-------------------------------------------------------
 Please be advised that use constitutes consent to
monitoring
```

```
(Elec Comm Priv Act, 18 USC 2701-2711)
**[vt220] InterNIC > whois
InterNIC WHOIS Version: 1.2 Wed, 18 Sep 96 09:49:50
**Whois: 199.182.20.0
Hill Associates (NET-HILLASSC)
17 Roosevelt Highway
Colchester, VT 05446

Netname: HILLASSC
Netnumber: 199.182.20.0

Coordinator:
Monaghan, Carol A. (CAM4) c.monaghan@HILL.COM
802-655-0940

Record last updated on 17-May-94.
**Whois: com-dom
   Commercial top-level domain (COM-DOM)
   Network Solutions, Inc.
   505 Huntmar Park Dr.
   Herndon, VA 22070

Domain Name: COM
Administrative Contact, Technical Contact, Zone Con-
tact:
Network Solutions, Inc. (HOSTMASTER)
hostmaster@INTERNIC.NET
(703) 742-4777 (FAX) (703) 742-4811
Record last updated on 02-Sep-94.
Record created on 01-Jan-85.

Domain servers in listed order:

A. ROOT-SERVERS.NET      198.41.0.4
H. ROOT-SERVERS.NET      128.63.2.53
B. ROOT-SERVERS.NET      128.9.0.107
C. ROOT-SERVERS.NET      192.33.4.12
D. ROOT-SERVERS.NET      128.8.10.90
```

```
E. ROOT-SERVERS.NET        192.203.230.10
I. ROOT-SERVERS.NET        192.36.148.17
F. ROOT-SERVERS.NET        192.5.5.241
G. ROOT-SERVERS.NET        192.112.36.4
```

```
**Would you like to see the known domains under this top-level domain? n

**Whois: exit

**[vt220] InterNIC > quit

Wed Sep 18 09:50:29 1996 EST

Connection #0 closed

C:>
```

5.2 KNOWBOT

KNOWBOT is an automated username database search tool that is related to WHOIS. The Knowbot Information Service (KIS), operated by the Corporation for National Research Initiatives (CNRI) in Reston, Virginia, provides a simple WHOIS-like interface that allows users to query several Internet user databases (White Pages services) all at one time. A single KIS query will automatically search the InterNIC, MILNET, MCImail, and PSI White Pages Pilot Project; other databases may also be included.

KNOWBOT may be accessed by TELNETing to host info.cnri.reston.va.us. The help command will supply sufficient information to get started. The sample dialogue below shows use of the query command to locate a user named "Steven Shepard"; this command automatically starts a search through the default set of Internet databases.

```
**C:> telnet info.cnri.reston.va.us

Knowbot Information Service

KIS Client (V2.0). Copyright CNRI 1990. All Rights Reserved.

KIS searches various Internet directory services

to find someone's street address, email address and phone number.

Type 'man' at the prompt for a complete reference with examples.

Type 'help' for a quick reference to commands.

Type 'news' for information about recent changes.

Please enter your email address in our guest book. . .
```

```
**(Your email address?) > s.shepard@hill.com
**> query shepard, steven
Trying whois at ds.internic.net. . .
The ds.internic.net whois server is being queried:
Nothing returned.
The rs.internic.net whois server is being queried:
Shepard, Steven (SS2192) 708-810-5215
Shepard, Steven (SS1302) axisteven@AOL.COM (954) 974-
4569 The nic.ddn.mil whois server is being queried:
Shepard, Steven (SS2192)
R.R. Donnelley & Sons
750 Warrenville Road
Lisle, IL 60532
Trying mcimail at cnri.reston.va.us. . .
Trying ripe at whois.ripe.net. . .
Trying whois at whois.lac.net. . .

No match found for .SHEPARD,STEVEN
**> quit
KIS exiting
Connection #0 closed
C:>
```

6 Information Servers

File transfer, remote login, and electronic mail remained the primary applications of the ARPANET/Internet until the early 1990s. But as the Internet user population shifted from hard-core computer researchers and academics to more casual users, easier-to-use tools were needed for the Net to become accepted as a useful resource. That means making things easier to find. This section will discuss some of the early tools that made it easier to locate and access information on the Internet.

6.1 Archie

Archie, developed in 1992 at the Computer Science Department at McGill University in Montreal, allows users to find software, data, and other information files that reside at anonymous FTP archive sites;

the name of the program, reportedly, is derived from the word "archive" and not from the comic book character. Archie tracks the contents of several thousand anonymous FTP sites containing millions of files. The archie server automatically updates the information from each registered site about once a month, providing relatively up-to-date information without unduly stressing the network. Archie, however, is not as popular as it once was and many sites have not updated their information; as the examples below show, many of the catalog listings are several years old.

Before using archie, you must identify a server address. The sites below all support archie; most (but not all) archie sites support the servers command which lists all known archie servers. Due to the popularity of archie at some sites and its high processing demands, many sites limit access to nonpeak hours and/or limit the number of simultaneous archie users. Available archie sites include the following:

archie.au	archie.rediris.es
archie.edvz.uni-linz.ac.at	archie.luth.se
archie.univie.ac.at	archie.switch.ch
archie.uqam.ca	archie.ncu.edu.tw
archie.funet.fi	archie.doc.ic.ac.uk
archie.th-darmstadt.de	archie.unl.edu
archie.ac.il	archie.internic.net
archie.unipi.it	archie.rutgers.edu
archie.wide.ad.jp	archie.ans.net
archie.kr	archie.sura.net
archie.sogang.ac.kr	

All archie sites can be accessed using archie client software. Some archie servers may be accessed using TELNET; when TELNETing to an archie site, login as archie (you must use lower case) and hit <ENTER> if a password is requested.

Once connected, the help command assists users in obtaining more information about using archie. Two more useful archie commands are prog, used to search for files in the database, and whatis, which searches for keywords in the program descriptions.

In the accompanying dialogue, the set maxhits command is used to limit the number of responses to any following prog commands; if this is not done, the user may get an enormous amount of information. In this example, the user issues a request to find entries related to "dilbert"; armed with this information, a user can use anonymous FTP to examine these directories and files.

The next request is for files with "tcp/ip" as a keyword descriptor. These responses can be used for subsequent prog commands.

Exit archie using the exit command. At this point, TELNET closes the connection and control returns to the local host.

Additional information about archie can be obtained by sending e-mail to Bunyip Information Systems (archie-info@bunyip.com). Client software is not required to use archie, but can make life a little easier; some such software can be downloaded using anonymous FTP from the /pub/archie/clients/ directory at ftp.sura.net (note that the newest program in this directory is dated June 1994). Most shareware and commercial archie clients hide the complexity described in this section; users usually connect to a preconfigured archie server merely by typing an archie command line.

```
**C:> telnet archie.unl.edu

SunOS UNIX (crcnis2)

**login: archie

**Password:

Welcome to the ARCHIE server at the University of
Nebraska-Lincoln

# Bunyip Information Systems, 1993

**unl-archie> help

These are the commands you can use in help:

      .  go up one level in the hierarchy

      ?  display a list of valid subtopics at the current level

<newline>

done, ^D, ^C quit from help entirely

<string> help on a topic or subtopic

Eg.

"help show"

will give you the help screen for the "show" command

"help set search"

Will give you the help information for the "search"
variable.

The command "manpage" will give you a complete copy
of the archie manual page.
```

**help. done

**unl-archie.>set maxhits 5

**unl-archie> prog dilbert

Search type: sub.

Your queue position: 2

Estimated time for completion: 00:20

Host ftp.wustl.edu (128.252.135.4)

Last updated 10:08 25 Dec 1993

Location: /multimedia/images/gif/unindexed/931118

FILE -rw-r-r- 9747 bytes 19:18 17 Nov 1993 dilbert.gif

**unl-archie> whatis tcp/ip

RFC 1065 McCloghrie, K.; Rose, M.T.

Structure and identification of management information for TCP/IP-based internets. 1988 August; 21 p. (Obsoleted by RFC 1155)

RFC 1066 McCloghrie, K.; Rose, M.T. Management Information Base for network management of TCP/IP-based internets. 1988 August; 90 p. (Obsoleted by RFC 1156)

RFC 1085 Rose, M.T. ISO presentation services on top of TCP/IP based internets. 1988 December; 32 p.

RFC 1095 Warrier, U.S.; Besaw, L. Common Management Information Services and Protocol over TCP/IP (CMOT). 1989 April; 67 p. (Obsoleted by RFC 1189)

RFC 1144 Jacobson, V. Compressing TCP/IP headers for low-speed serial links. 1990 February; 43 p.

RFC 1147 Stine, R.H., ed. FYI on a network management tool catalog: Tools for monitoring and debugging TCP/IP internets and interconnected devices. 1990 April; 126 p. (Also FYI 2)

RFC 1155 Rose, M.T.; McCloghrie, K. Structure and identification of management information for TCP/IP-based internets. 1990 May; 22 p. (Obsoletes RFC 1065)

RFC 1156 McCloghrie, K.; Rose, M.T. Management Information Base for network management of TCP/IP-based internets. 1990 May; 91 p. (Obsoletes RFC 1066)

```
RFC 1158 Rose, M.T., ed. Management Information Base
for network management of TCP/IP-based internets: MIB-
II. 1990 May; 133 p.

RFC 1180 Socolofsky, T.J.; Kale, C.J. TCP/IP tutor-
ial. 1991 January; 28 p.

RFC 1195 Callon, R.W. Use of OSI IS-IS for routing
in TCP/IP and dual environments. 1990 December; 65 p.

RFC 1213 McCloghrie, K.; Rose, M.T., eds. Management
Information Base for network management of TCP/IP-
based internets: MIB-II. 1991 March; 70 p. (Obsoletes
RFC 1158)

    log_tcp    Package to monitor tcp/ip connections
    ping       PD version of the ping(1) command. Send ICMP

ECHO requests to a host on the network (TCP/IP) to
see whether it's reachable or not

**unl-archie> exit

# Bye.

Connection #0 closed

C:>
```

6.2 Gopher

The Internet Gopher protocol was developed at the University of Minnesota's Microcomputer Center in 1991, as a distributed information search and retrieval tool for the Internet. Gopher is described in RFC 1436 [1]; the name derives from the University's mascot.

Gopher provides a tool so that publicly available information at a host can be organized in a hierarchical fashion using simple text descriptions, allowing files to be perused using a simple menu system. Gopher also allows a user to view a file on demand without requiring additional file transfer protocols. In addition, Gopher introduced the capability of linking sites on the Internet, so that each Gopher site can be used as a stepping stone to access other sites and reducing the amount of duplicate information and effort on the network.

Any Gopher site can be accessed using Gopher client software (or a WWW browser). In many cases, users can access Gopher by TELNETing to a valid Gopher location; if the site provides a remote Gopher client, the user will see a text-based, menu interface. The number of Gopher sites grew rapidly between 1991 and 1994, although growth tapered due to the introduction of the Web; in any case, most Gopher sites have a menu item that will allow you to identify other Gopher sites. If using TELNET, login with the username gopher (this must be in lowercase); no password is required.

In the sample dialogue below, the user attaches to the Gopher server at the Internet Network Information Center (InterNIC) by TELNETing to ds.internic.net. With the menu interface shown here, the user merely follows the prompts. Initially, the main menu will appear. Selecting item 3 causes Gopher to seize and display the "InterNIC Registration Services (NSI)" menu; move to the desired menu item by typing the item number or by moving the pointer (→) down to the desired entry using the DOWN-ARROW key on the keyboard, and then hitting ENTER. To quit the program at any time, press q (quit);? and u will provide help or go back up to the previous menu, respectively. Users may also search for strings within files using the / command or download the file being interrogated using the D command.

Menu item 1 within the first submenu (selected in the dialogue shown here) is titled "InterNIC Registration Archives." As its submenu implies, this is a place to obtain files containing the InterNIC's domain registration policies, domain data, registration forms, and other information related to registering names and domains on the Internet.

```
**SMCVAX$ telnet ds.internic.net
UNIX® System V Release 4.0 (ds2)
**login: gopher
.........................................................................
Welcome to the InterNIC Directory and
Database Server.
.........................................................................
Internet Gopher Information Client v2.1.3

  Gopher Information Client v2.1.3

  Home Gopher server: localhost

  •     >1. About InterNIC Directory and Database Ser-
vices/

  2. InterNIC Directory and Database Services (AT&T)/

  3. InterNIC Registration Services (NSI)/

  4. README

  Press ? for Help, q to Quit                 Page: 1/1

  **View item number: 3

  Internet Gopher Information Client v2.1.3

  InterNIC Registration Services (NSI)

  •     > 1. InterNIC Registration Archives/
```

```
  2. Whois Searches (InterNIC IP, ASN, DNS, and POC
Registry) <?>

  Press ? for Help, q to Quit, u to go up a menu
                                              Page: 1/1

  **View item number: 1

  Internet Gopher Information Client v2.1.3

  InterNIC Registration Archives

  •     > 1. archives/

  2. domain/

  3. netinfo/

  4. netprog/

  5. policy/

  6. pub/

  7. templates/

  Press ? for Help, q to Quit, u to go up a menu
                                              Page: 1/1

  **q

  **Really quit (y/n)? y

  Connection closed by Foreign Host

  SMCVAX$
```

6.3 VERONICA, JUGHEAD, and WAIS

The problem with being blessed with so much information from FTP, archie, Gopher, and other sources is exactly that—too much information. To make it easier for users to locate the system on which their desired information resides, a number of other tools have been created.

VERONICA (Very Easy Rodent–Oriented Netwide Index to Computerized Archives) was developed at the University of Nevada at Reno as an archie-like adjunct to Gopher. As the number of Gopher sites quickly grew after its introduction, it became increasingly harder to find information in gopherspace since Gopher was designed to search a single database at a time. VERONICA maintains an index of titles of Gopher items and performs a keyword search on all of the Gopher sites that it has knowledge of and access to, obviating the need for the user to perform a menu-by-menu, site-by-site search for information.

When a user selects an item from the menu of a VERONICA search, "sessions" are automatically established with the appropriate Gopher servers, and a list of data items is returned to the originating Gopher client in the form of a Gopher menu so that the user can access the files. VERONICA is available as an option on many Gopher servers.

Another Gopher-adjunct is JUGHEAD (Jonzy's Universal Gopher Hierarchy Excavation And Display). JUGHEAD supports key word searches and the use of logical operators (AND, OR, and NOT). The result of a JUGHEAD search is a display of all menu items which match the search string which are located in the University of Manchester and UMIST Information Server, working from a static database that is recreated every day. JUGHEAD is available from many Gopher sites, although VERONICA may be a better tool for global searches.

The Wide Area Information Server (WAIS, pronounced "ways") was initiated jointly by Apple Computer, Dow Jones, KMPG Peat Marwick, and Thinking Machines Corp. It is a set of free-ware, share-ware, and commercial software products for a wide variety of hardware/software platforms, which work together to help users find information on the Internet. WAIS provides a single interface through which a user can access many different information databases. The user interface allows a query to be formulated in English and the WAIS server will automatically choose the appropriate databases to search. Further information about WAIS can be obtained by reading the WAIS FAQ, from host rtfm.mit.edu in file /pub/usenet/news.answers/wais-faq.

7 The World Wide Web

The World Wide Web (WWW) is thought (erroneously) by many to be the same thing as the Internet. But the confusion, in many ways, is justified; by early 1996, the WWW accounted for over 40% of all the traffic on the Internet. In addition, the number of hosts on the Internet named www has grown from several hundred in mid-1994 to 17,000 in mid-1995 to 212,000 in mid-1996 to over 410,000 by early 1997. The Web has made information on the Internet accessible to users of all ages and computer-skill levels. It has provided a mechanism so that nearly anyone can become a content provider. According to some, growth in the number of WWW users is unparalleled by any other event in human history.

The WWW was developed in the early 1990s at the CERN Institute for Particle Physics in Geneva, Switzerland. The Web was designed to combine aspects of information retrieval with multimedia communications, unlike archie and Gopher, which were primarily used for the indexing of text-based files. The Web allows users to access information in many different types of formats, including text, sound, image, animation, and video. WWW treats all searchable Internet files as hypertext documents. Hypertext is a term which merely refers to text that contains pointers to other text, allowing a user reading one document to jump to another document for more information on a given topic, and then return to the same location in the original document. WWW hypermedia documents are able to employ images, sound, graphics, video, and animation in addition to text.

To access WWW servers, users must run client software called a browser. The browser and server use the Hypertext Transfer Protocol (HTTP) [3]. WWW documents are written in the Hypertext Markup Language (HTML) [2, 20], a simple text-based formatting language that is hardware and software platform-independent. Users point the browser at some location using a shorthand format called a Uniform Resource Locator (URL), which allows WWW servers to obtain files from any location on the public Internet using a variety of protocols, including HTTP, FTP, Gopher, and TELNET.

Mosaic, developed in 1994 at the National Center for Supercomputer Applications (NCSA) at the University of Illinois at Urbana–Champaign, was the first widely used browser. Because it was available at no cost over the Internet via anonymous FTP, and had a version for Windows, Mac, and UNIX systems, Mosaic was probably the single reason that the Web attracted so many users so quickly. The most commonly used browsers today include the Netscape Navigator (http://www.netscape.com), Microsoft's Internet Explorer (http://www.microsoft.com), and NCSA Mosaic (http://www.ncsa.uiuc.edu/SDG/Software/Mosaic/).

The WWW is ideally suited to a windows environment, or to other point-and-click graphical user interfaces. Nevertheless, several text-based Web browsers to exist, although their usefulness is limited if trying to obtain graphical images, or audio or video clips. One text-based Web browser is Lynx, and an example of its use is shown below. Items in square brackets in the sample dialogue are Lynx's way of indicating an image or other display that cannot be shown on an ASCII terminal.

```
**gck@zoo.uvm.edu> lynx www.hill.com

Getting http://www.hill.com/

Looking up www.hill.com.

Making HTTP connection to www.hill.com.Sending HTTP request.

HTTP request sent; waiting for response. Read 176 bytes of data.

512 of 2502 bytes of data.

1024 of 2502 bytes of data.
```

```
536

2048

502

Data transfer complete

Hill Associates

[INLINE] Hill Associates, Inc.

Leaders in Telecommunications Training and Education
Worldwide
```

Hill Associates is an international provider of voice and data telecommunications training and education. We cover the full breadth of the field, including telephony, computer networks, ISDN, X.25 and fast packet technologies (frame relay, SMDS, ATM), wireless, TCP/IP and the Internet, LANs and LAN interconnection, legacy networks, multimedia and virtual reality, broadband services, regulation, service strategies, and network security.

Hill Associates' products and services include instructor-led, computer-based (CBT), and hands-on workshop courses. Courseware distribution media include audio tape, video tape, CD-ROM, and 3.5" disks (PC).

Hill Associates products, services, and corporate information

- About Hill Associates
- HAI Products and Services Catalog
- Datacomm/2000-ED Series
- Contacting Hill Associates
- Employment Opportunities
- HAI Personnel Home Pages

On-line information resources from Hill Associates

- HAI Telecommunications Acronym List
- Articles, Books, and On-Line Presentations by HAI Staff
- GCK's Miscellaneous Sites List . . .

Hill Associates is host to the:

- IEEE Local Computer Networks Conference Home Page . . .
- Vermont Telecommunications Resource Center

```
Please send any comments or suggestions to the HAI
Webmaster. Come back again soon!
```

Information at this site © 1994-1997 Hill Associates.

Arrow keys: Up and Down to move. Right to follow a link; Left to go back.

H) elp O)ptions P)rint G)o M)ain screen Q)uit /=search [delete]=history list

**G

**URL to open: http://www.bbn.com

Getting http://www.bbn.com/

Looking up www.bbn.com.

Making HTTP connection to www.bbn.com.Sending HTTP request.

HTTP request sent; waiting for response. Read 119 bytes of data.

500

1000 bytes of data.

2

5

925

Data transfer complete

BBN On The World Wide Web

[LINK]

BBN Reports Fourth-Quarter and Year-End 1996 Results

[INLINE]

[ISMAP]

[ISMAP]

[LINK]

[INLINE]

Who Won Our Sweepstakes

How The Noc Solves Problems

Noc Noc Who's There

BBN Planet Network Map

[LINK] [LINK] [LINK] [LINK] [LINK] [LINK]

[LINK]

```
  Contact BBN Planet
  Directions to BBN
  Text only index of the BBN Web site
    |
  Corporate Disclaimer
  Send questions and comments about our site to
Webmaster@bbn.com
  © 1996 BBN Corporation
  Arrow keys: Up and Down to move. Right to follow a
link; Left to go back.
  H)elp O)ptions P)rint G)o M)ain screen Q)uit /=search
[delete]=history list
  **Q
  gck@zoo.uvm.edu>
```

7.1 Uniform resource locators

As more and more protocols have become available to identify files, archive and server sites, news lists, and other information resources on the Internet, it was inevitable that some shorthand would arise to make it easier to designate these sources. The common shorthand format is called the Uniform Resource Locator. The list below provides information on how the URL format should be interpreted for the protocols and resources that will be discussed in this document. A complete description of the URL format may be found in [4].

file://host/directory/file-name
Identifies a specific file, e.g., the file htmlasst in the edu directory at host ftp.cs.da would be denoted, using the full URL form:
<URL:file://ftp.cs.da/edu/htmlasst>.

ftp://user:password@host:port/directory/file-name
Identifies an FTP site, e.g.:
ftp://ftp.eff.org/pub/EFF/Policy/Crypto/*.

gopher://host:port/gopher-path
Identifies a Gopher site and menu path; a "00" at the start of the path indicates a directory and "11" indicates a file, e.g.:
gopher://info.umd.edu:901/00/info/Government/Factbook92.

http://host:port/directory/file-name?searchpart
Identifies a WWW server location, e.g.:
http://info.isoc.org/home.html.

mailto:e-mail address
Identifies an individual's Internet mail address, e.g.: mailto:s.shepard@hill.com.

telnet://user:password@host:port/
Identifies a TELNET location (the trailing "/" is optional), e.g.: telnet://envnet:henniker@envnet.gsfc.nasa.gov.

7.2 User directories on the web

While finding users on the Internet remains somewhat like alchemy if using the tools and utilities mentioned earlier, the Web has added a new dimension to finding people. Since 1995, many telephone companies have placed national white and yellow page telephone directories on-line, accessible via the World Wide Web.

For a while, it seemed that the easiest and most reliable approach to finding people's e-mail address on the Internet was to look up their telephone number on the Web, call them, and ask for their e-mail address! More recently, however, many third parties are augmenting the standard telephone directory with an e-mail directory. These services primarily rely on users voluntarily registering, resulting in incomplete databases because most users do not know about all of the services. Nevertheless, some of the personal directory services available via the Web with which e-mail addresses (and telephone numbers) can be found include Four11 Directory Services (http://www.Fourll.com/), Excite (http://www.excite.com/Reference/locators.html), and Yahoo! People Search (http://www.yahoo.com/search/people/).

In addition, the Knowbot Information Service (KIS), CNRI's automated username database search tool described earlier in this document, is also available on the Web, at http://info.cnri.reston.va.us/kis.html. Users can select several options for the KIS search, including the InterNIC, MILNET, MCImail, and Latin American InterNIC databases; UNIX finger and whois servers; and X.500 databases.

7.3 Other services accessible via the web

Many of the other utilities described earlier in this document can also be accessed via the WWW. In general, the Web browser acts as a viewer to a remote client rather than requiring specialized software on the user's system.

Several sites provide DNS information, obviating the need for a user to have a local DNS client such as NSLOOKUP. The hosts http://nsl.milepost.com/dns/ and http://shl.ro.com/~mprevost/netutils/dig.html are among the best DNS sites, allowing the user to access all DNS information. The site http://www.bankes.com/nslookup.htm

allows users to do multiple, sequential searches at a given domain. Other Web sites providing simple DNS name/address translation services include http://rhinoceros.cs.inf.shizuoka.ac.jp/dns.html, http://www.engin.umich.edu/htbin/DNSquery, http://www.lubin.pl/cgi-bin/ns/nsgate, and http://www.trytel.com/cgi-bin/weblookup. Ping is another service available on the Web. The http://shl.ro.com/~mprevost/netutils/ping.html page allows a user to select a host name, number of times to ping (1–10), and number of seconds between each ping (1–10), and returns a set of summary statistics. Other Web-based ping sites include http://www.net.cmu.edu/bin/ping (sends ten pings, and reports the times and min/max/avg summary statistics) and http://www.uia.ac.be/cc/ping.html (indicates whether the target host is alive or not).

Traceroute is also available on the Web. Unfortunately, these servers trace the route from their host to a host that the user chooses, rather than from the user's host to the target. Nevertheless, interesting route information can be found at http://www.net.cmu.edu/bin/traceroute. Traceroute service and a list of a number of other traceroute sites on the Web can be found at http://www.lublin.pl/cgi-bin/trace/traceroute.

Access to archie is also available via the WWW, where your browser acts as the graphical interface to an archie server. To find a list of archie servers, and to access them via the Web, point your browser at http://www.yahoo.com/Computers and Internet/Internet/FTP Sites/Searching/Archie/.

Finally, even Finger can be found on the World Wide Web; check out http://shl.ro.com/~mprevost/netutils/finger.html.

8 Discussion Lists and Newsgroups

Among the most useful features of the Internet are the discussion lists that have become available to allow individuals to discuss topics of mutual concern. Discussion list topics range from SCUBA diving and home brewing of beer to AIDS research and foreign policy. Several, naturally, deal specifically with the Internet, TCP/IP protocols, and the impact of new technologies.

Most of the discussion lists accessible from the Internet are unmoderated, meaning that anyone can send a message to the list's central repository and the message then will be automatically forwarded to all subscribers of the list. These lists provide very fast turn-around between submission of a message and delivery, but often result in a lot of messages (including inappropriate junk mail, or "spam"). A moderated list has an extra step; a human list moderator examines all messages before they are forwarded to ensure that the messages are appropriate to the list and not needlessly inflammatory!

Users should be warned that some lists generate a large number of messages each day. Before subscribing to too many lists, be sure that you are aware of local policies and/or charges governing access to discussion lists and e-mail storage.

8.1 Internet Discussion Lists

Mail can be sent to almost all Internet lists at an address with the following form: list name@host name.

The common convention when users want to subscribe, unsubscribe, or handle any other administrative matter is to send a message to the list administrator; do not send administrivia to the main list address! The list administrator can usually be found at: list name-REQUEST@host name

To subscribe to a list, it is often enough to place the word "subscribe" in the main body of the message, although a line with the format:

subscribe list_name your_full_name

will satisfy most mail servers. A similar message may be used to get off a list; just use the word "unsubscribe" followed by the list name. Not every list follows this convention, but it is a safe bet if you do not have better information!

8.2 LISTSERV

A large set of discussion groups is maintained using a program called LISTSERV. LISTSERV is a service provided widely on BITNET and EARN, although it is also available to Internet users. A LISTSERV User Guide can be found on the Web at http://www.earn.net/lug/notice.html. Mail can be sent to most LISTSERV lists at an address with the following form: list name@host name

The common convention when users want to subscribe, unsubscribe, or handle any other administrative matter is to send commands in a message to the LISTSERV server; do not send administrivia to the main list address! The list server can usually be found at: LISTSERV@host name

LISTSERV commands are placed in the main body of e-mail messages sent to an appropriate list server location. Once you have found a list of interest, you can send a message to the appropriate address with any appropriate command, such as:

subscribe list_name	your_full_name	Subscribe to a list

unsubscribe list_name	Unsubscribe from a list
help	Get help & a list of commands
index	Get a list of LISTSERV files
get file_name	Obtain a file from the server

8.3 Majordomo

Majordomo is another popular list server for Internet discussion lists. The Web site http://www.greatcircle.com/majordomo/ has a large amount of information about Majordomo.

Mail is sent to Majordomo lists using the same general address format as above: list name@host name

The common convention when users want to subscribe, unsubscribe, or handle any other administrative matter is to send a message to the Majordomo list server; do not send administrivia to the main list address! The Majordomo server can usually be found at: MAJORDOMO@host name

Majordomo commands are placed in the main body of e-mail messages sent to an appropriate list server location. Available commands include the following:

help Get help & a list of commands

subscribe list_name your_e-mail

Subscribe to a list (E-mail address is optional)

unsubscribe list_name your_e-mail

Unsubscribe from a list (E-mail address is optional)

info list Sends an introduction about the specified list

lists Get a list of lists served by this Majordomo server

8.4 Usenet

Usenet, also known as NETNEWS or Usenet news, is another information source with its own set of special interest mailing lists organized into newsgroups. Usenet originated on UNIX systems but has migrated to many other types of hosts. Usenet clients, called newsreaders, use the Network News Transfer Protocol [13] and are available for virtually any operating system; several web browsers, in fact, have this capability built in.

While Usenet newsgroups are usually accessible at Internet sites, a prospective Usenet client host must have appropriate newsreader software to be able to read news. Users will have to check with their local host or network administrator to find out what Usenet newsgroups are locally available, as well as the local policies for using them.

Usenet newsgroup names are hierarchical in nature. The first part of the name, called the hierarchy, provides an indication about the general subject area. There are two types of hierarchies, called mainstream and alternative; the total number of newsgroups is in the thousands. The news.announce.newusers newsgroup is a good place for new Usenet users to find a detailed introduction to the use of Usenet, as well as an introduction to its culture.

Usenet mainstream hierarchies are established by a process that requires the approval of a majority of Usenet members. Most sites that receive a NETNEWS feed receive all of these hierarchies, which include the following:

comp	Computers
misc	Miscellaneous
news	Network news
rec	Recreation
sci	Science.
soc	Social issues
talk	Various discussion lists

The alternative hierarchies include lists that may be set up at any site that has the server software and disk space. These lists are not formally part of Usenet and, therefore, may not be received by all sites getting NETNEWS. The alternative hierarchies include the following:

alt	Alternate miscellaneous discussion lists
bionet	Biology, medicine, and life sciences
bit	BITNET discussion lists
biz	Various business-related discussion lists
ddn	Defense Data Network
gnu	GNU lists
ieee	IEEE information
info	Various Internet and other networking information
k12	K-12 education
u3b	AT&T 3B computers
vmsnet	Digital's VMS operating system

8.5 Finding discussion lists and newsgroups

Armed with the rules for signing up for a discussion list or accessing a newsgroup, how does one find an appropriate list given one's interests?

There are tens of thousands of e-mail discussion lists on the Internet. One List of Lists may be found using anonymous FTP at ftp://sri.com/netinfo/interest-groups.txt; the List of Lists can be searched using a Web browser by going to http://catalog.com/vivian/interest-group-search.html. Other places to look are the Publicly Accessible Mailing Lists index at http://www.neosoft.com/internet/paml/byname.html and the LISZT Directory of E-Mail Discussion Groups at http://www.liszt.com. To obtain a list of LISTSERV lists, send e-mail to listserv@bitnic.cren.net with the command lists global in the body of the message. Alternatively, look on the Web at http://www.tile.net/tile/listserv/index.html. The Web site http://www.liszt.com has a Mailing Lists Database of lists served by LISTSERV and Majordomo.

There are also thousands of Usenet newsgroups. One Usenet archive can be found at gopher://rtfm.mit.edu/11/pub/usenet/news.answers; see the /active-newsgroups and /alt-hierarchies subdirectories. Usenet news may also be read at gopher://gopher.bham.ac.uk/11/Usenet. A good Usenet search facility can be found at DejaNews at http://www.dejanews.com/; messages can also be posted to Usenet newsgroups from this site.

Note that there is often some overlap between Usenet newsgroups and Internet discussion lists. Some individuals join both lists in these circumstances or, often, there is cross-posting of messages. Some Usenet newsgroup discussions are forwarded onto an Internet mailing list by an individual site to provide access to those users who do not have Usenet available.

9 Internet Documentation

To fully appreciate and understand what is going on within the Internet community, users might wish to obtain the occasional Internet specification. The main body of Internet documents are Request for Comments (RFCs), although a variety of RFC subsets have been defined for various specific purposes. The sections below will describe the RFCs and other documentation, and how to get them. The Internet standardization process is alluded to in the following sections. The Internet Engineering Task Force (IETF) is the guiding body for Internet standards; their Web site is http://www.ietf.org. The IETF operates under the auspices of the Internet Society (ISOC), which has a Web site at http://www.isoc.org. For complete, up-to-date information on obtaining Internet documentation, go to the InterNIC's Web site at http://ds.internic.net/ds/dspg0intdoc.html. The IETF's history and role in the Internet today is described in Kessler [15]. For information on the organizations involved in the IETF standards process, see RFC 2028 [11]. For information on the relationship between the IETF and ISOC, see RFC 2031 [12].

9.1 Request for comments (RFCs)

RFCs are the body of literature comprising Internet protocols, standards, research questions, hot topics, humor (especially those dated 1 April), and general information. Each RFC is uniquely issued a number which is never reused or reissued; if a document is revised, it is given a new RFC number and the old RFC is said to be obsoleted. Announcements are sent to the RFC-DIST mailing list whenever a new RFC is issued; anyone may join this list by sending e-mail to majordomo@zephyr.isi.edu with the line "subscribe rfc-dist" in the body of the message.

RFCs may be obtained through the mail (i.e., postal service), but it is easier and faster to get them on-line. One easy way to obtain RFCs on-line is to use RFC-INFO, an e-mail-based service to help users locate and retrieve RFCs and other Internet documents. To use the service, send e-mail to rfc-info@isi.edu and leave the Subject: field blank; commands that may go in the main body of the message include the following:

`help`	(Help file)
`help: ways_to_get_rfcs`	(Help file on how to get RFCs)
`RETRIEVE: RFC`	
`Doc-ID: RFCxxxx`	(Retrieve RFC xxxx; use all 4 digits)
`LIST: RFC`	(List all RFCs. . .)
`[options]`	(. . .[matching the following options])
`KEYWORDS: xxx`	(Title contains string "xxx")
`AUTHOR: xxx`	(Written by "xxx")
`ORGANIZATION: xxx`	(Issued by company "xxx")
`DATED-AFTER: mmm-dd-yyyy`	

```
DATED-BEFORE: mmm-dd-yyyy
OBSOLETES: RFCxxxx          (List RFCs obsoleting RFC
                             xxxx)
```

Another RFC e-mail server can be found at the InterNIC. To use this service, send an e-mail message to mailserv@ds.internic.net, leaving the Subject: field blank. In the main body of the message, use one or more of the following commands:

help	(Help file)
file /ftp/rfc/rfcNNNN.txt	(Text version of RFC NNNN)
file /ftp/rfc/rfcNNNN.ps	(Postscript version of RFC NNNN)
document-by-name rfcNNNN	(Text version of RFC NNNN)

To obtain an RFC via anonymous FTP, connect to one of the RFC repositories listed in Table 1 using FTP. After connecting, change to the appropriate RFC directory (as shown in Table F.1) using the cd command. To obtain a particular file, use the get command:

GET RFC-INDEX.TXT local_name	(RFC Index)
GET RFCxxxx.TXT local_name	(Text version of RFC xxxx)
GET RFCxxxx.PS local_name	(Postscript version of RFC xxxx)

TABLE F.1 Primary RFC Repositories

Host address	Directory
ds.internic.net	rfc
nis.nsf.net	internet/documents/rfc
nisc.jvnc.net	rfc
ftp.isi.edu	in-notes
wuarchive.wustl.edu	info/rfc
src.doc.ic.ac.uk	rfc
ftp.ncren.net	rfc
ftp.sesqui.net	pub/rfc
nis.garr.it	mirrors/RFC
funet.fi	rfc
munnari.oz.au	rfc

The RFC index, or a specific reference to an RFC, will indicate whether the RFC is available in ASCII text (.txt) or Postscript (.ps) format. By convention, all RFCs are available in ASCII while some are also available in Postscript where use of graphics and/or different fonts adds more information or clarity; an increasing number are also being converted to HTML. Be aware that the index file is very large, containing citations for over 2000 documents. Note that not all RFCs numbered below 698 (July 1975) are available on-line.

Finally, the InterNIC's Web site at http://ds.internic.net/ds/dspglint-doc.html contains the RFC index and a complete set of RFCs. More information about Web-based RFC servers can be found at http://www.isi.edu/rfc-editor/rfc-sources.html. The sample dialogue below, although highly abbreviated, shows a user obtaining RFC 1594 (Answers to Commonly asked "New Internet User" Questions) using e-mail and anonymous FTP.

```
**SMCVAX$ mail
**MAIL> send
**To: in% "rfc-info@isi.edu"
Subject:
Enter your message below. Press CTRL/Z when complete, CTRL/C to quit
**retrieve: rfc
**doc-id: rfc1594
**^Z
**MAIL> exit

**SMCVAX$ ftp ds.internic.net
**Username: anonymous
**Password:
**NIC.DDN.MIL> cd rfc
**NIC.DDN.MIL> get rfc1594.txt rfc-1594.txt
**NIC.DDN.MIL> exit
SMCVAX$
```

9.2 Internet standards

RFCs describe many aspects of the Internet. By the early 1990s, however, so many specifications of various protocols had been written that it was not always clear as to which documents represented standards for the Internet. For that reason, a subset of RFCs have been designated as STDs to identify them as Internet standards. Unlike RFC numbers that are never reused, STD numbers always refer to the latest version of the standard. UDP, for example, would be completely identified as "STD-6/RFC-768." Note that STD numbers refer to a standard, which is not necessarily a single document; STD 19, for example, is the NetBIOS Service Protocols standard comprising RFCs 1001 and 1002, and a complete citation for this standard would be "STD-19/RFC-001/RFC-1002."

The availability of new STDs is announced on the RFC-DIST mailing list. STD-1 [23] always refers to the latest list of "Internet Official Protocol Standards." The Internet standards process is described in RFC 2026 [5] and STD notes are explained in RFC 1311 [24].

STDs can be obtained as RFCs via anonymous FTP from any RFC repository. In addition, some RFC sites (such as ds.internic.net) provide an STD directory so that STD documents can be found in the path /STD/xx.TXT, where xx refers to the STD number.

STD documents may be obtained as RFCs using the methods described in Section 9.1. STDs may also be obtained via the RFC-INFO server using the RETRIEVE: STD and Doc-ID: STDxxxx commands. Also, check out the InterNIC's Web site at http://www.internic.net/std/ for the STD index and a complete set of STDs.

9.3 For Your Information documents

The For Your Information (FYI) series of RFCs provides Internet users with information about many topics related to the Internet. FYI topics range from historical to explanatory to tutorial, and are aimed at the wide spectrum of people that use the Internet. The FYI series includes answers to frequently asked questions by both beginning and seasoned users of the Internet, an annotated bibliography of Internet books, and an explanation of the domain name system.

Like the STDs, an FYI number always refers to the latest version of an FYI. FYI 4, for example, refers to the answers to commonly asked questions by new Internet users; its complete citation would be "FYI-4/RFC-1594." The FYI notes are explained in FYI 1 [18]. FYIs can be obtained as RFCs via anonymous FTP from any RFC repository. In addition, some RFC sites (such as ds.internic.net) provide an FYI directory so that FYI documents can be found in the path /FYI/xx.TXT, where xx refers to the FYI number.

FYI documents may be obtained as RFCs using the methods described in Section 9.1. FYIs may also be obtained via the RFC-INFO server using the RETRIEVE: FYI and Doc-ID: FYIxxxx commands. Also, check out the InterNIC's Web site at http://www.internic.net/fyi/ for the FYI index and a complete set of FYIs.

9.4 Best current practices

Standard track RFCs are formally part of the IETF standards process, subject to peer review, and intended to culminate in an official Internet Standard. Other RFCs are published on a less formal basis and are not part of the IETF process. To provide a mechanism of publishing relevant technical information which it endorsed, the IETF created a new series of RFCs, called the Best Current Practices (BCP) series. BCP topics include variances from the Internet standards process and IP address allocation in private networks. Like the STDs and FYIs, a BCP number always refers to the latest version of a BCP. BCP 5, for example, describes an IP address allocation plan for private networks; its complete citation would be "BCP-5/RFC-1918." The BCP process is explained in BCP 1 [25].

BCP documents may be obtained as RFCs using the methods described in Section 9.1. BCPs may also be obtained via the RFC-INFO server using the RETRIEVE: BCP and Doc-ID: BCPxxxx commands. Also, check out the RFC Editor's Web site at http://www.isi.edu/rfc-editor/ for the BCP index and a complete set of BCPs.

9.5 RARE technical reports

RARE, the Réseaux Associés pour la Recherche Européenne (Association of European Research Networks), has a charter to promote and participate in the creation of a high-quality European computer communications infrastructure for the support of research endeavors. RARE member networks use Open Systems Interconnection (OSI) protocols and TCP/IP. To promote a closer relationship between RARE and the IETF, RARE Technical Reports (RTRs) have also been published as RFCs since the summer of 1993.

RTR documents may be obtained as RFCs using the methods described in Section 9.1. RTRs may also be obtained via the RFC-INFO server using the RETRIEVE: RTR and Doc-ID: RTRxxxx commands. Also, check out the InterNIC's Web site at http://www.internic.net/rtr/ for the RTR index and a complete set of RTRs. Finally, RTRs may be obtained via anonymous FTP from ftp://ftp.rare.nl/rare/publications/rtr/.

10 Perusing the Internet

This guide is intended to provide the reader with a rudimentary ability to use the utilities that are provided by TCP/IP and the Internet. By now, it is clear that the user's knowledge, ability, and willingness to experiment are about the only limits to what can be accomplished.

There are several books that will help you get started finding sites on the Internet, including the INTERNET Yellow Pages [9]. But much more timely and up-to-date information can be found on the Internet itself, using such search tools as Yahoo! (http://www.yahoo.com), Excite (http://www.excite.com), Lycos (http://www.lycos.com), WebCrawler (http://www.webcrawler.com), and AltaVista (http://altavista.digital.com).

There are several other sources that cite locations from which one can access specific information about a wide range of subjects using such tools as FTP, Telnet, Gopher, and WWW. One of the best periodic lists and archives is through the Scout Report, a weekly publication by the InterNIC's Net Scout Services Project at the University of Wisconsin's Computer Science Department. To receive the Scout Report by e-mail each week, join the mailing list by sending email to listserv@lists.internic.net; place the line subscribe scout-report your_full_name in the body of the message to receive the text version or use subscribe scout-report-html your_full_name to receive the report in HTML. The Scout Report is also available on the Web at http://www.cs.wisc.edu/scout/report and http://rs.internic.net/scout/report, or via anonymous FTP at ftp://rs.internic.net/scout/.

Another list is Yanoff's Internet Services List, which may be found at http://www.spectracom.com/islist/ or ftp://ftp.csd.uwm.edu/pub/inet.services.txt. Gary Kessler, one of the co-author's of this document, maintains his own eclectic Miscellaneous Sites List at http://www.together.net/~kessler/gck site.html. If you are looking for Internet-specific information, one good starting point is http://www.yahoo.com/Computers and Internet/Internet/. The InterNIC is another valuable resource, with their Scout Report and Scout Toolkit (http://rs.internic.net/scout/toolkit).

There is also a fair amount of rudimentary tutorial information available on the Internet. The InterNIC cosponsors "The 15 Minute Series" (http://rs.internic.net/nic-support/15min/), a collection of free, modular, and extensible training materials on specific Internet topics. ROADMAP96 (http://www.ua.edu/~crispen/roadmap.html) is a free, 27-lesson Internet training workshop over e-mail.

More books and specialized articles came out about the Internet in 1993 and 1994 than in all previous years (squared!), and that trend has seemed to continue into 1995, 1996, and beyond. Three books are worth notable mention because they do not directly relate to finding your way around, or finding things on, the Internet. Hafner and Lyon [8] have written *Where Wizards Stay Up Late: The Origins of the Internet,* a history of the development of the Advanced Research Projects Agency (ARPA), packet switching, and the ARPANET, focusing primarily on the 1960s and 1970s. While culminating with the APRANET's 25th Anniversary in 1994, its main thrusts are on the groups building the ARPANET backbone (largely BBN) and the host-to-host application and communication protocols (largely the Network

Working Group). Salus' book, *Casting The Net: From ARPANET to INTERNET and beyond. . .* [28], goes into the development of the network from the perspective of the people, protocols, applications, and networks. Including a set of "diversions," his book is a bit more whimsical than Hafner & Lyon's. Finally, Carl Malamud has written a delightful book called *Exploring the Internet: A Technical Travelogue* [17], chronicling not the history of the Internet as much as a subset of the people currently active in building and defining it. This book will not teach you how to perform an anonymous FTP file transfer nor how to use Gopher, but provides insights about our network (and Carl's gastro-pathology) that no mere statistics can convey.

11 Acronyms and Abbreviations

ASCII	American Standard Code for Information Interchange
BCP	Best Current Practices
BITNET	Because It's Time Network
DDN	Defense Data Network
DNS	Domain Name System
EARN	European Academic Research Network
FAQ	Frequently Asked Questions list
FTP	File Transfer Protocol
FYI	For Your Information series of RFCs
HTML	Hypertext Markup Language
HTTP	Hypertext Transport Protocol
ICMP	Internet Control Message Protocol
IP	Internet Protocol
ISO	International Organization for Standardization
NetBIOS	Network Basic Input/Output System
NIC	Network Information Center
NICNAME	Network Information Center name service
NSF	National Science Foundation
NSFNET	National Science Foundation Network
RFC	Request For Comments
RARE	Réseaux Associés pour la Recherche Européenne
RTR	RARE Technical Reports
STD	Internet Standards series of RFCs
TCP	Transmission Control Protocol
TTL	Time-To-Live
UDP	User Datagram Protocol

URL	Uniform Resource Locator
WAIS	Wide Area Information Server
WWW	World Wide Web

12 Security Considerations

Security issues are not discussed in this memo.

13 Acknowledgments

Our thanks are given to all sites that we accessed or whose system resources we used in preparation for this document. We also appreciate the comments and suggestions from our students and members of the Internet community, particularly after the last version of this document was circulated, including Mark Delany and the rest of the gang at the Australian Public Access Network Association, Margaret Hall (BBN), John Martin (RARE), Tom Maufer (3Com), Carol Monaghan (Hill Associates), Michael Patton (BBN), N. Todd Pritsky (Hill Associates), and Brian Williams. Special thanks are due to Joyce Reynolds for her continued encouragement and direction.

14 References

1. Anklesaria, F., M. McCahill, P. Lindner, D. Johnson, D. Torrey, and B. Alberti, "The Internet Gopher Protocol," RFC 1436, University of Minnesota, March 1993.
2. Berners-Lee, T., and D. Connolly, "Hypertext Markup Language—2.0," RFC 1866, MIT/W3C, November 1995.
3. _____, R. Fielding, and H. Frystyk, "Hypertext Transfer Protocol—HTTP/1.0," RFC 1945, MIT/LCS, UC Irvine, May 1996.
4. _____, L. Masinter, and M. McCahill, Editors, "Uniform Resource Locators (URL)," RFC 1738, CERN, Xerox Corp., University of Minnesota, December 1994.
5. Bradner, S. "The Internet Standards Process-Revision 3," RFC 2026, Harvard University, October 1996.
6. Comer, D. *Internetworking with TCP/IP, Vol. I: Principles, Protocols, and Architecture,* 3/e. Englewood Cliffs, NJ: Prentice-Hall, 1995.
7. Feit, S. *TCP/IP: Architecture, Protocols, and Implementation with IPv6 and IP Security*, 2/e. New York: McGraw-Hill, 1997.
8. Hafner, K., and M. Lyon. *Where Wizards Stay Up Late: The Origins of the Internet.* New York: Simon & Schuster, 1997.
9. Hahn, H., and R. Stout. *The Internet Yellow Pages,* 3/e. Berkeley, CA: Osborne McGraw-Hill, 1996.
10. Harrenstien, K., M. Stahl, and E. Feinler, "NICNAME/WHOIS," RFC 954, SRI, October 1985.
11. Hovey, R., and S. Bradner. "The Organizations Involved in the IETF Standards Process," RFC 2028, Digital, Harvard University, October 1996.
12. Huizer, E. "IETF-ISOC Relationship," RFC 2031, SEC, October 1996.
13. Kantor, B., and P. Lapsley. "Network News Transfer Protocol," RFC 977, UC San Diego, UC Berkeley, February 1986.
14. Kessler, G.C. "An Overview of TCP/IP Protocols and the Internet." URL: http://www.hill.com/library/tcpip.html. Last accessed: 17 February 1997.
15. __________. "IETF-History, Background, and Role in Today's Internet." URL: http://www.hill.com/library/ietf hx.html. Last accessed: 17 February 1997.

16. ________. "Running Your Own DNS." Network VAR, July 1996. (See also URL: http://www.hill.com/library/dns.html. Last accessed: 17 February 1997.)
17. Malamud, C. *Exploring the Internet: A Technical Travelogue.* Englewood Cliffs, NJ: PTR Prentice Hall, 1992.
18. Malkin, G.S., and J.K. Reynolds, "F.Y.I. on F.Y.I.: Introduction to the F.Y.I. notes," FYI 1/RFC 1150, Proteon, USC/Information Sciences Institute, March 1990.
19. Mockapetris, P., "Domain Names—Concepts and Facilities," STD 13/RFC 1034, USC/Information Sciences Institute, November 1987.
20. National Center for Supercomputer Applications (NCSA). "A Beginner's Guide to HTML." URL: http://www.ncsa.uiuc.edu/General/Internet/WWW/HTMLPrimer.html. Last accessed: 2 February 1997.
21. Postel, J., "Domain Name System Structure and Delegation," USC/Information Sciences Institute, RFC 1591, March 1994.
22. ______, "Internet Control Message Protocol," USC/Information Sciences Institute, RFC 792, September 1981.
23. _______, Editor, "Internet Official Protocol Standards," STD 1/RFC 2000, Internet Architecture Board, February 1997.
24. ______, "Introduction to the STD Notes," RFC 1311, USC/Information Sciences Institute, March 1992.
25. ______, T. Li, and Y. Rekhter, "Best Current Practices," BCP 1/RFC 1818, USC/Information Sciences Institute, Cisco Systems, August 1995.
26. ______, and J. Reynolds, "File Transfer Protocol (FTP)," STD 9/RFC 959, USC/Information Sciences Institute, October 1985.
27. _____, and J. Reynolds, "TELNET Protocol Specification," STD 8/RFC 854, USC/Information Sciences Institute, May 1983.
28. Salus, P.H. *Casting The Net: From ARPANET to INTERNET and beyond . . .* Reading, MA: Addison-Wesley, 1995.
29. Socolofsky, T.J., and C.J. Kale, "TCP/IP Tutorial," RFC 1180, Spider Systems Ltd., January 1991.
30. Stevens, W.R. *TCP/IP Illustrated, Volume 1: The Protocols.* Reading, MA: Addison-Wesley, 1994.
31. Williamson, S., "Transition and Modernization of the Internet Registration Service," RFC 1400, Network Solutions Inc., March 1993.
32. Zimmerman, D., "The Finger User Information Protocol," RFC 1288, Rutgers University, December 1991.

15 Authors' Address

Gary C. Kessler
Hill Associates
17 Roosevelt Highway
Colchester, VT 05446
Phone: 11 802-655-8659
Fax: 11 802-655-7974
E-mail: kumquat@hill.com

Steven D. Shepard
Hill Associates
17 Roosevelt Highway
Colchester, VT 05446
Phone: 11 802-655-8646
Fax: 11 802-655-7974
E-mail: s.shepard@hill.com

Glossary

10 Mbps 10 Megabytes per second; standard Ethernet operating speed. (Also called bandwidth.)

10Base2 An Ethernet network implemented on RG58 thin coaxial cable, commonly known as Thin Ethernet or ThinNet.

10Base5 The original Ethernet standard as defined by DEC™, Xerox, and Intel, which uses thick yellow cable.

10BaseF A general classification of standards for fiber-optic Ethernet.

10BaseFB The portion of the 10BaseF standard that defines the requirements for synchronous data transmission over a fiber backbone.

10BaseFL The portion of the 10BaseF standard that defines a fiber-optic link between concentrator and station.

10BaseT An Ethernet network implemented on twisted-pair cabling.

100BaseT A proposed 100-Mbps Ethernet standard, based on CSMA/CD technology.

100BaseT4 IEEE standard for the physical media of Fast Ethernet products. 100BaseT4 standard products run on four-pair category three, four, or five cable.

100BaseTX IEEE standard for the physical media of Fast Ethernet products. Products that conform to the 100BaseTX standard run on two pair category five cable. Dayna Fast Ethernet products support 100BaseTX.

100BaseVG A proposed 100-Mbps LAN topology over UTP as suggested by HP.

100 Mbps 100 Megabits per second; Fast Ethernet operating speed. (Also called bandwidth.)

23B+D See PRI.

286, 386, 486 See Microprocessor.

2B+D ISDN basic-rate service presentation of two 64-Kbps B channels and one 16-Kbps D channel. See BRI, B Channel, and D Channel.

3174 A controller used in IBM 3270 mainframe environments. Provides connectivity between printers and terminals and the mainframe processor.

3270, 3270 Information Display System A once-common IBM mainframe data-entry and display system that consists of control units, display stations, printers, and other equipment. At one time the de facto standard for business computing, it is being replaced by PCs connected to LANs.

5250 A once-common IBM midrange data-entry and display system that consists of control units, display stations, printers, and other equipment. At one time the de facto standard for business computing, it is being replaced by PCs connected to LANs.

5294 A remote controller for IBM 5250 environments.

A

A and B signaling A form of in-band signaling used in T1 transmissions.

A/D converter A device that converts analog signals to digital.

Access method (1) In IBM environments, a host program managing the movement of data between the main storage and an input/output device of a computer system. BTAM, TCAM, and VTAM are common data communications access methods. (2) In LAN technology, a means to allow stations to make use of the network's transmission medium; classified as shared access (which is further divided into explicit access or contended access) or discrete access methods.

Accunet A family of AT&T high-speed digital facilities; includes packet-switched network, DDS, and formatted T1.

ACK (Acknowledgement) A control character used (with NAK) in BSC communications protocol to indicate that the previous transmission block was correctly received and that the receiver is ready to accept the next block. Also used as a ready reply in other communications protocols, such as the Hewlett-Packard ENQ/ACK protocol and the ETX/ACK method of flow control.

Acoustic coupler A modem device that connects a terminal or computer to the handset of a telephone.

ACS (Advanced Communications Service) A proposed packet-switched network from AT&T.

Active link A logical communications circuit or call that is set up only for the duration of the communications. A call-setup and call-clearing procedure is initiated for every connection.

Active matrix A type of LCD panel that has three transistors (red, green, blue) for each pixel. Produces brighter, sharper color than passive-matrix displays.

Adaptive equalizer An equalizer that adjusts to meet varying line conditions; mose operate automatically.

ADB (Apple Desktop Bus) A low-speed serial bus used on Apple Macintosh computers to connect input devices (such as the mouse or keyboard) to the CPU.

ADDCCP (Advanced Data Communications Control Procedures) The standard communications protocol for the U.S. government.

Address A unique sequence of bits, a character, or a group of characters that identifies a network station, user, or application; a unique ID for the destination of data; used for polling and routing purposes.

Advanced Digital Network (ADN) Usually refers to 56 Kbps leased line.

ADSL Asymmetric Digital Subscriber Line.

AFP (AppleTalk Filing Protocol) The AppleTalk protocol for remote file access. It works at the presentation level.

AGC (Automatic Gain Control) The capability of data-line devices such as faxes, modems, line drivers, and multiplexors to automatically compensate (within a given range) for a received signal that is too weak or too strong.

Agent A component of network- and desktop-management software, such as SNMP, that gathers information from MIBs.

Aggregate input rate The sum of all data rates of the terminals or computer ports connected to a multiplexor or concentrator. Burst aggregate input rate efers to the instantaneous maximum.

AI (Artificial Intelligence) Software that can do things normally associated only with human intelligence, such as logical deduction, guesswork, and composition.

Air Interface The standard operating system of a wireless network.

AIX The IBM version of UNIX.

Alphanumeric Consisting of letters and numbers.

Alternate routing In PABX technology, a method of completing calls by using another path when the primary circuit is unavailable, out of service, or busy.

AM (Amplitude Modulation) One of three basic ways (see also FM and phase modulation) to add information to a sine-wave signal; the magnitude of the sine wave, or carrier, is modified in accordance with the information to be transmitted.

AMI (Alternate Mark Inversion) A method of transmitting binary digits. In this method, successive marks, equal in amplitude, normally alternate between positive and negative polarity. A space designates zero amplitude. Also called a bipolar signal.

Amplifier Electronic component used to boost, or amplify, signals. Performance, also called gain, is measured in decibels.

Amplitude distortion An unwanted change in signal amplitude, usually caused by nonlinear elements in the communications path.

Amplitude modulation Changing the voltage level or amplitude of a carrier frequency to transmit digital and analog information.

AMPS (Advanced Mobile Phone Service) The standard operating standard for analog cellular phone services.

Analog A transmission mode in which data is represented by a continuously varying electrical signal. Contrast with digital.

Analog data Data that varies continuously on interval.

Analog loopback A diagnostic test that forms the loop at the modem's telephone line interface (see loopback).

Analog signal A continuously varying electromagnetic wave.

ANSI (American National Standards Institute) The principal-standards development organization in the U.S. The U.S.'s member body to the ISO, ANSI is a nonprofit, independent body that is supported by trade organizations, professional societies, and industry.

Answering tone A signal sent by the called modem (the "answer" modem) to the calling modem (the "originate" modem) on public telephone networks to indicate the called modem's readiness to accept data.

API (Application Program Interface) A set of formalized software calls and routines that can be referenced by an application program to access underlying network services.

APC (Adaptive Power Control) A feature that increases the cellular device's battery's talk and standby times by decreasing the power consumption.

APPC (Advanced Peer-to-Peer Communications) An API developed by IBM for its SNA networks. APPC features high-level program interaction capabilities on a peer-to-peer basis, and allows communications between computers without the intervention of a mainframe. Also called LU6.2.

Applet A small Java program that can be embedded in an HTML page.

Application Layer The highest of the seven-layer OSI model structure; contains all user or application programs. In the IBM SNA, it is the end-user layer.

Application software Programs that perform useful functions in the processing or manipulation of data; includes database managers, word processors, spreadsheets, and other programs that enable the useful manipulation of data. Compare with utility software.

Archie An Internet query system that tracks the contents of over 800 anonymous FTP archive sites containing over one million files.

Architecture The manner in which a system (such as a network or a computer) or program is structured. See also closed architecture, distributed architecture, and open architecture.

ARCNET (Attached Resource Computer Network) A LAN from Datapoint Corporation that interconnects personal computers via coaxial, twisted-pair, or fiber-optic cable. It transmits at 2 megabits per second and uses a token-passing access method and distributed-star topology.

ARP (Address Resolution Protocol) A Transmission Control Protocol/Internet Protocol (TCP/IP) process that maps IP addresses to Ethernet addresses; required by TCP/IP for use with Ethernet.

ARQ (Automatic Request for Retransmission) A communications feature whereby the receiver asks the transmitter to resend a block or frame, generally because of errors detected by the receiver.

ASCII (American Standard Code for Information Interchange, pronounced "ask key") (1) A 7-bit binary data code used in communications with most minicomputers and personal computers. ASCII has 128 possible character combinations. The extra bit in its 8-bit byte can be used as a parity bit or for special symbols. (2) Also called TTY. A protocol that uses the ASCII code set. It provides very little error checking. Transmission is asynchronous: beginning with a start bit, followed by a number of data bits (usually five to eight), and a stop bit.

Asymmetric References data transmission where the *upstream* and *downstream* speeds are different. Typically, the downstream speeds are much greater than the upstream speeds. Contrast with *symmetric*.

Asynchronous character A binary character, used in asynchronous transmission, that contains equal-length bits—including a start bit and one or more stop bits, which define the beginning and end of the character.

Asynchronous modem A modem that uses asynchronous transmission, and therefore does not require timing synchronization with its attached DTE or the remote modem; also used to describe a modem that converts asynchronous inputs from the DTE to synchronous signals for modem-to-modem transmission.

Asynchronous transmission A transmission method in which time intervals between transmitted characters may be of unequal length. Transmission is controlled by start and stop bits on each character, rather than by clocking as in synchronous transmission. Contrast with synchronous transmission.

ATDM (Asynchronous Time-Division Multiplexor) A TDM that multiplexes asynchronous signals by oversampling; also infrequently used to mean concentrator.

ATM (Asynchronous Transfer Mode) A specification from the ISDN standards for providing cell-relay services; a high-bandwidth networking standard.

Attenuation The decrease of signal strength as it travels through the cable or device, measured in decibels; opposite of gain.

AUI (Attachment Unit Interface) The cable between the transceiver (mounted on the backbone Ethernet cable) and the network interface card in a PC or other network mode.

Authentication The function of ensuring that the receiver can positively identify the sender. Authentication is the process of forcing users to prove their identity before they can gain access to network resources.

Auto-answer The ability of a device to respond to an incoming call on a dial-up telephone line and establish a data connection with a remote device without operator intervention.

Autodial The ability of a device to initiate a call over the switched telephone network and establish a connection without operator intervention.

Automatic baud-rate detection The process by which a receiving device determines the speed, code level, and stop bits of incoming data by examining the first character. ABR allows receiving devices to accept data from a variety of transmitting devices operating at different data rates, without being configured for each specific data rate in advance.

Availability A measure of equipment, system, or network performance, usually expressed in percent; the ratio of operating time to the sum of operating time plus down time. Based on MTBF and MTTR.

AWG (American Wire Gauge) A method of defining the cross-section area of a wire.

B

B Channel (Bearer channel) A 64-Kbps ISDN user channel that carries digital data, PCM-encoded digital voice, or a mixture of lower speed data traffic (digital data or digitized voice at a fraction of 64 Kbps).

Backbone (1) In 10Base5 Ethernet, the main network cable to which nodes are attached through transceivers. (2) In packet-switched networks, the major transmission path for a PDN.

Backbone network A transmission facility designed to interconnect low-speed distribution channels or clusters of dispersed user devices.

Background printing A feature that allows you to keep using your computer while it is sending a document to the printer.

Balanced A type of line in which both wires are electrically equal.

Balun (BALanced/UNbalanced) An impedance-matching transformer device used to connect balanced twisted-pair cabling with unbalanced coaxial or other cabling systems. Sometimes found in the IBM cabling systems.

Band splitter A multiplexor (commonly an FDM or TDM) designed to divide the composite bandwidth into several independent, narrower bandwidth channels, each suitable for data transmission at a fraction of the total composite data rate.

Bandwidth The data-carrying capacity of a communications channel; measured (in Hertz) as the difference between the highest and lowest frequencies of the channel. Bandwidth varies depending on the transmission method.

Barrel distortion A video-display defect in which the image appears to bulge outward from the middle of the screen.

Baseband A transmission method, typically for shorter distances, in which the entire bandwidth of the cable is required to transmit a single digital signal. (Compare broadband). Digital signals are put onto the cable without modulation and are transmitted one at a time, making baseband a simpler and cheaper way to transmit data. Simultaneous transmissions can be achieved through time-division multiplexing.

Batch processing A data-processing technique in which input data is accumulated off-line and processed in batches. Compare with interactive processing.

Baud Unit of signaling speed. The speed in baud is the number of line changes (in frequency, amplitude, etc.) or events per second. At low speeds, each event represents only one bit condition, and baud equals bps. As speed increases, each event represents more than one bit, and baud rate does not truly equal bps. But in common usage, baud rate and bps are often used interchangeably.

Baudot Data-transmission code in which five bits represent one character. Use of shift letters/figures enables 64 alphanumeric characters to be represented. Baudot is used in many teleprinter systems with one start bit and 1.5 stop bits added.

BBS (Bulletin-Board Service) An on-line service from which users can get information or files, or exchange messages, via modem.

BCD (Binary-Coded Decimal) A digital system that uses binary codes to represent decimal digits.

Bel Equal to 10 decibels, see decibel.

Bell 103 An AT&T 0-to-300-bps modem providing asynchronous transmission with originate/answer capability; also often used to describe any Bell 103-compatible modem.

Bell 113 An AT&T 0-to-300-bps modem providing asynchronous transmission with originate or answer capability (but not both); also often used to describe any Bell 113-compatible modem.

Bell 201, 201 B, 201 C An AT&T 2400-bps modem providing synchronous transmission. Bell 201 B was designed for full-duplex leased-line or public telephone network applications; Bell 201 C was designed for half-duplex public telephone network applications. Also often used to describe any Bell 201-compatible modem.

Bell 202 An AT&T 1800-bps modem providing asynchronous transmission that requires 4-wire circuit for full-duplex operation; also an AT&T 1200-bps modem providing asynchronous transmission over 2-wire, half-duplex leased-lines or public telephone networks. Often used to describe any Bell 202-compatible modem.

Bell 208, 208 A, 208 B An AT&T 4800-bps modem providing synchronous transmission. Bell 208 A was designed for leased-line applications; Bell 208 B was designed for public telephone network applications. Also often used to describe any Bell 208-compatible modem.

Bell 209 An AT&T 9600-bps modem providing synchronous transmission over 4-wire leased lines; also often used to describe any Bell 209-compatible modem.

Bell 212, 212 A An AT&T 1200-bps full-duplex modem providing asynchronous transmission or asynchronous transmission for use on the public telephone network; also often used to describe any Bell 212-compatible modem.

Bell 43401 An AT&T publication that defines requirements for data transmission over telco-supplied circuits with DC continuity (that is, metallic circuits). Typically used to cover short distances with special line drivers. See local dataset.

Bell standards AT&T standards for modem use. These standards are numbered in the 200s: Bell 208, Bell 209, etc.

Bend loss Increased attenuation in a fiber that results from the fiber's being bent, or from minute distortions within the fiber.

BERT/BLERT (Bit Error Rate Test/Block Error Rate Test) Tests that measure the quality of a data transmission. Both tests compare received data with an established data pattern, then count the number of mismatches (errors).

Binary Digital system with two states, 1 and 0; contrast with octal, decimal, and hexadecimal.

BIOS (Basic Input/Output System) ROM code in ROM in an IBM PC-compatible system that supports low-level system functions related to memory, disks, the monitor, and other hardware.

Bipolar transmission Method of sending binary data in which negative and positive states alternate; used in digital transmission facilities such as DDS and T1.

Bipolar violation The bit stream contains a pulse of the same polarity as the previous pulse and which is not part of a B8ZS code.

bis The second version of a CCITT standard.

Bisynchronous transmission (BSC) Often abbreviated "bisync." A byte- or character-oriented IBM communications protocol that has become an industry standard. It uses a defined set of control characters for synchronized transmission of binary-coded data between stations in a data communications system. Bisync communications require that both the sending and receiving devices be synchronized before transmission of data begins.

Bit (Binary digIT) The smallest unit of information in a binary system; a bit can have a zero or a one value.

Bit duration The time it takes one encoded bit to pass a point on the transmission medium; in serial transmission, a relative unit of time measurement, used for comparison of delay times (propagation delay, access latency, etc.) where the data rate of a transmission channel, typically high-speed, can vary.

Bit-oriented protocol A communications protocol that uses individual bits within the byte as control codes.

Blackout A sudden, unexpected loss of all electrical power, typically lasting for many minutes or even hours.

Block-multiplexor channel In IBM systems, a multiplexor channel that interleaves bytes of data; also called byte-interleaved channel. Contrast with selector channel.

Blocking (1) The process of grouping data into transmission blocks. (2) The inability of a PABX to service connection requests, usually because its switching matrix can only handle a limited number of connections simultaneously. Blocking occurs if a call request from a user cannot be handled because of an insufficient number of paths through the switching matrix; blocking thus prevents free stations from communicating.

BNC A bayonet-locking connector for miniature coax; BNC is said to be short for Bayonet-Neill-Concelman. Contrast with TNC.

BOC (Bell Operating Company) One of the local telephone companies resulting from the divestiture of AT&T.

BONDING Protocol An industry-standard protocol for B-channel aggregation. "BONDING" is an acronym for the Bandwidth on Demand Interoperability Group, which developed the protocol.

Booting Loading a computer's memory with the necessary information so it can function. A "cold boot" occurs when a computer's power is turned on, and there is nothing in memory.

bps (bits per second) The basic unit of measurement for serial data transmission capacity; Kbps for kilo (thousands of) bits per second; Mbps for mega (millions of) bits per second; Gbps for giga (billions of) bits per second; Tbps for tera (trillions of) bits per second.

BPS (Backup Power Supply) A device that switches to an alternate power source when the main source fails. Compare UPS.

Break (1) The point on a defective cable at which a signal stops, because of a damaged or severed circuit. (2) Any open-cable/circuit condition. (3) A space (or spacing) condition that exists longer than one character time (typical length is 110 milliseconds). Often used by a receiving terminal to interrupt (break) the sending device's transmission, to request disconnection, or to terminate computer output.

Breakout box (BOB) A testing device that permits the user to monitor the status of the various signals between two communicating devices, and to cross and tie interface leads, using jumper wires.

BRI (Basic Rate Interface) An ISDN service referred to as 2B+D. BRI provides two 64-Kbps digital channels to your desktop. It is capable of simultaneously transmitting or receiving any digital signal—voice, video, or data. ISDN Terminal Adapters replace modems as the customer-premise connection to this service, enabling you to make direct connections of data terminals and telephones.

Bridge A device that connects two or more LANs (often called subnetworks) that are running the same protocols and cabling. This arrangement creates an extended network, in which any two workstations on the linked LANs can share data. Bridges use only the bottom two layers of the OSI model. Compare router, gateway.

Broadband A transmission method that uses a bandwidth greater than a voice-grade channel's, and potentially capable of much higher transmission rates; also called wideband. In broadcast transmission, multiple channels access a medium (usually coaxial cable) that has a large bandwidth, using radio-frequency modems. Each channel occupies (is modulated to) a different frequency slot on the cable, and is demodulated to its original frequency at the receiving end. Cable television is an example, with up to 50 channels occupying one coaxial cable.

Broadcast A method of transmitting messages to two or more stations at the same time, such as over a bus-type local area network or by satellite; a protocol mechanism that supports group and universal addressing. Any simultaneous transmission to many receiving locations. One example is a message sent over a multipoint line to all terminals that share the line.

Broadcast storm Network messages that overload the network capacity.

Brouter A device that combines the functions of a bridge and a router. Brouters can route one or more protocols, such as TCP/IP and XNS, and bridge all other traffic. Contrasts with bridge and gateway.

Brownout A sudden, unexpected reduction in electrical power, usually lasting just a few seconds, but long enough to cause computer equipment to fail from insufficient power levels.

Browser A client program (software) that is used to look at various kinds of Internet resources. Examples are Netscape Navigator and Microsoft Internet Explorer.

BSC See bisynchronous transmission.

BTAM (Basic Telecommunications Access Method) An IBM software routine; the basic access method for 3270 data communications control.

Buffer A device that temporarily stores data from a faster device, then sends it out to a slave device. The faster device can thus move onto another task while the slower device is still accepting data. A buffer between your PC and your printer lets you get back to work quickly after you send a file to be printed. Buffers are also called spoolers; see spool.

Buffering The process of temporarily storing data in a software program or in RAM, to allow transmission devices to accommodate differences in data transmission rates.

Burroughs Poll Select (BPS) A protocol similar to BSC, but able to run on both synchronous and asynchronous links.

Bus topology A networking setup in which a single cable, such as thin Ethernet, is used to connect one computer to another like a daisy chain to carry data over a network. (See star topology.)

Byte A unit of information, used mainly in referring to data transfer, semiconductor capacity, and data storage; also referred to as a character; a group of eight (sometimes seven) bits used to represent a character.

Byte-interleaved channel See block-multiplexor channel.

Byte-oriented protocol Similar to bit-oriented protocol, but control information may be coded in fields one-byte long.

C

Cable taps Devices that connect the transceiver to the main cable (in an Ethernet network).

Cable modem A device that allows your computer high-speed access to data (such as information on the Internet) through a cable television network.

Cable segment A cable segment is a section of network cable separated by hubs, routers, and/or bridges to create a subnet.

Cache The portion of a computer's RAM reserved to act as a temporary memory for the last items read from a disk.

Call A request for connection or the connection that results from such a request.

Call accounting In packet-switched networks, the process of accumulating data on individual calls; a call-accounting record is a report of such data; it usually includes start and end times, NTN or NUI, and number of data segments and packets transmitted for each individual call.

Call request packet In packet-switched networks, the packet sent by the originating DTE showing requested NTN or NUI, network facilities, and call user data.

Call user data In packet-switched networks, the user information transmitted in a call-request packet to the destination DTE.

Called channel In LAN technology and packet-switched networks, a channel that can receive but cannot originate calls.

Called/calling channel In packet-switched networks, a channel that can both originate and receive calls.

Calling channel In packet-switched networks, a channel that can originate but not receive calls.

Camp-on, camp-on-busy A PABX or cable-based facility that allows users to wait on line (in queue) if the requested resource is busy. It connects the users in queue—on a first-come, first-served basis—when the requested resource becomes available.

Capacitance The tendency of a metallic transmission medium to store energy. Capacitance causes distortion of the data signal.

Card module A printed-circuit board that plugs into an equipment chassis.

Carpal tunnel syndrome (CTS) An ailment caused by activities such as typing that can involve chronic stress to the wrists.

Carrier (a) An analog signal whose frequency, amplitude, phase, or amplitude and phase have been altered to allow it to represent data. (b) A company that provides cellular phone service. Per FCC regulations, there are two competing cellular service carriers in each market, called A and B side carriers.

Carrier Detect See CD (Carrier Detect).

Carrier sense Signal provided by the Physical Layer to the Data-Link Layer of Ethernet to indicate that one or more stations are currently transmitting on the channel.

Carrier signal A tone or radio signal modulated by data, usually for long-distance transmission.

Cascade To connect a multiple-port device to another identical device, thus increasing the number of ports available.

Cascading Connecting hubs together with 10BaseT cable. Sometimes requires a crossover cable.

Category 3 A cabling standard for UTP horizontal wiring. Used for 10-Mbps 10BaseT Ethernet networks or 4-Mbps token-ring networks.

Category 4 A cabling standard for UTP wiring with a bandwidth of 20 MHz. Commonly used for 16-Mbps token-ring networks.

Category 5 A higher grade of unshielded twisted-pair cabling required for networking applications such as 100 Mbps Fast Ethernet. Most commonly prewired in buildings wired within the last five years.

Category 5E

Category 6 (proposed)

CATV (Community Antenna Television) One of the most common facilities found on broadband networks; standards exist for allocating channels on a CATV system.

CCITT (Comité Consultatif Internationale de Télégraphique et Téléphonique) An international consultative committee that set world-wide communications standards (such as V.21, V.22, and X.25). Replaced by the ITU-TSS.

CCITTX.25 An international standard defining packet-switched communication protocol for a public or private network. The recommendation is prepared by the Comité Consultatif Internationale Télégraphique et Téléphonique (CCITT). Along with other CCITT recommendations, the X.25 Recommendation defines the physical-, data-link-, and network-layer protocols necessary to interface with X.25 networks. The CCITT X.25 Recommendation is supported by most X.25 equipment vendors, but a new CCITT X.25 Recommendation is published every four years.

CCTV (Closed-Circuit Television) Television transmitted on customer-owned cables. One of the services often found on broadband networks.

CCU (Communications Control Unit) In IBM 3270 systems, a communications computer, often a minicomputer, associated with a host mainframe computer. It may perform communications protocol control, message handling, code conversion, error control, and application functions.

CD (Carrier Detect) An RS-232 control signal (on pin 8) which indicates that the local modem is receiving a signal from the remote modem. Also called Received Line Signal Detector (RLSD) and Data Carrier Detect (DCD).

CD (Compact Disc) Small plastic disc on which digital information is recorded. The data thus recorded can be played back using a laser. The packing density on the disc is very high, enabling hundreds of megabytes of data to be stored.

CDMA (Code Division Multiple Access) A digital communications standard that allows carriers to provide PCS services. A competing technology to TDMA and GSM.

CDPD (Cellular Digital Packet Data) An industry standard for data communication at 19,200 bps across unused portions of analog cellular voice channels.

CD-R (Recordable CD) CD format compatible with CD-ROM that can be written to once and read many times.

CD-ROM (Compact Disc Read-Only Memory) A compact disc with data prerecorded, normally used in large database-type applications such as directory, reference, or data retrieval.

Cell (1) A unit of storage for power or data. (2) A regional transmission area in cellular systems.

Cell Site The radio tower and transceiver located at the center of each cell.

Cellular Based on the use of cells. Cellular telephone and data networks divide an area into regional cells, each of which has its own central transmission facilities; that way, every point in the area is within range of some transmission facility.

Central office The building in which common telephone carriers terminate customer circuits (also known as Central Exchange).

Centronics parallel The de facto standard for personal computer printers. This 36-pin parallel interface allows the connection of printers and other devices to a computer.

CGA (Color Graphics Array) A graphics standard for IBM PC, PC/XT, and AT(R) computers or compatibles, introduced by IBM in 1981. CGA can display a maximum of 16 colors in two graphics modes: High Resolution, which has 640 × 200 pixels in black and white, and Low Resolution, which has 320 × 200 pixels in four colors.

Channel (1) As used in CCITT standards, a means of 1-way transmission. Compare with circuit. (2) As used in tariffs and in common usage, a path for electrical transmission between two or more points without common carrier-provided terminal equipment, such as a local connection to DTE. Also called circuit, line, data link, path, or facility. (3) In an IBM host system, a high-speed data link connecting the CPU and its peripheral devices. See also block-multiplexor channel and selector channel.

Channel bank Equipment in a telephone central office or customer premises that performs multiplexing of lower speed digital channels into a higher speed composite channel. The channel bank also detects and transmits signaling information for each channel, thus by transmitting framing information so that time slots allocated to each channel can be identified by the receiver.

Channel loopback A diagnostic test that forms the loop at the multiplexor's channel interface (see loopback).

Character A standard 8-bit unit representing a symbol, letter, number, or punctuation mark; generally means the same as byte.

Character set A collection of characters, such as ASCII or EBCDIC, used to represent data in a system. Usually includes special symbols and control functions. Often synonymous with code.

Character-oriented Describing a communications protocol or a transmission procedure that carries control information encoded in fields of one or more bytes; compare with bit-oriented and byte-oriented.

Checksum The total of a group of data items or a segment of data that is used for error-checking purposes. Both numeric and alphabetic fields can be used in calculating a checksum, since the binary content of the data can be added. If the checksum calculated does not match the original checksum, something in the data has changed. Checksums can detect single-bit errors and some multiple-bit errors.

Circuit (1) In data communications, a means of 2-way communications between two points, consisting of transmit and receive channels. (2) In electronic design, one or more components that act together to perform one or more functions.

Circuit-switching A technique in which physical circuits (as opposed to virtual circuits) are transferred (switched) to complete connections. Contrast with packet-switched network.

CISC (Complex Instruction Set Computing) A microprocessor architecture in which a large number of machine instructions are available. Compare RISC.

Cladding In fiber-optic cable, a colored low refractive index material that surrounds the core and provides optical insulation and protection to the core.

Clamping voltages The "sustained" voltage held by a clamp circuit at a desired level.

Client A device or entity in a distributed computing architecture that requests services and information.

Client/Server A network computing system in which individual computers (clients), use a central computer (server) for such services as file storage, printing, and communications. (See peer-to-peer.)

Clock An oscillator-generated signal that provides a timing reference for a transmission link; used to control the timing of functions such as sampling interval, signaling rate, and duration of signal elements. An enclosed digital network typically has only one master clock.

Clone (Cloning) A fraudulent method of duplicating electronic serial and mobile identification numbers.

Closed architecture An architecture that is compatible only with hardware and software from a single vendor. Contrast with open architecture.

Cluster A collection of terminals or other devices in a single location.

Cluster controller A device that manages the input/output operations of a group (cluster) of terminals (usually dumb) or workstations.

Clustered star A topology of several stars linked together.

CMOS (Complementary Metal Oxide Semiconductor) A type of microprocessor that demands very little power.

Coax cable Thin or thick coax cable used in Ethernet networking, usually in a bus topology or backbone use.

Code A set of unambiguous rules that specify the way data is represented, such as ASCII or EBCDIC.

Codec A device that encodes and decodes signals.

Code level The number of bits used to represent characters.

Coding and decoding equipment In LAN technology, PABX equipment or circuits that digitally code and decode voice signals.

Cold boot See booting.

Collision (1) In LAN technology, two stations attempting to use the same transmission medium at the same time. (2) In a half-duplex system, the result of both ends trying to transmit at the same time.

Collision detection In LAN technology, the act of detecting when a collision has occurred; typically occurs when a workstation does not receive an acknowledgement from a receiving station. Collision detection is an integral part of the CSMA/CD access method.

Command Port The console used to control and monitor a network or system; also, the interface to which the console is connected.

Common carrier A private utility company that furnishes voice- or data-communications services to the general public.

Communication server An intelligent device (a computer) providing communications functions; an intelligent, specially configured node on a LAN, designed to enable remote communications access and exit for LAN users.

Communications adapter A device attached to a System 3X host that allows communications over RS-232 lines to remote devices.

Communications protocol The rules governing the exchange of information between devices on a data link (such as SDLC, LAT, TCP/IP, and X.25).

Compaction See compression.

Compandor A single device that combines the functions of a compressor and expandor.

Composite The side signal of a concentrator or multiplexor that includes all the multiplexed data.

Composite link The line or circuit connecting a pair of multiplexors or concentrators; the circuit carrying multiplexed data.

Composite loopback A diagnostic test that forms the loop at the line side (output) or a multiplexor; see loopback.

Composite video The video format most widely used by TVs, VCRs, and camcorders. It carries all color and sync components of the picture on one signal, so its picture quality is relatively poor. Compare RGB, RGBHV, and RGBS.

Compression A technique used to increase the number of bits per second sent over a data link by replacing often-repeated characters, strings, and command sequences with electronic code. When this compressed data reaches the remote end of the transmission link, the coded data is replaced with the actual data. Also called "compaction."

Compressor A device that performs analog compression. See also compandor.

Compromise equalizer An equalizer set for best overall operation for a given range of line conditions; often fixed, but may be manually adjustable.

Concentration Collection of data at an intermediate point from several low- and medium-speed lines for transmission across one high-speed line.

Concentrator Any communications device that allows a shared transmission medium to accommodate more data sources than there are channels currently available within the transmission medium.

Conditioning Extra-cost options that users may apply to leased, or dedicated, voice-grade telephone lines in which line impedance is carefully balanced; allows improved line performance with regard to frequency response and delay distortion. It will generally allow for higher quality and/or higher speed data transmission. In increasing order of resultant line quality and cost, conditioning may be C1, C2, C4, or D1 and D2.

Connect time (1) A measure of system usage: the interval during which the user was on-line for a session. (2) The interval during which a request for a connection is being completed.

Connection (1) An established data-communications path. (2) The process of establishing that path. (3) A point of attachment for that path.

Connectivity The ability of one device to exchange data with another.

Connector A device that holds two parts of a circuit together, so that they make electrical contact.

Console The device used by the operator, system manager, or maintenance technician to monitor or control computer, system, or network performance.

Contended access In LAN technology, a shared access method that allows stations to use the medium on a first-come, first-served basis. Contrast with explicit access.

Contention A first-come, first-served method of access used in public telecommunication or PBX systems in which multiple devices must access a limited number of communication ports.

Continuously variable Capable of having one of an infinite number of values, differing from each other by an arbitrary small amount; usually used to describe analog signals or analog transmission.

Control character A nonprinting character used to start, stop, or modify a function. CR is an example of a control character.

Control signal A modem interface signal used to announce, start, stop, or modify a function; for example, CD is an RS-232 control signal that announces the presence of a carrier.

Control unit In an IBM host system, equipment coordinating the operation of an input/output device and the CPU. See CCU and TCU.

Convergence A measurement of how closely paths of the three color guns in a color monitor or TV match one another. The closer the convergence, the better the picture.

Cookie The most common meaning of "cookie" on the Internet refers to a piece of information sent by a Web Server to a Web Browser that the browser is expected to save and to send back to the server whenever the browser makes additional requests from the server.

Coprocessor A chip that is specific to one particular kind of computation, such as mathematics or video. The coprocessor takes over when the CPU is given the kind of task the coprocessor specializes in.

Core The central region of an optical waveguide through which light is transmitted; typically 8 to 12 microns in diameter for single-mode fiber, and 50 to 100 microns for multimode fiber.

CPI (Computer-PABX Interface) In LAN technology, a voice/data PABX standard (supported by DEC) for using T1 transmission that involves 56-Kbps channels, representing a move toward an open architecture. Compare with DMI.

CPU (Central Processing Unit) The heart of a computer; the component that does the computing. In a PC, it might be contained in a single microprocessor; in a minicomputer, on one or several printed circuit boards; in a mainframe, on many printed circuit boards.

CPS (characters per second) The number of characters transmitted per second.

CR (Carriage Return) An ASCII or EBCDIC control character used to position the print mechanism at the left margin on a printer—or the cursor at the left margin on a display terminal.

CRC (Cyclic Redundancy Check) A basic error-checking mechanism for link-layer data transmissions; a characteristic link-layer feature of (typically) bit-oriented data-communications protocols. The data integrity of a received frame or packet is checked via a polynomial algorithm based on the content of the frame, and then matched with the result that is performed by the sender and included in a (most often 16-bit) field appended to the frame.

Cross-bar switch In older PABXes, a switch having multiple vertical paths, multiple horizontal paths, and electromagnetically operated mechanical means for connecting any vertical path with any horizontal path. Modern PABXes often use an electronic version of the cross-bar switch.

Cross-pinned or crossover cable A cable configured to allow two DTE devices or two DCE devices to communicate. Also called a null modem or modem eliminator.

Cross talk The unwanted transfer of a signal from one circuit to another.

CRT (Cathode-Ray Tube) A specialized vacuum tube that forms pictures on a screen by means of a beam of electrons. The monitor on a computer or terminal contains a cathode-ray tube, but the term "CRT" is often used to refer to the entire terminal.

CSA (Canadian Standards Association) A Canadian agency that sets standards and tests electrical devices for safety.

CSMA (Carrier Sense Multiple Access) A LAN contention protocol by which workstations connected to the same channel are able to sense transmission activity on that channel and so defer their own transmission while the channel is active.

CSMA/CA (Carrier Sense Multiple Access/Collision Avoidance) A LAN access method in which contention between two or more stations is avoided. When a station is prepared to transmit, it first listens to see if the channel is clear.

CSMA/CD (Carrier Sense Multiple Access/Collision Detection) A LAN access method in which contention between two or more stations is resolved by collision detection. When two stations transmit at the same time, they both stop and signal that a collision has occurred. Each then tries again after waiting a predetermined time period, usually several microseconds.

CSU (Channel Service Unit) A digital DCE used to terminate digital circuits (such as DDS or T1 lines) at the customer site. It conditions the line, ensures network compliance with FCC rules, and responds to loopback commands from the central office; it also ensures proper ones density in transmitted bit streams and corrects bipolar violations.

CTF (Cheyenne Tape Format) A LAN-backup standard developed by Cheyenne Software Inc.

CTS (Clear To Send) An RS-232 modem interface control signal used in a switched half-duplex circuit to notify the computer that it has line control. In a full-duplex circuit, this signal is constant.

CUG (Closed User Group) In public data networks, a selected collection of terminal users that do not accept calls from sources not in their group, and often are restricted from sending messages outside the group.

Current loop A transmission technique that detects the presence or absence of current, rather than voltage levels, as ones and zeros. It has traditionally been used in teletypewriter networks that use batteries as the transmission power source. The most common interface found on a current-loop device is unipolar, full duplex, passive 20 mA. Other, less-common versions of current loop are bipolar, analog, half duplex. Many different operating-current and voltage levels are used throughout the world. The active device in a current loop supplies the current for the loop. Passive devices receive current rather than supply it, and are always attached to an active device.

Cursor A movable underline, rectangular-shaped block of light, or an alternating block of reversed video on the screen of a display device, usually indicating where the next character will be entered.

Customer premises equipment A general term for the telephones, computers, private branch exchanges, and other hardware located on the end user's side of the network boundary, established by the Computer Inquiry 11 action of the Federal Communications Commission.

Cyberspace The aggregate "landscape" or environment consisting of the memory and electronic pathways of all computers.

Cyclic redundancy check See CRC.

D

D bit, delivery confirmation bit In an X.25 packet-switched network, used to request end-to-end acknowledgement.

D Channel (Delta channel) A 16-Kbps ISDN channel that carries signaling information to control circuit-switched calls on associated B channels at the user interface. May also be used for packet-switching or low-speed (100 bps) telemetry when there is no signaling information waiting.

D4 framing A T1 12-frame format in which the 193rd bit is used for framing and signaling information; ESF is an equivalent but newer 24-frame technology.

D/A converter A device that changes digital pulses into analog signals.

DAA (Data Access Arrangement) DCE furnished or approved by a common carrier that permits privately-owned DCE or DTE to be attached to the common carrier's network; all modems now built for the public telephone network have integral DAAs.

Daisychaining An arrangement of devices connected in a series, one after the other. Any signal transmitted to the devices goes to the first device, and from the first to the second, and so on.

Data collection Procedure in which data from various sources is accumulated at one location (in a file or queue) before being processed.

Data communications The processes, equipment, or facilities used to transport signals from one data-processing device at one location to another data-processing device at another location.

Data encryption standard (DES) See DES.

Data integrity A measure of data-communications performance, indicating a sparsity (or, ideally, the absence) of undetected errors.

Data link Any serial data-communications transmission path, generally between two adjacent nodes or devices and without intermediate switching nodes. A data link includes the physical transmission medium, the protocol, and associated devices and programs, so it is both a physical and a logical link.

Data-Link Layer Layer Two in the OSI model; the network processing entity that establishes, maintains, and releases data-link connections between (adjacent) elements in a network; controls access to the physical medium (Layer One).

Data offset A field of the TCP head that reveals the distance (expressed in units of 32 bits) from the start of a TCP head to the end of the head (and thus, beginning of the accompanying data).

Data PABX, data-only PABX A PABX used solely for data; a device whose main purpose is to furnish connectivity—to set up and break connections on demand—between computers, terminals, and peripherals.

Data packet In X.25, a block of data that transports full-duplex information via an X.25 switched virtual circuit (SVC) or permanent virtual circuit (PVC). X.25 data packets can contain up to 1024 bytes of user data but the most common size is 128 bytes (the X.25 default).

Data rate, data-signaling rate A measure of how quickly data is transmitted, expressed in bps. Also commonly, but often incorrectly, expressed in baud. Synonymous with speed.

Data set A synonym for modem.

Data stream The collection of characters and data bits transmitted through a channel.

Data transfer rate The average number of bits, characters, or blocks per unit of time transferred from a data source to a data sink.

Database A large, ordered collection of information.

Datagram A finite-length packet with sufficient information to be independently routed from source to destination; datagram transmission typically does not involve establishing an end-to-end session, and may or may not entail confirmation/acknowledgement of delivery.

Data-over-voice An FDM technique which combines data and voice on the same line by assigning a portion of the unused bandwidth to the data; usually implemented on the twisted-pair cables used for in-house telephone-system wiring.

Datascope A diagnostic tool for monitoring and capturing data transmissions.

Daughterboard A board that mounts on, and connects to, the motherboard. Sometimes used for memory upgrades.

dB (decibel) A comparative (logarithmic) measure of signal power (strength or level): +10 dB (or +1 Bel) represents a gain of 10:1; –3 dB represents a 50 percent loss of power. Contrast with dBm.

DB9, DB15, DB15HD, DB25, DB37, DB50 Common names for D-shaped connectors used in data communications. The number indicates the number of possible pins in sockets in the connector.

dBm An absolute measure of signal power, where 0 dBm is equal to one milliwatt. Contrast with dB.

DCD (Data Carrier Detect) See CD (Carrier Detect).

DCE (Data Communications Equipment) The equipment that enables a DTE to communicate over a telephone line or data circuit. The DCE establishes, maintains, and terminates a connection, and performs the conversions necessary for communications. In RS-232, designation as either DCE or DTE determines the signaling role in handshaking.

DCE (Distributed Computing Environment) A set of software-interoperability standards developed and promoted by the OSF.

DDCMP (Digital Data Communications Message Protocol) A communications protocol used in DEC computer-to-computer communications.

DDD (Direct Distance Dialing) A telephone service in North America that enables a subscriber to call other subscribers outside the local area without operator assistance.

DDS (Dataphone Digital Service) A private-line digital service offered within one LATA by local telephone companies and between LATAs by AT&T Communications, with data rates typically at 2.4, 4.8, 9.6, and 56 Kbps; part of the services listed by AT&T under the Accunet family.

DDS-SC (Dataphone Digital Service with Secondary Channel); also called DDS II A tariffed private-line service offered by AT&T and certain BOCs that allows 64-Kbps clear-channel data, with a secondary channel that provides end-to-end supervisory, diagnostic, and control functions.

De facto Commonly accepted, as in "de facto standard."

De jure Specified by law, as in a "de jure standard."

Decibel A measure of the intensity of sound. Logarithmic, i.e., 5 is 10 times louder than 4.

Decimal A digital system that has ten states, 0 through 9.

DECnet™ The Digital Equipment Corporation proprietary network architecture that works across all of the company's machines; uses a peer-to-peer methodology.

Dedicated line A dedicated circuit, a nonswitched channel; see leased line.

Dedicated server A computer on a network that is assigned to function only as a resource server and cannot be used as a client.

Delay In communications, the time between two events; see propagation delay, response time.

Demodulation The conversion of a signal from analog to its original digital form.

Demultiplexing The process of breaking a composite signal into its component channels; the reverse of statistical multiplexing.

DES (Data Encryption Standard) A scheme approved by the National Bureau of Standards that encrypts data for security purposes. DES is the data-communications encryption standard specified by Federal Information Processing Systems (FIPS) Publication 46.

Desktop publishing (DTP) The creation of publication-quality documents on personal computers.

Destination address The address to which a packet is sent.

Destination field A field in a message header that contains the address of the station to which a message is being directed.

Destination group Same as rotary.

Destination port The TCP or UDP header field containing a two-octet value identifying the destination upper-level service or program.

Device Networking equipment such as a hub, switch, bridge, router, etc.

Device-independent A system that does not depend on properties in the device.

Diagnostics Programs or procedures used to test a piece of equipment, a communications link or network, or any similar system.

Dial network A network that is shared among many users, any one of whom can establish communication between desired points by use of a dial or push button telephone.

Dial-up line, dial-in line, dial line A temporary data connection activated by establishing a direct-dialed telephone link between two modems. Compare with leased line.

Dibit A group of two bits. In 4-phase modulation such as DPSK, each possible value of a dibit is encoded as a unique carrier-phase shift; the four possible values of a dibit are 00, 01, 10, and 11.

Differential modulation A type of modulation in which the absolute state of the carrier for the current signal element is dependent on the state after the previous signal element. See DPSK.

Digital Referring to communications procedures, techniques, and equipment by which informaition is encoded as either a binary one (1) or zero (0); the representation of information in discrete binary form, discontinuous in time; compare with analog.

Digital data Information transmitted in a coded form from a computer, represented by discrete signal elements.

Digital loopback A technique for testing the digital processing circuitry of a communications device. It may be initiated locally, or remotely via a telecommunications circuit. The device being tested decodes and then reencodes a received test message, then echoes it back to the sender. The results then are compared with the original message.

Digital pipe The line between the central office and the ISDN subscriber used to carry the communications channels. The capacity of the pipe and the number of channels carried varies from service to service.

Digital service High-speed digital data-transmission services offered for lease by telecommunication service providers. Services include ISDN, Frame Relay, T1, and dedicated or switched 56-Kbps transmission lines.

Digital signal A signal that is composed of two energy levels (on and off, or positive and negative current). Digital signals are used by computers to transmit data. The pattern of the energy-level change represents individual bits of information

DIN (1) The Deutsches Institut für Normung (German Institute for Standardization). (2) A type of rounded connector standardized by this organization.

DIP (Dual In-Line Package) switch A switch for opening and closing leads between two points. Often used to configure a device.

Directed broadcast A broadcast packet sent by a network node to all nodes on all directly attached networks.

Direct memory access (DMA) The ability of devices to access memory without CPU intervention.

Discrete access In LAN technology, an access method used in star LANs. Each station has a separate (discrete) connection through which it makes use of the LAN's switching capability. Contrast with shared access.

Discretely variable Capable of one of a limited number of values; usually used to describe digital signals or digital data transmission. Contrast with continuously variable.

Disk An electromagnetic storage medium for digital data.

Disk duplexing A fault-tolerant technique that writes simultaneously to two hard disks using different controllers.

Disk mirroring A fault-tolerant technique that writes data simultaneously to two hard disks using the same controller.

Disk server A device providing shared access to information on mass storage devices.

Distortion The unwanted changes in signal or signal shape that occur during transmission between two points.

Distributed architecture LAN architecture that uses a shared communications medium, used on bus or ring LANs; uses shared access methods.

Distributed computing An arrangement that puts more of the computing power at individual workstations, rather than in one central processing plant. Modern LANs are good examples of distributed computing. Contrast with distributed processing.

Distributed processing An arrangement that allows separate CPUs to share work on the same application program, thus allowing each CPU to perform a certain task. Often used to mean distributed computing.

Distribution frame A wallmounted structure for terminating telephone wiring, usually the permanent wires from or at the telephone central office, where cross-connections are readily made to extensions. Also called "distribution block."

DLC (Data Link Control) The set of rules (protocols) used by two nodes or stations on a network to perform an orderly exchange of information. A data link includes the physical transmission medium, the protocol, and associated devices and programs, so it is both a physical and a logical link.

DMA (Direct Memory Access) A method of moving data from a storage device directly to RAM, without using the CPU's resources.

DMI (Desktop Management Interface) A standard created by the DMTF for managing PCs from network-management platforms such as SNMP.

DMI (Digital Multiplexed Interface) A voice/data PABX standard (supported by AT&T) for using T1 transmission that involves 64-Kbps channels, representing a move toward an open architecture via ISDN. Compare with CPI.

DMTF (Desktop Management Task Force) An industry group promoting standardization of desktop-management software.

DNA (Digital Network Architecture) The Digital Equipment Corporation layered data-communications protocol.

DNIC (Data Network Identification Code) A four-digit number assigned to public data networks and to specific services within those networks.

Domain Logical grouping, usually geographical, of registered objects in the Clearinghouse. It is also one part of the fully qualified three-part Clearinghouse name.

Domain name The unique name that identifies an internet site.

DOS (Disk Operating System) (1) An operating system (set of programs) that instructs a disk-based computing system to manage resources and operate related equipment. (2) Often used as a synonym for MS-DOS and compatible operating systems.

Dot pitch The distance between sets of red, green, and blue dots on a color monitor's screen. The lower this measurement, the closer the dots are and the better the image on the monitor.

Downline loading The process of sending configuration parameters, operating software, or related data from a central source to individual stations.

Download To load data from a remote device, such as a server or on-line service, into a local device, such as a PC.

Downstream Used to describe traffic on a network from the provider to the endpoint (your computer). Downloading a Web page is downstream traffic. See also *upstream*.

Downtime The period during which computer or network resources are unavailable to users because of a system or component failure.

dpi (dots per inch) A measurement of resolution, usually used with printers and scanners.

DPMS (Display Power Management Signaling) Standard A monitor-industry standard for managing monitor's power consumption.

DPSK (Differential Phase Shift Keying) The modulation technique used in Bell 201 modems; see dibit.

Draft proposal An ISO standards document that has been registered and numbered but not yet given final approval.

DRAM RAM in which the settings of memory locations are continuously refreshed, even when they are not being read from or written to. Used when memory access is expected to be frequent.

Driver A software module that, under control of the processor, manages an I/O port to an external device, such as a serial RS-232C port to a modem.

Drop A connection point between a communicating device and a communications network; a single connection (or node) on a multipoint line.

Drop cable In LANs, a cable that connects the main network cable, or bus, and the data terminal equipment (DTE).

DSL Digital Subscriber Line. Digital *modems* attached to regular telephone wires (twisted-pair copper wiring) that transmit up to 1.5 *Mbps downstream* (to the subscriber) and up to 512 *Kbps upstream,* depending on service purchased. The speed examples listed above are for G.Lite DSL; other types of DSL service will have differing speeds.

DSP (Digital Signal Processing) chip A microprocessor optimized to handle large amounts of real-time data from an analog input, such as sound or video.

DSR (Data Set Ready) An RS-232 modem interface control signal that indicates the modem is ready to transmit data.

DSU (Data Service Unit) A component of customer premise equipment used to interface to a digital circuit (DDS or T1); combined with a channel service unit (CSU), it converts a customer's data stream to bipolar format for transmission. (Also see CSU, ISU.)

DTE (Data Terminal Equipment) User devices, such as terminals and computers, that connect to data communications equipment (DCE), such as modems; they either generate or receive the data carried by the network. In RS-232C connections, designation as either DTE or DCE determines the signaling role in handshaking; in a CCITT X.25 interface, it determines the device or equipment that manages the interface at the user's premises. See DCE.

DTMF (Dual-Tone Multiple-Frequency) The audio signaling frequency generated by touch-tone push button telephones.

DTR (Data Terminal Ready) An RS-232 modem interface control signal (sent from the DTE to the modem on pin 20) that indicates that DTE is ready for data transmission, and requests that the modem be connected to the telephone circuit.

Dumb terminal A display terminal with no processing capabilities. For processing, the terminal is entirely dependent on the main computer with which it communicates.

Duplex (1) In communications circuits, the ability to transmit and receive at the same time; also referred to as full duplex. Half-duplex circuits can receive only or transmit only. (2) In terminals, a choice between displaying locally generated characters and echoed characters.

Dynamic bandwidth allocation A feature of some network nodes that allows additional transmission circuits to be temporarily added to the basic allocation during a high burst of traffic.

Dynamic routing strategy In packet-switching networks, the process of directing messages through a network. When a route is disabled or too busy, an alternate route is automatically chosen.

E

E1 A digital data circuit that runs at 2.048 Mbps.

E911 A FCC mandated system that allows law enforcement to locate a caller.

EBCDIC (Extended Binary Coded Decimal Interchange Code) An 8-bit character code used primarily in IBM equipment; the code provides for 256 different bit patterns. Compare with ASCII.

Echo The distortion created when a signal is reflected back to the originating station.

Echo cancellation The technique used in modems to filter out unwanted signals.

Echo suppressor A device used by telcos or PTTs that blocks the receive side of the line during the time that the transmit side is in use.

Echoplex A method of checking data integrity by returning characters to the sending station for verification of data integrity.

EEPROM (Electrically Erasable Programmable Read-Only Memory) Similar to EPROM, except that chips can be erased with electrical signals (rather than ultraviolet light) while installed in a device, and then reloaded with firmware. Often used to store system configuration data.

EES (Escrowed Encryption System) A security system proposed by the U.S. Department of Justice for the U.S. government's data communication. It involves inserting into all new federal computers a special encryption chip whose output would be reasonably secure but could be tapped by law-enforcement agencies.

EGA (Enhanced Graphics Adapter) A video standard for IBM PC, PC/XT, and AT computers or compatibles. EGA retains compatibility with CGA, but has more graphics modes. Its resolution, at 640 × 350 pixels, is better than that of CGA.

EIA (Electronic Industries Association) A standards organization in the U.S. specializing in the electrical and functional characteristics of interface equipment. Its recommended standards are marked with the RS- prefix (RS-232, RS-485, etc.).

EIA/TIA-568 A wiring standard for commercial buildings supported by both the EIA and the TIA, as well as by ANSI. It describes every aspect of cable installation for both voice and data communication.

EISA (Extended Industry Standard Architecture) Bus A 32-bit adaptation of the 8- and 16-bit buses originally developed by IBM. It was designed for use with Intel microprocessors. The EISA bus was a joint development of COMPAQ and other PC manufacturers as an alternative to the proprietary IBM Micro Channel bus.

Electronic mail, E-mail Messages sent between subscribers electronically via a public or private data communications system.

EMC (Electromagnetic Compatibility) A directive that specifies the acceptable limits for electromagnetic emissions from an electronic device, and how much electromagnetic interference the device should tolerate.

EMI (Electromagnetic Interference) Unwanted electromagnetic emissions, generated by lightning or by electronic or electrical devices, that degrade the performance of another electronic device. Interference may be reduced by shielding. Maximum acceptable levels of EMI from electronic devices are detailed by the FCC.

EMI/RFI (Electromagnetic Interference/Radio Frequency Interference) Filtering Background noise that could alter or destroy data transmission.

Empty-slot ring In LAN technology, a ring LAN in which a free packet circulates through every station; a bit in the packet's header indicates whether it contains any messages (if it contains messages, it also contains source and destination addresses).

Emulation The imitation of all or part of one device, terminal, or computer by another, so that the emulating device accepts the same data, performs the same functions, and appears to other network devices as if it were the emulated device.

Encapsulation The technique used by layered protocols in which a layer adds header information to the protocol data unit from the layer above.

Encoding/decoding The process of organizing information into a format suitable for transmission, and then reconverting it after transmission; for pulse-code-modulated voice transmission, the generation of digital signals to represent quantified samples, and the subsequent reverse process.

Encryption The function of ensuring that data in transit may only be read by the intended recipient. Encryption disguises/scrambles the contents of a message as it travels over a network, making it unintelligible to hackers who may wish to monitor or copy it. Encryption uses a mathematical algorithm and a digital key (series of bits) based on the algorithm to code a message at one end of a transmission and then decode it at the other end.

Energy Star A U.S. government program that mandates strict limits on power consumption by electronic equipment sold to U.S. government offices.

ENQ (Enquiry) A control character (Control-E in ASCII) used as a request to obtain identification or status.

ENQ/ACK protocol A Hewlett-Packard communications protocol. The HP3000 computer follows each transmission block with ENQ to determine if the destination terminal is ready to receive more data; the destination terminal indicates its readiness by responding with ACK.

Enterprise network A network (usually large) that connects all sites of a particular organization.

EPROM (Erasable Programmable Read-Only Memory) A nonvolatile semiconductor PROM that can be erased for reuse by exposing it to intense ultraviolet light.

Equalization The process of compensating for line distortions.

Equalizer A device used by modems to compensate for distortions caused by telephone line conditions.

Erasable storage Storage whose contents can be modified (Random Access Memory, or RAM), as contrasted with read-only storage (Read-Only Memory, or ROM).

Ergonomics The science of how "natural" or "easy" things are to use. An "ergonomic design," as of a keyboard, monitor, or mouse, should cause less stress and strain than nonergonomic designs.

Error control An arrangement that combines error detection and error correction.

Error correction An arrangement that restores data integrity in received data, either by manipulating the received data or by requesting retransmission from the source. See ARQ.

Error detection An arrangement that senses flaws in received data by examining parity bits, verifying block check characters, or using other techniques.

Error rate A measure of data integrity, given as the fraction of bits that are flawed. Often expressed as a negative power of 10, as in 10-6 (a rate of one error in every one million bits).

ESF (Extended Superframe) A DS1 framing format in which the 193rd bit of the bit stream is a shared bit. This means there are 24 framing bits. The 193rd bit follows a 24-frame sequence instead of 12 (as in D4 superframe). This is the same as using an 8 K bit data channel to frame the signal and identify which frames carry signaling information.

ESMR (Enhanced SMR) A digital radio service using base stations to provide wireless voice and data communication. Alternative to cellular. See FHMA, MIPS.

ESN (Electronic Serial Number) The 32-bit binary number assigned by the manufacturer that uniquely identifies a cell phone.

Essential facilities In packet-switched networks, standard network facilities, which are on all networks.

Ethernet A network standard first developed by Xerox, and refined by DEC and Intel. Ethernet interconnects personal computers and transmits at 10 megabits per second. It uses a bus topology that can connect up to 1024 PCs and workstations within each main branch. Ethernet is codified as the IEEE-802.3 standard.

ETX (End of Text) A control character used to indicate the conclusion of a message; it immediately precedes the block check character in transmission blocks.

Even parity A dumb-terminal data-verification method in which each character must have an even number of one bits.

Exchange A unit established by a common carrier for the administration of communications services in a specified geographical area such as a city. It consists of one or more central offices together with the equipment used in providing the communications services. Frequently used as a synonym for central office.

Expanded memory A PC's address memory (any amount above 640 K) that conforms to the LIM (Lotus, Intel, and Microsoft) memory specification.

Expandor A device that reverses the effect of analog compression. See also compandor.

Explicit access In LAN technology, a shared access method that allows stations to use the transmission medium individually for a specific time period; every station is guaranteed a turn, but every station must also wait for its turn. Contrast with contended access.

Extended memory A PC's address memory above the IBM boundary that may be accessed directly by some operating environments; a page frame is not required as with expanded memory. Must be converted to expanded memory by software drivers to be of any use.

Extender A device that increases the distance over which data can be sent from a PC to a printer.

F

Facility (1) A feature or capability offered by a system, item of hardware, or software. (2) In telco environments, line and equipment used to furnish a completed circuit. (3) In packet-switched networks.

Facsimile (fax) The transmission of page images by a system that is concerned with patterns of light and dark rather than with specific characters. Older systems use analog signals; newer devices use digital signals and may interact with computers and other digital devices.

Fallback The ability to default to an alternate line or device, should the primary unit fail.

Fanout A device that splits one port into a number of identical ports.

FAQ Frequently Asked Questions. FAQs are documents that list and answer the most commonly asked questions on a given subject.

Fast Ethernet Any 100-Mbps Ethernet-based networking scheme.

Fast select In packet-switched networks, a calling method that allows the user to expedite the transmission of a limited amount of information (usually 128 bytes). The information is sent along with the call-request packet; therefore, the information arrives faster than in other call methods (which send the information in the packets that follow the call-request packet).

Fault The point of failure in a malfunctioning or inoperative device.

Fault management One of the five basic categories of network management defined by the International Standards Organization (ISO). Fault management is used for the detection, isolation, and correction of faults on the network.

Fault tolerance The ability to continue operating normally if one component of the system fails.

FCC (Federal Communications Commission) A board of commissioners appointed by the President under the Communications Act of 1934, with the authority to regulate all interstate telecommunications originating in the United States, including transmission over phone lines.

FCS (Frame Check Sequence) In a bit-oriented protocol, a 16-bit field that contains transmission-error checking information, usually appended at the end of a frame.

FDDI (Fiber Distributed Data Interface) An American National Standards Institute (ANSI) specified standard for fiber-optic links with data rates up to 100 Mbps. The standard specifies: multimode fiber; 50/125, 62.5/125, or 85/125 core cladding; an LED or laser light source; and 2 km for unrepeated data transmission at 40 Mbps.

FDM See frequency-division multiplexor.

FEP (Front-End Processor) A dedicated computer linked to one or more host computers or multiuser minicomputers; performs data communications functions and serves to off-load the attached computers of network processing; in IBM SNA networks, an IBM 3704, 3705, 3725, or 3745 communications controller.

FF (form feed) An ASCII or EBCDIC printer control character used to skip to the top of the next page (or form).

FHMA (Frequency-Hopping Multiple-Access) A multiple-access spread-spectrum ESMR technology developed by the Israeli military.

Fiber loss The attenuation of the light signal in optical-fiber transmission.

Fiber optics A technology that uses light as a digital information carrier. The transmission medium is made up of small strands of glass, each of which provides a path for light rays that carry the data signal. Fiber-optic technology offers large bandwidth, very high security, and immunity to electrical interference. The glass-based transmission facilities also occupy far less space than other high-bandwidth media, which is a major advantage in crowded underground ducts.

Field A group of bits that describes a specified characteristic; displayed on a reserved area of a CRT or located in a specific part of a record.

FIFO (First-In, First-Out) A method of coordinating the sequential flow of data through a buffer.

FIGS (Figures Shift) (1) A physical shift in a terminal using Baudot Code that enables the printing of numbers and symbols. (2) The character that causes the shift.

File A collection of related data records.

File server In local area networks, a node dedicated to providing file and mass data storage services to the other stations on the network.

File-server protocol In LAN technology, a communications protocol that allows application programs to share files.

Filter An arrangement of electronic components designed to pass signals in one or several frequency bands and to attenuate signals in other frequency bands.

Filtering In LAN technology, discarding packets that do not meet the criteria for forwarding.

Finger An Internet software tool for locating people on other Internet sites.

Fire wall A combination of hardware and software that separates a LAN into two or more parts for security purposes.

Firewire A high-speed serial peripheral interface invented by Apple to replace SCSI.

Firmware A computer program or software stored permanently in PROM or ROM or semipermanently in EPROM.

Flag In communications, a bit pattern of six consecutive 1 bits (character representation is 01111110) used in many bit-oriented protocols to mark the beginning (and often the end) of a frame. Flag is also used as a general term that indicates when a certain condition or response should be initiated.

Flash ROM EEPROM that can be reprogrammed while it is still in the computer after being erased by a brief "flash"—a deliberately induced electric surge.

Flicker The flashing of a video screen as the light beam that makes up the image follows its raster pattern.

Flow control The procedure for regulating the flow of data between two devices; prevents the loss of data once a device's buffer has reached capacity.

FM (Frequency Modulation) One of three basic ways (see AM and phase modulation) to add information to a sine-wave signal: the frequency of the sine wave, or carrier, is modified in accordance with the information to be transmitted.

FOIRL (Fiber Optic Inter Repeater Link) A standard that defines a fiber-optic link between two repeaters in an IEEE-802.3 network.

Font A set of characters that share the same design style.

Footprint The space a device occupies on a desk or in a workplace.

Forward channel Channel on which cell sites communicate with mobile telephone sets.

Forwarding In LAN technology, sending a packet from one LAN to another through a bridge.

Four-wire circuit A circuit containing two pairs of wires, one pair for the transmit channel and one pair for the receive channel, for simultaneous (full duplex) two-way transmission. Contrast with two-wire channel.

Fox message A diagnostic test message that uses all the letters (and that sometimes includes numerals): "THE QUICK BROWN FOX JUMPED OVER A LAZY DOG'S BACK 1234567890." (In French, "VOYEZ LE BRICK GEANT QUE J'EXAMINE PRES DU WHARF.") Often run continuously during system testing and fault isolation.

Fractional T1 A service aimed at customers who do not need or cannot afford all 24 channels of a full T1 line. Fractional T1 service offers the use of one or more channels. The customers, then, pay only for the channels they use.

Fragment An IP datagram that contains a part of the data sent by a transport module as a discrete request. IP produces fragments when the transport request is too large to be held in a single datagram.

Fragmentation The process of breaking a datagram into smaller pieces and attaching new Internet heads to form smaller datagrams. Fragmentation allows the transfer of a datagram over a subnetwork that has a maximum packet size too small for the complete datagram. IP requires that receivers be able to reassemble datagrams.

Fragmentation offset A field in the IP head marking the relative position of a datagram fragment within the larger original datagram.

Frame (1) Same as transmission block. (2) The sequence of bits and bytes in a transmission block. (3) The overhead bits and bytes that surround the information bits in a transmission block.

Frame Relay A packet network service, relying on the data integrity inherent in digital transmissions to speed up transmission. Unlike old X.25 networks, frame relay "assumes" the data is correct and starts checking as soon as it receives the header, in a half-dozen error-checking steps. Frame-relay services are offered with T1 and DDS connections.

Framing A control procedure used with multiplexed digital channels, such as T1 carriers, whereby bits are inserted so that the receiver can identify the time slots that are allocated to each subchannel; framing bits may also carry alarm signals indicating specific alarm conditions.

Frequency The number of cycles of a signal per second.

Frequency converter In broadband cable systems, the device that translates between the transmitting and receiving frequencies.

Frequency-division multiplexing (FDM) A technique for combining many signals on one circuit by separating them in frequency.

Frequency-division multiplexor A device that divides the available transmission frequency range into narrower bands, each of which is used for a separate channel.

Frequency reuse Technique used in cellular communication whereby frequencies are used repeatedly in multiple cells to increase system capacity.

Front-end processor See FEP.

FSK (Frequency Shift Keying) An FM technique in which one frequency represents a mark and a second frequency represents a space.

FTAM (File Transfer, Access, and Management) An OSI application utility that provides transparent access to files stored on dissimilar systems.

FTP (File Transfer Protocol) An upper-level TCP/IP service that allows copying of files across a network.

Full duplex (FDX) Simultaneous, two-way, independent transmission in both directions (4-wire). Compare with half duplex.

Full-duplex Ethernet (FDE) An implementation of Ethernet that allows communication in both directions simultaneously and can greatly improve data rates.

G

G703 The general standard for interfacing to digital high-speed circuits. This standard now includes specifications for both the 1.544 Mbps and 2.048 Mbps data rates; however, G703 is normally referred to for 2.048 Mbps applications.

G704 The signaling specification for connection to the digital network.

G.Lite DSL Also referred to as Universal DSL, G.992.2, or G.Lite. A standard type of *DSL,* which is being developed jointly by a group of telecommunications and computer companies (including Compaq) known as the Universal ADSL Working Group, or UAWG. G.Lite allows a DSL modem to operate concurrently with normal telephone service on a single phone line.

Gain Increased signal power, usually the result of amplification; opposite of attenuation.

Garbage An informal term for corrupted data.

Gateway A hardware–software combination that connects two LANs (or a LAN and a host computer) that run different protocols—for example, a TCP/IP LAN and an SNA mainframe. The gateway provides the protocol conversion. Gateways, because they operate on the top three layers of the OSI model, are much more complex than bridges. See bridge.

GFI (Group Format Identifier) In X.25 packet-switched networks, the first 4 bits in a packet header; contains the Q bit, D bit, and modulus value.

GIF Graphical Interchange Format. A common format for image files, especially suitable for images containing large areas of the same color.

Gigabit One billion bits.

Gigabyte 1024 megabytes or 232 bytes.

Gopher A browsing tool on the Internet.

GOSIP (Government OSI Profile) The U.S. government's version of the OSI protocols. GOSIP compliance is typically a requirement in government networking purchase.

Graded-Index Fiber An optical fiber whose core, composed of concentric rings of glass, has a nonuniform index of refraction. The refractive indexes of the rings decrease from the center axis out, in order to increase bandwidth.

Ground An electrical connection or common conductor that, at some point, connects to the earth.

Ground loop A current across a conductor, created by a difference in potential between two grounded points, as in two buildings connected by a run of RS-232 (or other data) cable. Energy waiting to be used could travel across the RS-232 (or other data) line and damage sensitive circuitry in computer equipment.

Group I fax An analog facsimile device that transmits or receives a standard page in four to six minutes. Group I machines are no longer being manufactured and are rarely in use in today's market.

Group II fax An analog device that transmits or receives a page in two or three minutes. These systems offer some data-compression techniques for faster transmission and can be compatible with Group I devices. Like the Group I units, Group II models are not actively marketed today.

Group III fax The standard for current facsimile devices. Most facsimile systems marketed today are digital devices offering operating speeds of one minute or less. When equipped with automatic speed recognition, these systems can be compatible with Group I and II units, although a number of the lower cost models on the market today are strictly Group III compatible. Machines that can recognize speed automatically can select the fastest speed available when sending to or receiving from Group I or Group II devices.

Group addressing In transmission, the use of an identification field that is common to two or more stations; any environment where multiple stations recognize a common identifier, and either one station or a group of stations responds.

Groupware Software designed for network use by a group of users working on a related project.

GSM (Global Standard for Mobile) A digital communications standard used in over 60 countries.

GSTN (General Switched Telephone Network) Same as PSTN.

Guard band The unused bandwidth separating channels to prevent cross talk in an FDM system.

GUI (Graphical User Interface, pronounced "gooey") An operating system or environment which displays program choices and options on the screen as icons, or picture symbols. Users enter commands by pointing at icons with a mouse or other pointing device and clicking a button on the pointing device. Invented at the Xerox Palo Alto Research Center in the 1970s, the GUI first became popular on the Apple Macintosh. Some other GUIs are Microsoft Windows, OS/2, and X-Windows for UNIX.

H

Hacker A person preoccupied with trying to break into computer systems or networks. Originally, any amateur computer enthusiast.

Half duplex (HDX) Transmission in either direction, but not in both directions simultaneously (2-wire). Compare with full duplex.

Handset The part of a telephone containing the mouthpiece and receiver.

Hand-off The act of transferring a cellular call from one tower to another during the course of the call. The hand-off is not usually noticed by the cellular user.

Halftone The reproduction of continuous-tone artwork, such as a photograph, through a crossline or contact screen, which convert the image into dots of various sizes.

Handshaking Exchange of predetermined signals between two devices, establishing a connection or providing flow control. Usually part of a communications protocol.

Hardware Physical equipment, as opposed to the computer program or method of use; for example, mechanical, magnetic, electrical, or electronic devices. Contrast with firmware or software.

Hardware address The address intrinsic to a device attached to a network.

Hayes AT Command Set The de facto standards for controlling modem dialing operations and setting operational parameters. These instructions were developed to activate features and processes on intelligent modems.

HDLC (High-Level Data-Link Control) The international standard communication protocol defined by the ISO.

Head end In a broadband transmission network, a group of active and/or passive components that translate one range of frequencies (Transmit) to a different frequency band (Receive); allows devices on a single-cable network to send and receive signals without interference.

Header The control information added to the beginning of a message; contains the destination address, source address, and message number.

Heartbeat See SQE.

Hercules Graphics Card (HGC) An early video-adapter card and de facto video standard for monochrome graphics on IBM PC and compatible computers.

Hertz (Hz) A measure of frequency or bandwidth; 1 Hz equals one cycle per second.

Heterogeneous In local area networking, a communication system capable of accommodating multiple protocols and machine types.

Hexadecimal A digital system that has sixteen states, 0 through 9, followed by A through F. Any 8-bit byte can be represented by two hexadecimal digits.

Hierarchical System A computer/communication system organized around a central processor, to which all other elements are subordinate in operation and data flow.

Hit As used in reference to the World Wide Web, "hit" means a single request from a Web browser for a single item on a Web server.

Home page The main Web page for a business, organization, person or simply the main page out of a collection of Web pages.

Homogeneous In local area networking, a communication system in which all participants use the same protocols.

Hop A term used when counting components and wiring segments in an Ethernet network to determine whether Ethernet compliance has been met.

Host computer The central computer (or one of a collection of computers) in a data-communications system, which provides the primary data-processing functions such as computation, database access, or special programs or programming languages; often shortened to "host".

Hotkey A single key or a simple key combination that can be used to activate a computer program or a function in a program.

Hot-swapping The ability to add and/or remove PC cards without restarting the computer to use the cards.

HP-GL (Hewlett-Packard Graphics Language) A control language for Hewlett-Packard and compatible plotters, and for printing vector graphics or other printers.

HP-UX The Hewlett-Packard version of UNIX.

HSM (Hierarchical Storage Management) A model for network data-storage management.

HTML HyperText Markup Language. The coding used to create hypertext documents for use on the World Wide Web.

HTTP HyperText Transport Protocol. The protocol for moving hypertext files across the Internet.

Hub The core of a star-topology network or cabling system. Also referred to as a "repeater" or "concentrator," its primary function is to receive and send signals along the network between the nodes connected to it.

Hunt group Same as rotary.

Hybrid network A LAN that consists of a number of topologies and access methods—for example, a network that includes both a token ring and a CSMA/CD bus.

Hypertext Text that contains links to other documents—words or phrases in the document that can be chosen by the reader and which cause another document to be retrieved and displayed.

I

I/O Input/output.

I/O bound A condition where the operation of the I/O port is the limiting factor in program execution.

IC (Integrated Circuit) A multifunction semiconductor device.

ICMP (Internet Control Message Protocol) The TCP/IP process that provides the set of functions used for network layer management and control.

Idling signal A signal used to indicate when data is not being sent.

IEC (International Electrotechnical Commission) An organization that cooperates with the ISO for technology standards.

IEEE (Institute of Electrical and Electronic Engineers) An international society of professional engineers that issues widely used networking standards.

IEEE-802.2 A data-link layer standard used with IEEE-802.3, IEEE-802.4, and IEEE-802.5.

IEEE-802.3 The IEEE standard for Ethernet; a physical-layer standard that uses the CSMA/CD access method on a bus-topology LAN.

IEEE-802.3 10BaseT An IEEE standard describing 10-megabit per second twisted-pair Ethernet wiring using baseband signaling. This system requires a wiring hub.

IEEE-802.3 10Broad36 This IEEE specification describes a long-distance type of Ethernet cabling with 10-megabit per second signaling rate, a broadband signaling technique, and a maximum cable-segment distance of 3600 meters.

IEEE-802.4 A physical-layer standard that uses the token-passing access method on a bus-topology LAN. Nearly identical to MAP.

IEEE-802.5 A physical-layer standard that uses the token-passing access method on a ring-topology LAN.

IEEE-802.6 This IEEE standard for metropolitan area networks (MANs) describes what is called a Distributed Queue Dual Bus (DQDB). The DQDB topology includes two parallel runs of cable—typically fiber-optic cable—linking each node (typically a router for a LAN segment) using signaling rates in the range of 100 megabits per second.

IEEE-802.12 The draft standard for 100BaseVG networking.

IEEE Project 802 An IEEE team that developed the IEEE-802 family of LAN standards.

IEEE-488 Standard commonly used for HP plotters and in industrial applications. Also known as Hewlett-Packard Interface Bus (HPIB) or General Purpose Interface Bus (GPIB).

IETF (Internet Engineering Task Force) The organization responsible for managing and maintaining the Internet.

IGRP (Interior Gateway Routing Protocol) An internetworking protocol for routers developed by Cisco Systems Inc.

Impedance The resistance to the flow of alternating current in a circuit.

In the transmission block The first bit of the LRC is set to produce an odd (or even) number of first bits that are set to 1; the second through eighth bits are set similarly. Also called horizontal parity check.

Indeo A video-coding algorithm developed by Intel and used in the Intel PCS standard.

Information bit A data bit, as opposed to an overhead bit.

Insertion loss A power loss that results from inserting a component into a previously continuous path or creating a splice in it.

Inside wiring In telephone deregulation, wiring of the customer's premises; the wiring inside of a building.

Intelligence, intelligent A term for equipment (or a system or network) which has a built-in processing power (often furnished by a microprocessor) that allows it to perform sophisticated tasks in accordance with its firmware.

Intelligent port selector Same as Data PABX.

Intelligent TDM Same as concentrator.

Intelligent terminal A programmable terminal (for example, a computer or PC).

Interactive processing Time-dependent (real-time) data communications; a user enters data and then awaits a response from the destination before continuing. Compare with batch processing.

Interchange circuit In any interface, a circuit with an associated pin assignment on the interface connector that is assigned a data, timing, or control function.

Interface A shared boundary; a physical point of demarcation between two devices, where the electrical signals, connectors, timing, and handshaking are defined; the procedures, codes, and protocols that enable two entities to interact for an exchange of information (e.g., RS-232).

Interface adapter See terminal adapter.

Interface converter A device that allows communication between two systems with incompatible electrical signals, connectors, and handshaking.

International standard An ISO standards document that has been approved in final balloting.

International Telecommunications Union The telecommunications agency of the United Nations, established to provide standardized communications procedures and practices, including frequency allocation and radio regulations, on a worldwide basis.

Internet (1) Any large network made up of several smaller networks. (2) Capitalized, the international network of the networks that connects educational, scientific, and commercial institutions.

Internetwork A large multisegment network or "internet." Two or more networks connected by routers and bridges. Networks in an internetwork share information and services.

Internetwork router In LAN technology, a device used for communications between subnetworks; only messages for the corrected subnetwork are transmitted by this device. Internetwork routers function at the network layer of the OSI model.

Interrupt In LAN technology, the method of access used in certain bus and ring networks in which workstations can "interrupt" a server to get service. The server resumes its prior operation after handling the station's request.

Intranet Similar to the "Internet," this is a local, internal network, usually set up with "locations" and "pages" in the e-mail system, it is used for a variety of intercompany purposes including posting company policy, announcing job openings, listing engineering projects and schedules, meeting announcements, posting sales, benefits coordination, etc.

IP (Internet Protocol) The protocol used in gateways to connect networks at the OSI Network Level (Layer 3) and above.

IPDS (Intelligent Printer Data Stream) A page-description printer protocol that allows a complete page of text and graphics to be formatted and stored in the printer's memory.

IPX A communication protocol in Novell NetWare that creates, maintains, and terminates connections between network devices, such as workstations and servers.

IRL (Inter-Repeater Link) An Ethernet segment connecting two repeaters and not containing network stations.

ISA (Industry Standard Architecture) A bus standard developed by IBM for expansion cards in the first IBM PC. The original bus supported a data path only 8-bits wide. IBM subsequently developed a 16-bit version for its AT class computers. The 16-bit AT ISA bus supports both 8-and 16-bit cards. The 8-bit bus is commonly called the PC/XT bus, and the 16-bit bus is called the AT bus.

ISDN (Integrated Services Digital Network) A CCITT standard for a network that accommodates a variety of mixed digital-transmission services; the access channels are basic rate (144 Kbps) and primary rate (1.544 Mbps).

ISO (International Standards Organization) An organization that promotes the development of standards for computers.

Isochronous A form of data transmission in which individual characters are always separated by a whole number of bit-length intervals. The clocking information is encoded in the data stream. Contrast with asynchronous transmission, in which the characters may be separated by random-length intervals.

ISP Internet Service Provider. An organization offering and providing Internet services to the public and having its own computer servers to provide the services offered.

ISU (Integrated Service Unit) A single device that combines the functions of both a CSU and a DSU.

ITI (Interactive Terminal Interface) In packet-switched networks, a PAD that supports network access by asynchronous terminals.

ITU-TSSI (International Telegraphic Union Telecommunication Standards Sector) The replacement organization for the CCITT.

J

Jabber In LAN technology, continuous random data (garbage); normally used to describe the action of a station (whose circuitry or logic has failed) that locks up the network with its incessant transmission.

Jam signal A signal generated by a card to ensure that other cards know that a packet collision has taken place.

Java Java is a network-oriented programming language invented by Sun Microsystems that is specifically designed for writing programs that can be safely downloaded to your computer through the Internet and immediately run without fear of viruses or other harm to your computer.

JDK Java Development Kit. A software development package from Sun Microsystems that implements the basic set of tools needed to write, test, and debug Java applications and applets.

Jitter The slight movement of a transmission signal in time or phase; jitter can introduce errors and cause the loss of synchronization in high-speed synchronous communications.

JPEG (Joint Photographic Experts Group) A standard for lossy compression of graphic-image files.

Joystick A pointing device whose upright lever is used to manipulate a pointer on the computer screen.

Jukebox A data-storage device that can automatically switch between multiple removable disks or tapes.

Jumper A patch cable or wire used to establish a circuit, often temporarily, for testing or diagnostics.

K

k or K Kilo; (1) one thousand (for example, Kbps). (2) Kilobytes.

KB Kilobytes.

Kbps Kilobits per second; standard measurement of data rate and transmission capacity. One Kbps equals 1000 bits per second.

Keep-alive A process wherein one side of a connection will periodically send a packet to the remote side just to make sure the connection is still up.

Kernel The heart of an operating system, containing the basic scheduling and interrupt handling, but not the higher level services, such as the file system.

kHz Abbreviation for kilohertz, a unit of frequency equal to 1000 cycles per second.

Kilobyte A standard quantity measurement for disk and diskette storage and semiconductor circuit capacity: one kilobyte of memory equals 1024 bytes (8-bit characters) of computer memory.

L

LAN (Local Area Network) A data-communications system confined to a limited geographical area (up to 6 miles or about 10 kilometers) with moderate- to high-data rates (100 Kbps to 50 Mbps). The area served may consist of a single building, a cluster of buildings, or a campus-type of arrangement. The network uses some type of switching technology and does not use common-carrier circuits, although it may have gateways or bridges to other public or private networks.

LAP (Link Access Procedure) The data-link level protocol specified in the CCITT X.25 interface standard; LAP has been supplemented with LAPB and LAPD.

LAPB (Line Access Procedure, Balanced) In X.25 packet-switched networks, a link initialization procedure that establishes and maintains communications between the DTE and DCE; LAPB involves the T1 timer and N2 count parameters. All PDNs now support LAPB.

LAPD (Link Access Procedure-D) Link-level protocol devised for ISDN connections, differing from LAPB (LAP Balanced) in its framing sequence.

Laptop A small portable computer. Sometimes distinguished as larger than a notebook, but sometimes used as a synonym for notebook.

LAT (Local Area Transport) A protocol unique to Digital Equipment Corporation products, for virtual terminal access across an Ethernet network.

LATA (Local Access and Transport Area) A U.S. geographical subdivision used to define local (as opposed to long distance) telephone service.

Latency The time interval between when a network station seeks access to a transmission channel and when access is granted or received; equivalent to waiting time.

Layer In the OSI reference model, one of seven basic layers, referring to a collection of related network processing functions; one level of a hierarchy of functions.

LCD (Liquid Crystal Display) A display that uses liquid crystals to form patterns. Compare CRT.

LDM (Limited-Distance Modem) See line driver.

Leased line A telephone line reserved for the exclusive use of a leasing customer without interexchange switching arrangements. A leased line may be point-to-point or multipoint. (Also called a private line.)

LED (Light-Emitting Diode) A semiconductor device that accepts electrical signals and converts the energy to a light signal; with lasers, the main light source for optical-fiber transmission; used mainly with multimode fiber.

Level (1) Magnitude, as in signal level or power level. (2) Used as a synonym for layer.

Level 3, 4, 5 cable See Category 3 cable, Category 4 cable, and Category 5 cable.

LF (Line Feed) An ASCII or EBCDIC control character used to move to the next line on a printer or display terminal.

Lightpen See selector lightpen.

Line discipline Archaic term for communications protocol.

Line driver A DCE device that amplifies a data signal for transmission over cable for short distances beyond the RS-232 limit of 50 feet, and even up to several miles. The line driver also conditions the signal, by reshaping distorted pulses. Also called "limited-distance modem" (LDM) or "short-haul modem" (SHM).

Line turnaround The reversing of transmission direction when a half-duplex circuit is used.

Linearity The property of a transmission medium or of an item of equipment that allows it to carry signals without introducing distortion.

Link A communications circuit or transmission path connecting two points.

Link Layer Layer Two of the OSI reference model; also known as the Data-Link Layer (preferred usage).

LLC (Logical Link Control) A protocol developed by the IEEE-802 committee for data-link layer transmission control; the upper sublayer of the IEEE Layer 2 (OSI) protocol that complements the MAC protocol; IEEE standard 802.2; includes end-system addressing and error checking.

Loaded line A telephone line equipped with loading coils to add inductance in order to minimize amplitude distortion.

Local area network See LAN.

Local attachment In IBM environments, the connection of a peripheral device or control unit directly to a host channel.

Local bus A type of bus with a very short signal path between main processor and I/O processor(s). Used for functions like high-resolution video that demand high I/O speeds.

Local channel loopback A channel loopback test that forms the loop at the input (channel side) to the local multiplexor.

Local composite loopback A composite loopback test that forms the loop at the output (composite side) of the local multiplexor.

Local dataset See line driver.

Local digital loopback A digital loopback test that forms the loop at the DTE side (digital input) of the local modem.

Local exchange, local central office The exchange or central office in which the subscriber's lines terminate.

Local line, local loop A channel connecting the subscriber's equipment to the line-terminating equipment in the central office. Usually a metallic circuit (either 2-wire or 4-wire).

LocalTalk A 230-Kbps network from Apple computer used with Macintosh systems.

Logical channel, logical connection See virtual circuit.

Logical channel number In packet-switched networks, a number assigned when a virtual call is placed; up to 4095 independent logical channels may exist on a single link.

Loopback A diagnostic procedure used for transmission devices. A test message is sent to a device being tested. The message is then sent back to the originator and compared with the original transmission. Loopback testing may be performed with a locally attached device or conducted remotely over a communications circuit.

Loss Reduction in signal strength, expressed in decibels; also, attenuation; opposite of gain.

Lossy Describes compression methods that involve some loss of data.

LRC (Longitudinal Redundancy Check) An error detection method in which the BCC consists of bits calculated on the basis of odd or even parity for all the characters.

LU6.2 In SNA (Systems Network Architecture), a set of protocols that provides peer-to-peer communications between applications. See APPC.

M

M Mega; designation for one million (for example, Mbps).

m Milli; designation for one thousandth.

M bit The More Data mark in an X.25 packet that allows the DTE or DCE to indicate a sequence of more than one packet.

MAC (Media Access Control) A media-specific access control protocol within IEEE-802 specifications; currently includes variations for token-ring, token-bus, and CSMA/CD; the lower sublayer of the IEEE's link layer (OSI), which complements the Logical Link Control (LLC).

Macintosh A family of desktop computers manufactured by Apple. They were the first popular computers to use a GUI.

Macro A sophisticated command made up of a sequence of simple commands or primitives.

Magnetic medium Any data-storage medium and related technology, including disks, diskettes, and tapes, in which different patterns of magnetization are used to represent bit values.

Mainframe A large-scale computer system that can house comprehensive software and several peripherals and also handle multiple users, usually in the hundreds.

MAN (Municipal Area Network) An extended network or cluster of networks serving a city, an academic or business campus, or any site featuring several widely separated buildings.

Manchester encoding Digital encoding technique (specified for the IEEE-802.3 Ethernet baseband network standard) in which each bit period is divided into two complementary halves; a negative-to-positive voltage transition in the middle of the bit period designates a binary 1, while a positive-to-negative transition represents a 0. The encoding technique also allows for self-clocking: That is, the receiving device can recover the transmitted clock from the incoming data stream.

MAP (Manufacturing Automation Protocol) A suite of networking protocols that track the seven layers of the OSI model. Originated by General Motors.

Mapping In network operations, the logical association of one set of values, such as addresses on one network, with quantities or values of another set, such as devices on another network (name-address mapping, internetwork route mapping, protocol-to-protocol mapping).

Mark Presence of signal. In telegraph communication, a mark represents the closed condition or current flowing. A mark impulse is equivalent to a binary 1. Compare with space.

Master station (1) In multipoint circuits, the unit which controls/polls the nodes. (2) In point-to-point circuits, the unit that controls the slave station. (3) In LAN technology, the unit on a token-passing ring that allows recovery from error conditions, such as lost, busy, or duplicate tokens; a monitor station.

MAU (Multistation Access Unit) A wiring concentrator used in token-ring LANs.

MB See megabyte.

Mbps Millions of bits per second (bps).

MCGA (Multicolor Graphics Array) A video standard created specifically for the PS/2 Models 25 and 30. MCGA does not work with EGA software.

MDA (Monochrome Display Adapter) An old video standard for IBM PC, PC/XT, and AT computers or compatibles. MDA handled only text data.

Mean time between failures See MTBF.

Mean time to repair See MTTR.

Media Networking wiring such as 10BaseT and 100BaseT UTP cable, and coax cable.

Media filter A device that enables transition from one type of cable transmission scheme to another. Used specifically to run token-ring over UTP.

Media management The ability to manage and the process of managing different media (coaxial cable, twisted-pair cable, optical-fiber cable) used within the same network. Media management involves cable performance monitoring, cable break detection, planning for cable routes, etc.

Medium Anything used for the propagation or transmission of signals, usually in the form of electrons or modulated radio, light, or acoustic waves; examples include optical fiber, cable, wire, dielectric slab, water, air, or free space.

Megabit (Mb, Mbit) 1,000,000 bits; used in describing data transfer rates as a function of time (Mbps). See bit.

Megabyte (Mbyte, MB, Meg, or M) 1,048,576 bytes, equal to 1024 kilobytes; basic unit of measurement of mass storage. See byte.

Megahertz (MHz) A unit of frequency equal to 1,000,000 cycles per second.

Menu The list of available software functions for selection by the operator, displayed on the computer screen once a software program has been entered.

Message A complete transmission; used as a synonym for packet, but a message is often made up of several packets.

Metropolitan Service Area (MSA) A cellular service area covering a large city.

MHS (Message Handling System) The standard defined by the CCITT as X.400 and by the ISO as Message Oriented Text Interchange Standard (MOTIS).

MIB (Management Information Base) A small database that a device keeps about itself and from which it provides information to management software such as SNMP.

MIN (Mobile Identification Number) The ten-digit number given by the area code and telephone number.

Micro Channel A proprietary 16- or 32-bit bus developed by IBM for its PS/2 computers' internal expansion cards; also offered by others. Compare with EISA bus.

Microcomputer A desktop computer system. A computer that uses a microprocessor for its CPU. A personal computer (PC).

Microprocessor An electronic integrated circuit, typically a single-chip package, capable of receiving and executing coded instructions. IBM PCs and compatibles are based on Intel microprocessors—the 8088, 80286, 80386, 80486, Pentium, and so on. Apple Macintosh computers are based on the Motorola 68000 series. Other Apple and IBM computers use RISC-based processors (see RISC and PowerPC).

Microsoft LAN Manager A network operating system codeveloped by Microsoft and 3Com. LAN Manager offers a wide range of network management and control capabilities.

Microwaves Very short radio waves used for unbounded transmission. A microwave is any radio wave above 890 MHz per second.

MIDI (Musical Instrument Digital Interface) A standard interface for connecting computers to real-time audio signals as produced by digital instruments.

Midsplit A type of broadband cable system in which the available frequencies are split into two groups: one for transmission and one for reception. This requires a frequency converter.

MIF (Minimum Internetworking Functionality) A general principle within the

ISO standards that calls for minimum LAN station complexity when interconnecting with resources outside the LAN.

Minicomputer A small- or medium-scale computer, also called a "mini," usually operated with interactive dumb terminals. A mini can operate as a single powerful workstation or as a multiuser system.

Mini-MAP (Mini-Manufacturing Automation Protocol) A version of MAP consisting of only physical, link, and application layers, intended for lower-cost process-control networks. With Mini-MAP, a device with a token can request a response from an addressed device; unlike under standard MAP protocol, the addressed Mini-MAP device need not wait for the token to respond.

MIME Multipurpose Internet Mail Extensions. The standard for attaching nontext files to standard Internet mail messages.

MIPS (Million Instructions Per Second) A general comparison gauge of a computer's raw processing power.

Mirror A mirror site is an exact copy of material on a server at another location.

Mission-critical Necessary for the functioning of an organization; said of those system resources whose failure could seriously impair the ability of the system's owner to function.

MMJ (Modular Molded Jack) A modular connection with six wires, used in DEC systems. The locking tab is on the right-hand side, rather than in the center as in RJ-11 connectors.

MNP (Microcom Networking Protocol) Networking protocols that include standards for error correction and data compression.

MO (Magneto-Optical) disk, drive A storage technology that stores information on a magnetic medium at optical densities. MO disks are rewritable.

Mobile net A network composed at least partially of portable devices that communicate with other network devices using unbounded transmission techniques.

Mobile Telephone Switching Office (MTSO) Also called the switch, the MTSO is the link between the cellular phone and the landline phone system. The MTSO is the "brain" of the cellular system.

Modem (MOdulator-DEModulator) A device used to convert serial digital data from a transmitting terminal to an analog signal suitable for transmission over a telephone channel, or to reconvert the transmitted analog signal to serial digital data for acceptance by a receiving terminal.

Modem eliminator A device used to connect a local terminal and a computer port in lieu of the pair of modems that they would ordinarily connect to; allows DCE-to-DTE data and control signal connections otherwise not easily achieved by standard cables or connectors in a synchronous environment.

Modular A design technique that permits a device or system to be assembled from interchangeable components; the system or device can be expanded or modified simply by adding another module.

Modulation Varying the characteristics of a wave-frequency, amplitude, or phase to match those of another wave. This is done by varying one or more of the signal's basic characteristics: frequency, amplitude, or phase.

Modulo A term used to express the maximum number of states for a counter; this term is used to describe several packet-switched network parameters, such as packet number (usually set to modulo 8, counted from 0 to 7). When the maximum count is exceeded, the counter is reset to 0.

Monitor station On ring networks, the unit responsible for removing damaged packets and making sure that the ring is intact.

Motherboard The main board in a computer that holds the CPU chip, the ROM, the RAM, and sometimes coprocessors.

Mount The making of a portion of a file server's file system accessible to other hosts.

Mouse A pointing device that is moved on a flat surface to manipulate a pointer on the computer screen. Most commonly used with a GUI.

MPC (Multimedia PC Council) standards Industry standards for the interoperability of multimedia components, such as CD-ROM drives and video and sound cards.

MPEG (Motion Picture Experts Group) A standard for lossy compression of full-motion video.

MPR standards Standards for radiation from monitors, set by the Swedish National Board for Measurement and Testing (abbreviated "MPR" in Swedish).

MSAU See MAU.

MS-DOS (Microsoft Disk Operating System) A microcomputer operating system developed for the IBM PC and, hence, a de facto industry standard.

MSI (Medium-Scale Integration) A term used to describe a multifunction semiconductor device with a medium density (up to 100 circuits) of electronic circuitry contained on a single silicon chip.

MSL (Maximum Segment Lifetime) The maximum amount of time an IP datagram can exist within the Internet. The default MSL is set at two minutes.

MSS (Maximum Segment Size) The maximum size of the TCP segments to be exchanged by TCP peers. TCP modules who want to accommodate system concerns for buffering or subnet constraints on maximum datagram size can negotiate the default of 65,536 to a smaller value.

MTBF (Mean Time Between Failures) A stated or published period of time for which a user may expect a device to operate before a failure occurs.

MTTR (Mean Time To Repair) The average time between failure of a device or system and its repair.

MTU (Maximum Transmission Unit) A subnetwork dependent value that indicates the largest datagram that a subnetwork can handle.

Multicast bit A bit in the Ethernet addressing structure used to indicate a broadcast message (a message to be sent to all stations).

Multidrop line In 3X systems, a single communications circuit that interconnects many stations, each of which contains terminal devices.

Multimedia Technology that makes use of (1) several different media—video, audio, text, and so on—in one presentation or application. (2) In LAN applications, the use of mixed types of cable such as thin coax, UTP, and fiber, in LAN applications.

Multimode fiber An optical fiber designed to carry multiple signals, distinguished by frequency or phase, at the same time; compare with single-mode fiber.

Multiplatform Able to run on more than one type of computer.

Multiple routing The process of sending a message to more than one recipient, usually when all destinations are specified in the header of the message.

Multiplexor A device used for division of a transmission facility into two or more subchannels, either by splitting the frequency band into narrower bands (see frequency-division multiplexor) or by allotting a common channel to several different transmitting devices one at a time (see time-division multiplexor).

Multipoint line A single communications line or circuit connecting several stations, thus supporting terminals in several different locations. Compare with point-to-point line.

Multitasking The concurrent execution of two or more tasks or applications by a computer.

Multithreaded Capable of executing more than one thread simultaneously; said of some multitasking systems.

Mux Short for multiplexor.

N

N connector A threaded connector for coax; N is named after Paul Neill. See also BNC and TNC.

N2 count In X.25 packet-switched networks, a count for the allowable number of retransmissions.

NAK (Negative acknowledgement) (1) In BSC communications protocol, a control character used to indicate that the previous transmission block was in error and the receiver is ready to accept retransmission of the erroneous transmission block. Contrast with ACK. (2) In multipoint systems, the not-ready reply to a poll.

NAM (Number Assignment Module) The electronic components in a wireless system that stores the telephone number and ESN of a phone.

NAMPS (Narrowband Advanced Mobile Phone System) The combining of cellular voice processing and digital signaling.

Nanosecond One billionth of a second.

Narrowband PCS The newest generation of paging network with services including two-way paging and the acknowledgement of messages being received.

NBS (National Bureau of Standards) A U.S. government agency that produces Federal Information Processing Standards (FIPS) for all U.S. government agencies except the Department of Defense. Membership includes other U.S. government agencies and network users.

NCC (Network Control Center) Any centralized network diagnostic and management station or site, such as that of a packet-switching network.

NDIS (Network Driver Interface Specification) A standard established by Microsoft for writing hardware-independent drivers.

NEC (National Electrical Code) A collection of standards for electrical safety established by the National Fire Protection Association.

Netbeui A set of network-transport protocols developed by Microsoft.

NetBIOS (Network Basic Input/Output System) Software developed by IBM; provides the interface between a PC's operating system, the I/O bus, and the network; a de facto network standard.

NetView IBM mainframe network-management software that integrates the functions of several earlier IBM network-management products.

NetWare A LAN operating system from Novell.

Network An interconnection of computer systems, terminals, or data-communications facilities.

Network access control In LAN technology, circuits that dictate when individual workstations may transmit messages.

Network adapter A piece of hardware that is used to connect a computer to a network. A network adapter may be a PCI card (see network interface card) or it may connect to a computer externally, via a Universal Serial Bus or a parallel port.

Network address Host number, network number, and socket number of an entity, constituting its address on the Internet.

Network architecture A set of design principles, including the organization of functions and the description of data formats and procedures, used as the basis for the design and implementation of a network.

Network interface controller (Network Interface card, or NIC) Electronic circuitry that connects a workstation to a network; usually in the form of a card that fits into one of the expansion slots inside a personal computer. It works with the network software and computer operating system to transmit and receive messages on the network.

Network Layer Layer 3 in the OSI model; the logical network entity that services the transport layer; responsible for ensuring that data passed to it from the transport layer is routed and delivered through the network.

Network operating system (NOS) The software used to connect devices, share resources, transfer files, and perform network activity. Usually, there are two parts to a network operating system: server and workstation (requester).

Network topology The physical and logical relationship of nodes in a network; the schematic arrangement of the links and nodes of a network; networks typically have a star, ring, tree, or bus topology, or some hybrid combination.

Network utilities Programs that handle routine procedures such as troubleshooting.

Neural network A set of linked microprocessors that can form associations and learn like the neurons in a human brain.

Neutral current loop Same as single-current version of current loop.

NEXT (Near-end cross talk) A measurement of interference between the conductors of a cable when used in a LAN application.

NFS (Network File Server) An extension of TCP/IP that allows files on remote nodes of a network to appear locally connected.

Nibble, nybble The first or last half of an 8-bit byte (4 bits).

NNTP (Network News Transport Protocol) An extension of the TCP/IP protocol that provides a network news transport service.

Node A termination point for two or more communications links. The node can serve as the control location for forwarding data among the elements of a network or multiple networks, as well as performing other networking and, in some cases, local processing functions. A node is usually connected to the backbone network and serves end points and/or other nodes.

Noise Random electrical signals, generated by circuit components or by natural disturbances, that corrupt the data by introducing errors.

Nondedicated A term used to describe a file server that can be used simultaneously as a workstation. File server functions run in the background when it is used as a work station.

Nonpersistent In LAN technology, a term used to describe a CSMA LAN in which the stations involved in a collision do not try to retransmit immediately, even if the network is quiet.

Nonproprietary A specification or implementation that was not created by a single private organization.

Nonseed router A nonseed router waits to receive routing information (the routing maintenance table) from other routers on the network before it begins routing packets.

Nonvolatile storage A storage medium whose contents are not lost when power is removed.

Notebook A small portable computer, usually about the size of a standard letter-size notebook.

NRZ A binary coding technique that does not return to a neutral state after each bit is transmitted.

NTN (Network Terminal Number) Number identifying the logical location of a DTE connected to a network. The NTN may contain a subaddress used by the DTE rather than by the network to identify equipment or circuits attached to it. It can be up to 10-digits long.

NTSC (National Television System Committee) A standard for color broadcasting developed in the 1950s for use mainly in North America and some of South America.

NuBus A high-speed bus used in the Macintosh family of computers. NuBus is structured so that you can put a card into any slot you want and there is no conflict over priority between cards.

NUI (Network User Identification) In X.25 packet-switched networks, a combination of the network user's address and the corresponding password; replaces the NTN in newer networks.

Null character A character (with all bits set to mark) used to allow time for a printer's mechanical actions, such as return of carriage and form feeding, so that the printer will be ready to print the next data character.

Null modem See cross-pinned cable.

NVRAM (nonvolatile RAM) RAM whose contents are not lost when the power is turned off.

O

Object Any defined system resource—data, software, or hardware. In systems that treat their resources as objects, different applications can view or manipulate an object, but the object still maintains a consistent identity.

Object-oriented Treating resources as objects; see object.

Object-oriented programming (OOP) A type of computer programming based in the creation and use of a set of objects and the development of relationships among the objects.

OCR (Optical Character Recognition) The computer recognition of printed characters. Data can be entered by scanning a printed page.

Octal A digital system with eight states, 0 through 7.

Octet In packet-switched networks, a grouping of 8 bits.

ODI The Novell open data-link interface standard for hardware-independent drivers.

Off hook In a telephone environment, activated; by extension, a modem automatically answering a call on the dial network is said to go "off hook."

Off-line The condition in which a user, terminal, or other device is not connected to a computer or is not actively transmitting via a network; opposite of on-line.

Office automation A term used to describe the process of making wide use of data-processing and data-communications technology—electronic mail, word processing, file and peripheral sharing, and electronic publishing—in the office environment, usually involving the installation of LAN.

On hook In a telephone environment, deactivated; by extension, a modem not in use is said to be "on hook." Contrast with off hook.

On-line The condition in which a user, terminal, or other device is actively connected with the facilities of a communications network or computer; opposite of off-line.

On-line computer A computer used for on-line processing.

On-line processing A method of processing data in which data is input directly from its point of origin and output directly to its point of use.

Open architecture An architecture that is compatible with hardware and software from any of many vendors. Contrast with closed architecture.

Open-endedness The quality of system design that permits future extensions and modifications with little, if any, backward impact.

Operating system The software of a computer that controls the execution of programs, typically handling the functions of input/output control, resource scheduling, and data management. OS/2, MS-DOS, and VM/370 are examples of operating systems.

Optical disk A very high density information storage medium that uses light to read and write digital information. (See also WORM.)

Opto-isolator A small device that converts electrical signals to light signals, transmits the light signals across a short gap, and converts them back to electrical signals on the other side. Because no electricity flows across the gap, an opto-isolator gives your circuit immunity to electromagnetic interference and ground-loop problems.

OS/2 A multitasking operating system developed by IBM and Microsoft for use with Intel 80286 and later microprocessors.

OSINET A test network, sponsored by the National Bureau of Standards (NBS), designed to provide vendors of products based on the OSI model, a forum for interoperability testing.

OSI (Open Systems Interconnection) OSI reference model. A model for networks developed by the International Standards Organization, dividing the network functions into seven connected layers. Each layer builds on the services provided by those under it.

OSI Model Open Systems Interconnection Reference Model adopted by the ISO and CCITT.

OTMF (Open Tape Media Format) A LAN-backup standard developed by Legato Systems, Inc.

Overhead In communications, all information, such as control, routing, and error-checking characters, that is in addition to user-transmitted data; includes information that carries network status or operational instructions, network routing information, and retransmissions of user data messages that are received in error.

Oversampling A TDM technique where each bit from each channel is sampled more than once.

Overscan A video-display effect in which the image is enlarged so that its edges are off the screen.

Overspeed A condition in which the transmitting device runs slightly faster than the data presented for transmission.

P

PABX (Private Automatic Branch Exchange) A user-owned, automatic telephone exchange that accommodates the transmission of calls to and from the public telephone network.

Packet A sequence of data, with associated control information, that is switched and transmitted as a whole; refers mainly to the field structure and format defined with the CCITT X.25 recommendation.

Packet filter A feature of a bridge that compares each packet received with specifications set by the network administrator. If the packet matches the specifications, the bridge can either forward or reject it. Packet filters let the administrator limit protocol specific traffic to one network segment, isolate electronic mail domains, and perform many other traffic control features.

Packet header In packet-switched networks, the first three octets of an X.25 packet.

Packet type identifier In X.25 packet-switched networks, the third octet in the packet header, which identifies the packet's function and, if applicable, its sequence number.

Packet-switched network A data-communications network that transmits packets. Packets from different sources are interleaved and sent to their destination over virtual circuits. The term includes PDNs and cable-based LANs.

PAD (Packet Access Device) An interface between a terminal or computer and a packet-switching network.

PAL (Phase Alternation by Line) A standard for color broadcasting used mostly in Europe. This system avoids the color distortion that appears in the NTSC systems.

Parallel processing Concurrent or simultaneous execution of two or more processes, or programs, within the same processor, as contrasted with serial or sequential processing.

Parallel transmission Transmission mode in which a number of bits of information are sent simultaneously over separate lines (for example, eight bits over eight lines); usually unidirectional. Compare with serial transmission.

Parity In ASCII, a check of the total number of 1 bits (as opposed to 0's) in a character's binary representation. A final eighth bit is set so that the count, when transmitted, is always even or odd. This even or odd state can easily be checked at the receiving end; an incorrect parity bit can help reveal errors in the transmission.

Parity bit A bit that is set at 0 or 1 in a character to ensure that the total number of one bits in the data field is even or odd, as desired.

Parity checking An error-detection technique in which character bit patterns are forced into parity, so that the total number of one bits is always odd or always even. This is accomplished by the addition of a one or zero bit to each byte, as the byte is transmitted; at the other end of the transmission, the receiving device verifies the parity (odd or even) and the accuracy of the transmission.

Partition (1) To divide a disk or tape drive into independent volumes. (2) To divide a network into independent segments (often as an automatic response to a network error).

Pass-through The ability to gain access to one network element through another.

Passive device In current-loop applications, a device that must draw its current from connected equipment.

Passive head end A device that connects the two broadband cables of a duplicable system. It does not provide frequency translation.

Passive hub A device on a token-ring network that receives signals from one workstation for forwarding to a destination workstation.

Passive link A logical communications circuit or call that maintains a constant connection. A call-setup and call-clearing procedure is not necessary.

Passive matrix A type of LCD that has transistors for each row and column of pixels. Produces paler, dimmer color than active-matrix displays.

Password A word or group of characters a user has to enter to gain access to a computer or to files.

Patch panel A system of terminal blocks, patch cords, and backboards that let you cross-connect fields.

PBX (Private Branch Exchange) A manual, user-owned telephone exchange.

PCI (Peripheral Component Interconnect) bus A local-bus standard created by Intel.

PCL (Printer Control Language) A control language for Hewlett-Packard and compatible laser printers.

PCM (Pulse Code Modulation) A modulation technique used to convert analog voice signals into digital form. Used for voice multiplexing on T1 circuits.

PCMCIA (Personal Computer Memory Card International Association) Defines the format and interface for a credit card six device for use with portable or laptop computers. There are 3 types. Type 1, the thinnest at 3.3 mm, is normally used for memory type enhancements. Type 2, which is 5 mm thick, is for modem and LAN enhancements. Type 3, 10.5 mm thick, is for mass storage I/O.

PCS (Personal Communications System) A wireless data-communication and message network, such as the ARDIS PersonalMessaging service, that uses small terminals that can be carried on one's person.

PCS (Personal Conferencing Specification) A standard for desktop videoconferencing system developed by Intel. Uses the Indeo algorithm.

PDN (Packet Data Network) A network established and operated by a PTT, common carrier, or private operating company for the specific purpose of providing data communications services to the public. May be a packet-switched network or a digital network such as DDS.

Peer to Peer A network computing system in which all computers are treated as equals on the network. Individual computers may share hard drives, CD-ROM drives, and other storage devices with the other computers on the network.

Pentium See Microprocessor.

Peripheral Any separate device, such as a printer or modem, that connects to and is controlled by a computer.

Personal computer A microcomputer designed to be used by a single user.

PGA (Professional Graphics Adapter) A video standard for IBM PC/XT and AT or compatibles. It offers better resolution than EGA, but has been largely superseded by VGA.

Phase modulation One of three ways (see AM and FM) of modifying a sine wave signal to make it carry information. The sine wave or carrier has its phase changed in accordance with the information to be transmitted.

Photo CD A format developed by Kodak for storing and retrieving still photographs on CD.

Physical Layer Within the OSI model, the lowest level (Layer 1) of network processing, below the link layer; concerned with the electrical, mechanical, and handshaking procedures over the interface that connects a device to a transmission medium; an electrical interface.

Pincushioning A video-display defect in which the image appears to bend inward toward the middle of the screen.

Pinout A list or diagram that shows how the individual wires in a cable or connector are used.

Pipe A communications process within the operating system that acts as an interface between a computer's devices (keyboard, disk drives, memory, and so on) and an application program. A pipe simplifies the development of application programs by "buffering" a program from the intricacies of the hardware or the software that controls the hardware; the application developer writes code to a single pipe, not to several individual devices. A pipe is also used for program-to-program communications.

Pixel Picture element; the smallest unit of a graphics or video display. Its characteristics (color and intensity) are coded into an electrical signal for transmission.

Pixel array A two-dimensional array of samples that define the color everywhere in a rectangular region.

Plenum rated In cabling, approved by the NEC for installation in air spaces ("plenums"). Plenum-rated cable does not give off toxic fumes when it burns. Compare PVC.

Point-to-point line A single communications circuit connecting just two locations. Compare with multipoint.

Polarity Any condition in which there are two opposing voltage levels or changes, such as positive and negative.

Polling A communications technique that determines when a terminal is ready to send data. The computer continually interrogates all of its attached terminals in a round-robin sequence. A terminal acknowledges the poll when it has data to send.

Port A point of access into a computer, a network, or other electronic system; the physical or electrical interface through which one gains access; the interface between a process and a communications or transmission facility.

Port concentrator, port concentration A device that allows several terminals to share a single computer port; a concentrator link in which the port concentrator simplifies the software demultiplexing used in lieu of the demultiplexing normally performed by the computer-site concentrator.

Port selector See data PABX.

PostScript A programming language developed by Adobe, specifically designed to handle text and graphics and their placement on a page. Used primarily in laser printers and imagesetters. See also QuickDraw and PCL.

POTS (Plain Old Telephone Service) The basic service provided by the public telephone network, without any added facilities such as conditioning.

Power conditioning The process of maintaining uniform voltage on a power line.

Powerline networking This is an emerging technology which will allow a home network to utilize already existing AC wiring as the cable connection between computers and other devices in the network. Powerline networking will enable easy home automation, merging your home security system, lighting, and other environmental controls with your home network.

PowerPC (1) A series of microprocessor chips jointly produced by Apple, IBM, and Motorola. (2) Computers designed around these chips.

PPP (Point-to-Point Protocol) A protocol that allows a computer to use a regular telephone line and a modem to make TCP/IP connections.

Presentation Layer In the OSI model, the layer of processing that provides services to the Application Layer, allowing it to interpret the data exchanged, as well as to structure data messages for transmission in a specific display and control format.

PRI (Primary Rate Interface) An ISDN service that operates at a rate of 1.544 Mbps. This service provides 23 B channels at 64 Kbps and 1 D channel at 64 Kbps. The system uses 8 Kbps for framing bits. In Europe the service provides 30 B channels at 64 Kbps and 1 D channel at 64 Kbps, and uses 64 Kbps for framing bits. The service allows for simultaneous transmission over all B channels and the D channel.

Primitives Basic units of machine instruction.

Print driver A driver that manages an application's communication with a printer; see driver.

Print server A workstation with an attached printer, a spooler, and software to process and manage printing tasks for all users on the network.

Print spooler A program or device that acts as a buffer for data going to a printer; see buffer.

Printer-sharer A device that permits multiple users to print without the inconvenience of a mechanical switch.

Private line See leased line.

Profile In packet-switched networks, a set of parameter values, such as for a terminal, that can be defined and stored; the parameters can then be recalled and used as a group by identifying and selecting the appropriate profile.

Program A set of instructions for a computer; see software and firmware.

PROM (Programmable Read-Only Memory) A nonvolatile memory chip that allows a program to reside permanently in a piece of hardware; compare with volatile memory.

Propagation delay The time it takes a signal composed of electromagnetic energy to travel from one point to another over a transmission channel; usually most noticeable in communicating with satellites; normally, the speed-of-light delay.

Protocol The procedures used to control the orderly exchange of information between stations on a data link or on a data-communications network or system. Protocols specify standards in three areas: the code set, usually ASCII or EBCDIC; the transmission mode, usually asynchronous or synchronous; and the nondata exchanges of information by which the two devices establish contact and control, detect failures or errors, and initiate corrective action.

Protocol analyzer A diagnostic tool for displaying and analyzing communications protocols. With this tool, a user can test the performance of network data to make sure that the network and its hardware are operating within the network's specifications. LAN managers use protocol analyzers to plan network upgrades and expansions, and to perform network maintenance and troubleshooting tests.

Protocol converter A device that translates from one communications protocol into another, such as IBM SNA/SDLC to ASCII; compare with gateway.

Protocol layering A technique of structuring protocols so the protocol at a layer uses the protocol at the next lower layer without knowing the details of the operation.

PS/2 (Personal System/2) A family of IBM personal computers that succeed the AT line.

PSE (Packet Switch Exchange) A unit that performs packet switching in a network.

PSK (Phase Shift Keying) A phase modulation technique in which phase shifts represent signaling elements. Compare also with FSK.

PSTN (Public Switched Telephone Network) The telephone system over which calls may be dialed.

PTT (Post, Telephone, and Telegraph Authority) The government agency that functions as the communications common carrier and administrator in many areas of the world.

Public network A network operated by common carriers or telecommunications administrations for the provision of circuit-switched, packet-switched, and leased-line circuits to the public.

Public switched network Any switching communications system, such as Telex, TWX, or public telephone networks, that provides circuit switching to many customers.

Pulse dialing Older form of telephone dialing, utilizing breaks in DC current to indicate the number being dialed.

Pulse trap A device that monitors any RS-232 lead for changes in logic levels (high to low or low to high).

Punchdown block Wiring outlet/terminal into which unterminated cable is "punched down."

PVC (Permanent Virtual Circuit) In a packet-switched network, a fixed virtual circuit between two users; no call-setup or call-clearing procedures are necessary. The PDN equivalent of a leased line. Contrast with SVC.

PVC (Polyvinyl chloride) The type of plastic from which cable jackets and wire insulation are usually made. Produces toxic fumes when it burns. Compare with plenum rated.

Q

Q bit (Qualifier bit) In X.25 packet-switched networks, Bit 8 in first octet of packet header; it is used to indicate whether the packet contains control information.

QAM (Quadrature Amplitude Modulation) A modulation technique that combines phase modulation and AM techniques to increase the number of bits per baud.

Queue Any group of items, such as computer jobs or messages, waiting for service.

Queuing Sequencing of batch data sessions.

QuickDraw Programming routines that allow a Macintosh computer to display graphics on a screen. QuickDraw is also used for outputting text and images to printers not compatible with PostScript.

QuickTime A dynamic-data format developed by Apple to be used for animation. QuickTime files can be used in documents created by other applications. For instance, a QuickTime video clip can be pasted into a word-processing document. QuickTime VR is the new Apple standard for Virtual Reality (see Virtual Reality).

R

Rackmount Designed to be installed in a cabinet; usually 19" wide.

RAID (Redundant Array of Inexpensive Disks) A method of storing data on multiple hard-disk drives, for faster access, greater reliability, or both. There are six officially defined "levels," each designed for a specific kind of application.

RAM (Random-Access Memory) Semiconductor read/write volatile memory. Data stored is lost if power is turned off. It is where data is stored while being processed by the microprocessor.

Raster The rectangular pattern of scanning lines that produces the picture, or the illuminated face of the monitor without the picture.

RBOC (Regional Bell Operating Company) One of the regional Bell telephone companies that were created when AT&T was dismantled in 1984. Each RBOC may include several BOCs.

RD (Received Data) An RS-232 data signal (received by DTE from DCE on pin 3).

Real time An operating mode that allows immediate interaction with data as it is created, as in a process-control system or computer-aided design system.

Reboot To reset the computer. Rebooting clears RAM, and starts the booting process from the beginning; see booting.

Red alarm An alarm condition in T1 that indicates an out-of-frame condition for an incoming DS1 signal for 2 to 3 seconds. Also referred to as "loss of signal" alarm.

Redundancy (1) In data transmission, the portion of a message's gross information content that can be eliminated without losing essential information. (2) The technique of building in extra identical components, to be used as backups in case the primary components fail.

Refresh rate How quickly your screen redraws itself, usually measured in Hertz.

Remote access The ability of a computer in one location to reach a device that is some distance away, perhaps at another site.

Remote analog loopback An analog loopback test that forms the loop at the line side (analog output) of the remote modem.

Remote channel loopback A channel loopback test that forms the loop at the input (channel side) of the remote multiplexor.

Remote composite loopback A composite loopback test that forms the loop at the output (composite side) of the remote multiplexor.

Remote digital loopback A digital loopback test that forms the loop at the DTE side (digital input) of the remote modem.

Repeater In digital transmission, equipment that receives a pulse train, amplifies it, retimes it, and then reconstructs the signal for retransmission; in fiber optics, a device that decodes a low-power light signal, converts it to electrical energy, and then retransmits it via an LED or laser source. Also called a "regenerative repeater."

Resolution A measurement of the amount of detail in a graphic image. When referring to video, resolution is usually expressed in pixels; for example, standard VGA has a resolution of 640 × 480.

Resource class In LAN technology, a collection of computers or computer ports that offer similar facilities, such as the same application program; each can be identified by a symbolic name.

Response time (1) For interactive sessions, the elapsed time between the end of an inquiry and the beginning of a response. (2) For surge protectors, how fast the unit responds to block power surges; also called "clamp time."

Retransmissive star Also known as passive fiber star. In fiber-optic transmission, a passive component that permits the light signal on an input fiber to be retransmitted on multiple output fibers; formed by heating together a bundle of fibers to near the melting point; used mainly in fiber-based LANs.

Reverse channel (a) An answer back channel provided during half-duplex operation. It allows the receiving modem to send low speed acknowledgements to the transmitting modem without breaking the half-duplex mode. This is also used to arrange the turnaround between modems so that one ceases transmitting and the other can begin. (b) Channel on which a mobile telephone set communicates with a cell site.

RF (Radio-Frequency) Modulation The electromagnetic format in which broadcast and cable TV signals are transmitted.

RG58 A standard for Thin Ethernet coaxial cable; it specifies 50-ohm impedance.

RG62 A standard for ARCNET coaxial cable; it specifies 93-ohm impedance.

RGB Red, Green, Blue; a video standard in which the color signals for red, green, and blue are carried on separate lines, then combined to form a color video picture. Horizontal and vertical sync are imposed on one of the colors, usually green.

RGBHV Red, Green, Blue, Horizontal, Vertical; a video standard similar to RGB except that horizontal and vertical sync signals are each carried on a separate line.

RGBS Red, Green, Blue, Sync; a video standard similar to RGB except that horizontal and vertical sync signals are carried separately on one line.

RI (Ring Indicator) An RS-232 modem interface signal (sent from the modem to the DTE on pin 22) which indicates that an incoming call is present.

Ring topology A basic networking topology where all nodes are connected in a circle, with no terminated ends on the cable.

RIP (Routing Information Protocol) A protocol that is used by routers on a LAN to exchange routing information among participating routers.

RISC (Reduced Instruction Set Computing) Internal computing architecture in which processor instructions are pared down so that most can be performed in a single processor cycle, theoretically improving computing efficiency.

RJ-11 connector The standard telephone line connector that you plug into a jack in the wall to receive phone service.

RJ-45 connector The standard connector (plug) for *Ethernet* cables. Slightly wider than RJ-11.

RJE (Remote Job Entry) Submission of batch processing jobs through an input device (such as an IBM modem 3780) that has access to a computer through a data link.

RLSD (Received Line Signal Detector) See CD; also called "DCD."

Roaming Whenever a cellular phone is operating on a nonhome system, it is roaming. You are charged for roaming when you place or receive a call outside of your home area.

RO (Read Only) A teleprinter receiver without a transmitter.

ROM (Read-Only Memory) Memory chips that store data or software; firmware.

Rotary An arrangement of a group of lines, such as telephone or data PABX lines, that are identified by a single symbolic name or number; upon request, connection is made to the first free line.

Round-robin retraining A method of training in which the receiving modem asks for a training pattern by sending a training pattern.

Router A network device that examines the network addresses within a given protocol, determines the most efficient pathway to the destination, and routes the data accordingly. Operates at the Network Layer of the OSI model.

Routing The process of selecting the correct circuit path for a message.

RS-422, RS-423 EIA serial transmission standard that extends transmission speeds and distances beyond those of RS-232. RS-422 is a balanced system with a high level of noise immunity. RS-423 is an unbalanced version of RS-422.

RS-499 RS-449 specifies the pinning for RS-422 and RS-423 when a DB37 or DB9 connector is used.

RS-485 EIA serial interface standard for multipoint lines. Unlike its predecessors (RS-232, RS-422, etc.), RS-485 uses a tri-state driver. The third state the other interfaces lack is an off condition that allows equipment to communicate over shared wiring while another device sits quietly. It is tolerant to electrical noise, which makes it ideal for industrial users like manufacturing facilities and airports. And RS-485 supports long cabling distances up to 4000 feet (1219.2 m) at 100 Kbps.

RS-530 Specifies the pinning for balanced interfaces such as RS-422, when a DB25 connector is used. Designed to replace RS-449.

RSA A de facto standard for untappable public-key encryption developed by RSA Data Security Inc.

RTS (Ready To Send) An RS-232 signaling lead generated by the DTE. In a switched circuit (half duplex) the DTE raises RTS in a bid for line control. It then receives a verify signal (CTS) from the DCE when it is Clear To Send (CTS).

Run time For a BPS or UPS, the length of time that the unit can supply power from the batteries during AC power loss.

Rural Service Area (RSA) A cellular service area in less populated regions.

S

SAA (System Application Architecture) A set of standards developed by IBM that provides identical user interfaces for applications running on PCs, minicomputers, and mainframes.

Sag A persistent voltage shortfall. The undervoltage is not enough to cause a brownout but can damage sensitive equipment.

Scan converter A device that converts computer video to a TV-video format.

Scan line One of the many horizontal lines that make up the picture on a video screen.

Scanner A device that converts images or text on paper into data that can be manipulated by a computer.

SCSI (Small Computer Standard Interface; pronounced "scuzzy") A specification of mechanical, electrical, and functional standards for connecting small computers with intelligent peripherals such as hard disks, printers, and optical disks; each device on the bus has a unique identification.

ScTP (Screened Twisted-Pair) A type of 100-ohm cable, composed of 4 pairs of copper wire surrounded by a thin aluminum-foil shield, that eliminates EMI.

SDLC (Synchronous Data Link Control) The primary data-link protocol used in IBM SNA networks. It is a bit-oriented, synchronous protocol.

SECAM (Sequential Couleur à Mémoire) The line sequential color system used for television transmissions in France, some Middle East countries, eastern Europe, and Russia.

Sector A segment of a magnetic disk's track, typically occupied by one block of data.

Seed router A seed router supplies routing information (network numbers and ranges, zone names, etc.) to the network.

Segment A segment consists of one or more nodes. Segments are connected to subnets by hubs and repeaters.

Selector channel An input/output (I/O) channel designed to operate with only one I/O device at a time. Once the I/O device is selected, complete records are transferred in one-byte intervals. Compare with block-multiplexor channel.

Selector lightpen An instrument that can be attached to a display station as a special feature. When pointed at a portion of the display station's image on the screen and then activated, the selector lightpen identifies that portion of the displayed screen for subsequent processing.

Semaphore In LAN technology, a message transmitted when a user accesses a file or disk; it will deny other users access to the file or disk.

Sequencing Breaking down a data message for transmission; each "piece" is assigned a sequential number for reassembly.

Serial transmission The transmission of a character or byte of data one bit at a time. Contrast with parallel transmission.

Server A computer or processor that holds applications, files, or memory shared by users on a network.

Server-based network A network in which all client computers use a dedicated central server computer for network functions such as storage, security, and other resources.

Session A connection between two stations that allows them to communicate; the time period during which a user engages in a dialogue with an interactive computer; in the IBM SNA, the logical connection between two network-addressable units.

Session Layer Layer Five of the OSI reference model; provides protocols for assembling physical messages into logical messages.

Shared access In LAN technology, an access method that allows many stations to use the same (shared) transmission medium; contended access and explicit access are two kinds of shared access methods. Contrast with discrete access.

Shared data Files on the server that can be shared across the network.

Shared Ethernet Standard 10BaseT Ethernet method of sending data to a hub, which then rebroadcasts this data to every port on the network until it reaches its destination.

Shared resources Files, printers, peripherals and other services that can be shared across the network.

Shareware Software designed to be freely shared or paid for on the honor system.

Shielding The process of protecting a cable with a grounded metal surrounding, so that electrical signals from outside the cable cannot interfere with transmission inside the cable. Shielding lessens the chance that the information moving along the cable will interfere with adjacent cables.

Short An abnormal condition that occurs when there is an unwanted electrical connection between two wires. It results in a flow of excess current.

Short-haul modem See line driver.

SIDF (System Independent Data Format) A LAN-backup standard developed by Novell Inc.

Signal bounce When a bus topology network cable has not been properly terminated at each end of the cable, the signal from the network will travel from one end of the cable to the other. This is called signal bounce.

Signal-to-Cross talk Ratio (SCR) The ratio of the value of the signal to that of the cross talk. Calculated only for twisted pair, it is the difference between NEXT and attenuation on a cable and will vary on the same network.

Signal-to-noise ratio The relative strength of the desired signal compared to the strength of unwanted noise; usually measured in dB.

Sign-on character The first character sent on an ABR circuit; used to determine the data rate.

Simplex transmission Transmission in only one direction.

Single-mode fiber An optical fiber that supports only one mode of light propagation above the cutoff wavelength. Core diameters are usually between 5 and 10 microns, and claddings are usually ten times the core diameter. These fibers have a potential bandwidth of 50 to 100 GHz per kilometer.

Skipjack The EES data-encryption algorithm.

Slave station In point-to-point circuits, the unit controlled by the master station.

SLIP (Serial Line Internet Protocol) An older protocol for IP connections over telephone lines, RS-232 cables, or other serial lines. The Point-to-Point Protocol (PPP) is replacing it.

SMA A threaded screw-style fiber-optic connector.

Smart hub A type of twisted-pair concentrator used in either Ethernet or ARCNET networks. A smart hub has built-in network management facilities, usually in the form of programmed firmware that allow a network manager to control and plan network configuration, as well as to monitor network performance.

SMDS (Switched Multimegabit Digital Service) A service developed by Bell Communications Research (Bellcore) to address the market for high-speed public networking. SMDS, based on industry standards, provides for interoperability between the telecommunication carrier's network, and customer premise equipment from 1.544 to 45 Mbps.

SMR (Specialized Mobile Radio) An analog radio service using antennas to provide wireless voice communication. Alternative to cellular. See also ESMR.

SMRT (Single Message-unit Rate Timing) A U.S. telephone company tariff under which local service is measured and calls are timed in increments of 5 minutes or less—with a single message—unit charge applied to each increment.

SMTP (Simple Mail Transport Protocol) TCP/IP's electronic-mail subprotocol.

SMS (Short Messaging Service) A PCS phone feature that permits users to receive and transmit short text messages.

SNA (Systems Network Architecture) The IBM total description of the logical structure, formats, protocols, and operational sequences for transmitting information units between IBM software and hardware devices. Data communications system functions are separated into three discrete areas: the application layer, the function management layer, and the transmission subsystem layer. The structure of SNA allows the ultimate origins and destinations of information—that is, the end users—to be independent of, and unaffected by, the specific data communications system services and facilities used for information exchange.

SNMP (Simple Network Management Protocol) A protocol originally designed to be used in managing TCP/IP internets. SNMP is presently implemented on a wide variety of computers and networking equipment and may be used to manage many aspects of network and end-station operation.

Socket The concatenation of the IP Address and TCP port, which together specify a particular process or service within the Internet.

Software A computer program or set of programs held in some storage medium and loaded into read/write memory (RAM) for execution. Compare with firmware and hardware.

SOH (Start Of Header) A control character used to indicate the beginning of the header.

Solaris A UNIX-based operating system created by Sun Microsystems to run on Sun computers and workstations.

SONET A set of standards for data communication over fiber-optic cable at speeds between 51.84 Mbps and 13 Gbps.

Source address The address from which a packet originates.

Space Absence of signal. In telegraph communications, a space represents the open condition or no current flowing. A space impulse is equivalent to a binary 0.

Spanning Tree Algorithm An IEEE standard algorithm that allows loops to be configured in a bridged network to provide alternate data paths. Spanning Tree automatically makes sure only one data path is accessed at transmission time.

SPARC (Scalable Processor Architecture Reduced instruction set Computer) A powerful workstation similar to a reduced-instruction-set-computing (RISC) workstation.

Speed Same as data rate.

Speed dialing The process of using short sequences of digits to represent complete telephone numbers.

Speed plus A technique used to combine voice and data on the same line by assigning the top part of the normal voice bandwidth to data.

Spike A burst of extra voltage in a power line, lasting only a fraction of a second. Also called a "transient." Compare with surge.

Splice A connection of two cables where each pair in one cable is connected to the corresponding pair in the other.

Splitter A device that multiplies one input into a number of identical ports.

Spool (Simultaneous Peripheral Operation On Line) A program or piece of hardware that controls data going to an output device; also called "spooler." See buffer.

Spooling The queueing of documents for printing. (The first phase for printing in the printing protocol model.)

Spread spectrum The process of modulating a signal over a significantly larger bandwidth than is necessary for the given data rate for the purpose of lowering the bit error rate in the presence of strong interference signals.

SQE (Signal Quality Error) A test in Ethernet LANs; after the computer has sent a transmission onto the LAN, the transceiver sends a collision test signal back to the Ethernet card, so that the controller knows the collision detection circuitry is still operational. Also called "heartbeat."

SQL (Structured Query Language) A common database programming language.

SRAM (Static RAM) RAM in which the memory locations are set when they are written to, and then not refreshed until they are written again. Used when the memory access is expected to be infrequent.

SSI (Small-scale integration) A term used to describe a multifunction semiconductor device with a sparse density (10 circuits or fewer) of electronic circuitry contained on a single silicon chip.

ST A trademark for a bayonet-style fiber-optic connector invented by AT&T.

Standalone Self-contained; said of a modem or other device that works outside its host computer or a rack. The opposite of a card, which is installed inside the computer or a rack.

Starlan A local-area network design and specification, within the IEEE-802.3 standards, characterized by 1-Mbps baseband data transmission over two-pair twisted-pair wiring.

Star topology A networking setup used with 10BaseT cabling and a hub where each node on the network is connected to the hub like points of a star. (See bus topology.)

Start bit In asynchronous transmission, the first bit used to indicate the beginning of a character, normally, a space condition which serves to prepare the receiving equipment for the reception and registration of the character.

Start-stop transmission Asynchronous transmission such that a group of signals representing a character is preceded by a start bit and followed by a stop bit.

Station Any DTE that receives or transmits messages on a data link, including network nodes and user devices.

Statistical multiplexing A time-division multiplexing technique in which time slots are dynamically allocated on the basis of need. Slots are allocated to devices that have data to be transmitted.

Status-activity monitor (SAM) A testing device that provides full breakout-box functions in addition to LED monitoring for indicating low- or high-lead activity.

Step-index A type of optical fiber with a uniform refractive index at its core and a sharp decrease in the refractive index at its core/cladding interface.

Stop bit In asynchronous transmission, the last bit used to indicate the end of a character; normally a mark condition that serves to return the line to its idle or rest state.

Store and forward The most accurate data transferring technique that examines each packet of a transmission to verify accuracy, and ensuring bad or misaligned packets are eliminated, then sends it to its destination. When the network is busy, the packet is stored until the network is able to carry the traffic and packets are transmitted without error.

Straight-through pinning Cable configuration that matches DCE to DTE, pin for pin (Pin 1 with Pin 1, Pin 2 with Pin 2, and so on).

String A process consisting of a series of threads.

Strobe An electrical pulse used to call for the transfer of information.

STX (Start of Text) A control character used to indicate the beginning of a message; it immediately follows the header in transmission blocks.

Subnetwork One of the smaller networks linked by a bridge or router; see bridge, router.

Surge An oversupply of voltage from the power company, lasting as long as several seconds. A strong surge can damage electronic equipment.

SVC (Switched Virtual Circuit) In a packet-switched network, a temporary virtual circuit between two users.

S-VGA (Super VGA) A refinement of VGA that offers higher resolution, at least 800 × 600 pixel.

S-VHS (S-Video) A cost-effective method developed by TV engineers to reduce distortion that occurs in TV/monitor and VCR processing. S-Video gives more detail and less color distortion.

Swell A persistent voltage surplus. The overvoltage is not enough to cause a surge but can damage sensitive equipment.

Switch (1) Any device that makes or changes electrical connections in a circuit. (2) Informal for data PABX. (3) In packet-switched networks, the device used to direct packets. Usually located at one of the nodes on the network's backbone.

Switched Ethernet Unlike shared Ethernet, it provides a private connection between two nodes on a network, speeding up the rate at which data is sent along the network and eliminating collisions.

Switched 56 A dial-up version of DDS. It provides for 56-Kbps transmission via 4-wire circuits. Two-wire SW56 is also available from some carriers.

Switched line A communications link for which the physical path may vary with each usage, such as the public telephone network.

Switching matrix The electronic equivalent of a cross-bar switch.

Symbolic name A means used to identify a collection of stations (as in an access group) or computer ports (as in a resource class).

Symmetric References data transmission that is the same speed both *upstream* and *downstream*. Contrast with *asymmetric*.

SYN (Synchronous Idle) In synchronous transmission, a control character used to maintain synchronization and as a time fill in the absence of data. The sequence of two SYN characters in succession is used to maintain synchronization following each line turnaround.

Sync Short for synchronous or for synchronous transmission, or synchronization. See RGBS.

Synchronization, synchronizing The process of keeping the receiver "in step" with the transmitter; usually achieved by having a constant time interval between successive bits, by having a predefined sequence of overhead bits and information bits, and by having a clock.

Synchronous transmission Data transmission in which characters and bits are transmitted at a fixed rate, with the transmitter and receiver synchronized by a clock source usually generated by the DCE. This eliminates the need for individual start bits and stop bits surrounding each byte, thus providing greater efficiency. Compare with asynchronous transmission.

Sysop System Operator. Anyone responsible for the physical operations of a computer system or network resource.

System administrator A user with specific network privileges, who is responsible for setting up and maintaining the services and organization of network services.

T

T carrier A time-division multiplexed service, normally supplied by the telephone company, that usually operates digital-transmission facility, at an aggregate data rate of 1.544 Mbps and above.

T-connector A coaxial connector, shaped like a T that connects two thin Ethernet cables while supplying an additional connector for a network interface card.

T1 A digital carrier facility used to transmit a DS-1 formatted digital signal at 1.544 Mbps. A T1 carrier can transmit large volumes of information across great distances at high speeds at a (potentially) lower cost than that provided by traditional analog service. A T1 carrier uses time-division multiplexing to manipulate and move digital information. It consists of one 4-wire circuit providing 24 separate 64-Kbps logical channels; the aggregate data rate equals 1.544 Mbps.

T1 timer In X.25 packet-switched networks, a timer used to measure timeout intervals in link initialization and data exchanges.

T3 28 T1 lines in one; the aggregate data rate is 44.746 Mbps.

TA (Terminal adapter) An ISDN phone or a PC card that emulates the phone. Devices on the end of a basic rate interface line are known as terminals.

Tail circuit A feeder circuit or an access line to a network node. Usually used when a component (synchronous environment only) is not in close proximity to the DCE device, so that line drivers are used to drive the signal further. The connection between the two DCEs is referred to as the tail circuit.

Tap In cable-based LANs, a connection to the main transmission medium.

Tariff The published schedule of rates for specific equipments, facilities, or services offered by a common carrier; also, the vehicle by which regulatory agencies approve the rates. Thus, a contract between the customer and the common carrier.

TCAM (Telecommunications Access Method) An IBM software routine; the telecommunications access method for 3270 control.

TCP (Transmission Control Protocol) A specification for software that bundles and unbundles sent and received data into packets, manages the transmission of packets on a network and checks for errors.

TCP/IP (Transmission Control Protocol/Internet Protocol) A layered set of protocols that allows sharing of applications among PCs, hosts, or workstations in a high-speed communications environment. Because TCP/IP's protocols are standardized across all its layers, including those that provide terminal emulation and file transfer, different vendors' computing devices (all running TCP/IP) can exist on the same cable and communicate with one another across that cable. Corresponds to Layers 4 (Transport) and 3 (Network) of the OSI reference model.

TCU (Transmission Control Unit) A control unit (such as an IBM 2703) whose operations are controlled solely by programmed instructions from the computing system to which the unit is attached; no program is stored or executed in the unit. Contrast with communications control unit.

TD (Transmitted Data) An RS-232 data signal (sent from DTE to DCE on pin 2).

TDM (Time Division Multiplexor) A device that accepts multiple channels on a single transmission line by connecting terminals, one at a time, at regular intervals, interleaving bits (Bit TDM) or characters (Character TDM) from each terminal.

TDMA (Time-Division Multiple Access) A high-speed, burst mode of operation that can be used to interconnect LANs; first used as a multiplexing technique on shared communications satellites.

TDR (Time-Domain Reflectometry) A method of finding cable faults by sending out high-frequency pulses. If the signal reaches a problem point, it bounces back to the tester.

Telco A generic abbreviation for "telephone company"; also an abbreviation for "telephone central office."

Telecommuting Using telecommunications equipment to maintain contact with an office while working at home.

Telemetry Transmission of coded analog data, often real-time parameters, from a remote site.

Teleprinter A terminal without a CRT that consists of a keyboard and a printer.

Teleprocessing A form of information handling in which a data processing system utilizes communication facilities. (Originally, but no longer, an IBM trademark.) Synonymous with data communications.

Teletype A brand of teleprinter, made by Teletype Corporation.

TELEX (Teleprinter Exchange) A teleprinter dial network offered by Western Union and the International Record Carriers; uses Baudot code.

Telnet A virtual terminal service available through the TCP/IP protocol suite.

ter The third version of a CCITT recommendation.

Terabyte (TB) 1024 GB or 241 bytes.

Terminal (1) Any device capable of sending or receiving information over a communication channel. (2) A point at which information can enter or leave a communication network.

Terminal adapter (interface adapter) A device that converts existing POTS telephones and non-ISDN data terminals to the ISDN basic rate interface for circuit-switched voice and data communications.

Terminal control unit See cluster controller.

Terminal server A device that allows one or more terminals to connect to an Ethernet LAN.

Terminated line A circuit with a resistance at the far end equal to the characteristic impedance of the line, so no reflections or standing waves are present when a signal is entered at the near end.

Termination (1) Placement of a connector on a cable. (2) The maintenance of an electrical load at the end of a circuit; see terminated line.

Terminator A resistor at each end of an Ethernet cable that absorbs energy to prevent reflected energy back along the cable (signal bounce). It is usually attached to an electrical ground at one end.

Text (1) Information for human, as opposed to computer, comprehension, intended for presentation in a two-dimensional form. (2) In communications, transmitted characters forming the part of a message that carries information to be conveyed; in some protocols, the character sequence between start-of-text (STX) and end-of-text (ETX) control characters.

Thick Ethernet, Thicknet Standard 10Base5 Ethernet; see 10Base5.

Thin Ethernet An Ethernet LAN or IEEE-802.3 LAN that uses a smaller-than-normal diameter coax; often used to link IBM personal computers together. Operates at same frequency as Ethernet but at smaller distances. Also known informally as "Cheapernet" or "ThinNet."

Thin-film transistor (TFT) LCD See active-matrix.

ThinNet see Thin Ethernet.

Thread (1) A processor instruction or series of instructions that make up a complete operation or process and cannot be interrupted. (2) A series of interrelated messages in an on-line conversation.

Throughput The total useful information processed or communicated during a specified time period; expressed in bits per second, packets per second, or some similar measurement.

Throughput delay The length of time required to accept input and transmit it as output.

Timeout A set period of waiting before a system performs a designated action.

Time slot An assigned period of time or an assigned position in a sequence.

Time-sharing A method of computer operation that allows several interactive terminals to use one computer. Although the terminals are actually served in sequence, the high speed of the computer makes it appear as if all terminals were being served simultaneously.

Tip and Ring Traditional telephone terminology for "positive" and "negative." In old-style telephone switchboards, the "tip" wire was the one that connected to the tip of the plug; the "ring" wire was connected to a slip ring in the jack.

TNC A threaded connector for miniature coax; TNC is said to be short for threaded-Neill-Concelman. Contrast with BNC.

Token A continuously repeating frame, transmitted onto the network by the controlling computer; the frame that polls for network transmissions. See token-bus, token passing, token-ring.

Token-bus A LAN-access mechanism and topology in which all stations actively attached to the bus listen for a broadcast token or supervisory frame. Stations wishing to transmit must receive the token before doing so. The next physical station to transmit is not necessarily the next physical station on the bus, bus access is controlled by preassigned priority algorithms.

Token passing A LAN protocol in which the network continuously circulates a bit pattern known as a token. Only the workstation holding the token can put a message onto the LAN.

Token-ring A network access mechanism and topology in which a supervisory frame or token is passed from station to station in sequential order. Stations wishing to gain access to the network must wait for the token to arrive before transmitting data. In a token ring, the next logical station receiving the token is also the next physical station on the ring. The standard was developed by IBM and ratified as an IEEE standard, 802.5. Token-ring interconnects PCs via special twisted-wire cable in a star topology, connecting all computers to a central wiring hub.

TOP (Technical and Office Protocols) A Boeing version of the MAPsuite, aimed at office and engineering applications.

Topology The logical or physical arrangement of stations on a network in relation to one another. See bus, ring, star, and tree.

Total Length An IP head field containing the number of octets in an internet datagram, including both IP head and the data portion.

Track A ring-shaped portion of a magnetic disk's surface area, defined by its minimum and maximum distance from the center of the disk.

Trackball A pointing device, similar to a mouse, in which the rotation of a ball in a fixed housing controls the measurement of a pointer on the computer screen.

Training The process in which a receiving modem achieves equalization with a transmitting modem.

Training pattern The sequence of signals used in training.

Transaction (1) In communications, a message destined for an application program; a computer-processed task that accomplishes a particular action or result. (2) In interactive communications, an exchange between two devices, one of which is usually a computer. (3) In batch or remote job entry, a job or job step.

Transaction processing A real-time method of data processing in which individual tasks or items of data (transactions) are processed as they occur—with no primary editing or sorting.

Transceiver A hardware device that links a node with a baseband network cable, functioning as both a transmitter and receiver. Derived from transmitter/receiver.

Transformer An electromagnetic device that can change (or step up or step down) the voltage of alternating currents as used in power supplies. Transformers also provide DC isolation and are commonly used to terminate circuits connected to other devices.

Transient See spike.

Transmission The dispatching of a signal, message, or other form of intelligence by wire, radio, telegraphy, telephony, facsimile, or other means; a series of characters, messages, or blocks, including control information and user data; the signaling of data over communications channels.

Transmission block A sequence of continuous data characters or bytes transmitted as a unit, over which a coding procedure is usually applied for synchronization or error-control purposes.

Transmission code Any of the standard character sets used in information interchange. In order for two systems to communicate, they must agree on the transmission code. Examples are ASCII, EBCDIC, Baudot, and Unicode.

Transmission media In LAN technology, anything used to carry data in the form of electrical signals. Examples: cable, optical fiber.

Transmission mode The technique by which a device recognizes the beginning and end of a character clocking in synchronous transmission; start and stop bits in asynchronous.

Transport Layer In the OSI model, the network processing entity responsible, in conjunction with the underlying Network, Data Link, and Physical Layers, for the end-to-end control of transmitted data and the optimized use of network resources.

Tree A LAN topology that recognizes only one route between two nodes on the network. The map resembles a tree or the letter T.

Trojan horse A program or data that seems innocuous when it is loaded into a system or network but later facilitates an attack by a hacker or virus.

Trunk A dedicated aggregate telephone circuit connecting two switching centers, central offices, or data-concentration devices. The main network cable.

TTL (Transistor-to-Transistor Logic) A digital-logic family with well-defined operating characteristics; generally available in integrated-circuit form. Some TTL operating characteristics are current, input and output voltages, and drive characteristics.

TUV (Technischer Uberwachungs-Verein) A German electrical testing and certification organization similar to UL.

Twinaxial cable A cable that is similar to coaxial cable, but with two inner conductors instead of one. Used in IBM minicomputer and midrange systems: Systems 34, 36, and 38, and AS/400.

Twisted pair Two insulated copper conductors that are wound around each other, mainly to cancel the effects of electrical noise; typical of standard telephone wiring: Unshielded twisted pair contains no outside wraparound conductor.

Twisted-pair cable Cable made up of one or more twisted pairs. This kind of cable may or may not be shielded. Twisted-pair cable is easier to install, less expensive, and easier to change than coaxial cable, but its bandwidth is usually smaller.

Two-wire channel A circuit that indicates information signals in both directions for data carried by the same path.

TWX (Teletypewriter Exchange Service) A teletypewriter dial network owned by Western Union. ASCII-coded machines are used.

Tymnet A common carrier offering an X.25 PDN.

Type of Service An IP head field containing the transmission quality parameters, precedence level, reliability level, speed level, resource tradeoff (precedence versus reliability) and transmission mode (datagram versus stream).

Type 1 cable Two twisted pairs of 22 AWG solid conductors, enclosed in a braided cable shield; associated with the IBM cabling system.

Type 3 cable 22 or 24 AWG with at least two twists (unshielded) per linear foot. Four twisted pairs are recommended when installing a new wire. Associated with the IBM cabling system.

Type A Coax In IBM 3270 Systems, a serial transmission protocol operating at 2.35 Mbps, which provides for the transfer of data between a 3274 Control Unit and attached display stations or printers.

U

U Interface A standard basic rate interface using two copper wires.

UADSL Universal ADSL. See *G.Lite DSL.*

UART (Universal Asynchronous Receiver/Transmitter) A microprocessor that handles the actual electronic communication in an async modem.

UDP (User Datagram Protocol) The TCP/IP transaction protocol used for applications such as remote network management and name-service access; this lets users assign a name, such as "RVAX*2,S," to physical or numbered address.

UL (Underwriters Laboratories) A private organization that sets standards and tests electrical equipment for safety.

ULSI (Ultra-Large-Scale Integration) A term used to describe a multifunction semiconductor device with an ultra-high density (over 10,000 circuits) of electronic circuitry contained on a single silicon chip.

Underscan A video-display effect in which the image is shrunk so that it does not take up the whole screen.

UNIX An operating system originally designed by AT&T for communicating multiuser, 32-bit minicomputers; has come into wide commercial acceptance because of its predominance in academia and its programming versatility. Other versions include AIX and XENIX.

Unloaded line A line with no loading coils.

UPS (Uninterruptible Power Supply) A device that continues to supply electricity for a period of time after an outage. Compare BPS.

Upstream Used to describe traffic on a *network* from the endpoint (your computer) to the provider. Sending an e-mail message is upstream traffic. See also *downstream*.

URL Uniform Resource Locator. The address your browser uses to locate a specific site on the World Wide Web.

USOC (Universal Service Order Code) A set of phone-company standards for equipment, including connectors and interfaces.

Utility software Programs that make operation of a PC or a LAN more convenient, including programs to move disk files more easily, diagnostic programs, etc.

UTP cable Unshielded twisted-pair cable. This cable is used for creating *Ethernet* networks. See also *Category 5 cable.*

UUCP (UNIX-to-UNIX Copy Program) A transport protocol for dial-up e-mail access, remote command execution, and file transfers (on UNIX based systems).

UUENCODE Unix-to-Unix Encoding. A method for converting files from binary to ASCII so that they can be sent via e-mail.

V

V.10 A CCITT interface recommendation; electrically similar to RS-423 (unbalanced, high speed).

V.11 A CCITT interface recommendation; electrically similar to RS-422.

V.17 A CCITT 14,400-bps leased-line recommendation; used for 14,400-bps fax communication.

V.21 A CCITT 300-bps dial-line modem recommendation; similar to RS-422.

V.22 A CCITT 1200-bps dial or 2-wire leased-line modem recommendation; similar to Bell 212.

V.22 bis A CCITT 2400-bps dial or 2-wire leased-line modem recommendation.

V.23 A CCITT 600/1200-bps dial-line modem recommendation; similar to Bell 202.

V.24 A CCITT interface recommendation that defines interchange circuits; similar to and operationally compatible with RS-232.

V.25 A CCITT dial-line parallel-interface recommendation.

V.25 bis A CCITT dial-line serial-interface recommendation.

V.26 A CCITT 2400/1200-bps leased-line modem recommendation; similar to Bell 201 B.

V.26 bis A CCITT 2400/1200-bps dial-line modem recommendation; similar to Bell 201 C.

V.26 ter A CCITT 2400-bps dial or 2-wire leased-line modem recommendation.

V.27 A CCITT 4800-bps leased-line modem recommendation with manual equalizer; similar to Bell 208 A.

V.27 bis A CCITT 4800-bps dial-line modem recommendation with automatic equalizer.

V.27 ter A CCITT 4800-bps dial-line modem recommendation; similar to Bell 208 B; used for 4800-bps fax communication.

V.28 A CCITT interface recommendation that defines electrical characteristics for the interchange circuits defined by V.24; similar to and operationally compatible with RS-232.

V.29 A CCITT 9600-bps leased-line modem recommendation; similar to Bell 209; used for 9600-bps fax communication.

V.32 A CCITT 9600-bps dial or 2-wire leased-line modem recommendation.

V.32 bis A CCITT 14.4-Kbps dial or 2-wire leased-line modem recommendation.

V.32 terbo An industry standard for dial-up modem communication of 19.2 Kbps.

V.34 A CCITT modem standard (not yet ratified as this is written) for data transmission at 28.8 Kbps. The preliminary version is called "V.Fast."

V.35 A CCITT interface standard for high-speed communication. V.35 specifies a 34-pin connector and can transmit at speeds into the millions of bits per second. It cannot connect, physically or electrically, to any other interface without the aid of a converter.

V.36 A CCITT interface for 4-wire data communication at speeds of 48 Kbps and up; designed to replace V.35.

V.42 A CCITT standard for an error-correction protocol for modems using async-to-sync conversion; defines LAP-M protocol.

V.42 bis A CCITT standard for a protocol, built into a modem's firmware, that provides both error correction and data compression.

V.54 A CCITT standard for modem loopback tests.

V.110 Async rate adaption (for ISDN) at speeds up to 19.2 Kbps.

V.120 Async rate adaption (for ISDN) at speeds up to 64 Kbps; uses statistical multiplexing.

VAN (Value Added Network) A network whose services go beyond switching.

Vaporware A jocular term for software that is announced or advertised but never ships.

VDE (Verband Deutscher Electrotechniker) The German national electrical standards and testing agency.

VDT (Video Display Terminal) CRT terminal.

VDU (Video Display Unit) Same as VDT.

VESA (Video Electronic Standards Association) A video-industry group that promotes standardization.

V.Fast A preliminary version of V.34.

VGA (Video Graphics Array) A video standard for IBM PC and compatible computers. Standard VGA has a resolution of 640 × 480 and supports 16 colors.

Video card A PC interface card that processes video data.

Video teleconferencing The use of "television-type" transmissions to allow people at two or more locations to communicate as if they were in the same meeting.

Videotex A data-communication service that transmits on TV channels.

VINES (VIrtual NEtwork Software) Operating software for a LAN made by Banyan.

Virtual circuit In packet switching, a network facility that gives the appearance to the user of an actual end-to-end circuit; a dynamically variable network connection where sequential data packets may be routed differently during the course of a virtual connection. Virtual circuits enable transmission facilities to be shared by many users simultaneously.

Virtual machine The technique of making one computer behave like several.

Virtual memory A technique for using disk storage space to emulate random access memory (RAM).

Virtual reality (VR) An artificial environment created by computer. Virtual Reality attempts to provide an illusion of actually being in a place or of doing something. A Virtual Reality environment can be experienced using an headset and electronic gloves, or simply viewed on a monitor.

Virtual storage Storage space that may be viewed as addressable main storage, but is actually auxiliary storage (usually peripheral mass storage) mapped into read addresses; amount of virtual storage is limited by the addressing scheme of the computer.

Virus Any destructive self-replicating program.

VL-bus A VESA local-bus standard that allows fast video processing (on Intel processor-based PCs).

VLSI (Very Large-Scale Integration) A term used to describe a multifunction semiconductor device with a very high density (up to 10,000 circuits) of electronic circuitry contained on a single silicon chip.

Voice channel A transmission path usually limited to passing the bandwidth of the human voice.

Voice/data PABX A device that combines the functions of a voice PABX and a data PABX, often with emphasis on the voice facilities.

Voice-frequency The part of the audio frequency range that can transmit commercial-quality speech.

Voice-grade line A channel that is capable of carrying voice-frequency signals.

Voice PABX Voice-only PABX for voice circuits; a telephone exchange.

Volatile memory A storage medium that loses all data when power is removed.

Volume An individually identified data-storage device.

VRAM (Video RAM) RAM optimized for video processing.

VTAM (Virtual Telecommunications Access) An IBM software routine; the virtual access method for 3270 systems.

W

WAN (Wide-Area Network) A network that serves an area of hundreds or thousands of miles, using common carrier-provided lines; contrast with LAN.

WATS (Wide Area Telephone Service) A service provided by telephone companies in the U.S. that permits a customer to make calls to or from telephones in a specific nonlocal zone for a flat monthly charge. Also called "800 service."

Wavelength Distance between successive peaks of a sine wave.

Wideband A system in which multiple channels access a medium (usually coaxial cable) that has a large bandwidth, greater than that of a voice-grade channel; typically offers higher-speed data-transmission capability. Also see broadband.

Window A separate work area on a computer screen, usually in a GUI.

Windows™ A GUI environment developed by Microsoft. It permits users to run more than one application on a desktop computer simultaneously.

Wiring closet Central location for termination and routing of on-premises wiring systems.

Wiring hub A cabinet, usually mounted in a wiring closet, that holds connection modules for various kinds of cabling. The hub contains electronic circuits that retime and repeat the signals on the cable. The hub may also contain a microprocessor board that monitors and reports on network activity.

Workflow The "flow" of tasks as they pass from one worker to another, especially in a workgroup environment.

Workgroup A group of people working together on the same project, especially when they have their own dedicated network.

Workstation (1) Input/output equipment at which an operator works. (2) A high-end desktop computer using RISC processors, often running the UNIX or XENIX operating system.

World Wide Web A type of Internet browsing tool.

Worm A self-replicating program that consumes processor time but cannot destroy data, software, or other system resources.

WORM (Write Once, Read Many) A common type of optical disk drive. The disk can be written to only once; after that, the data is permanently stored there.

WWW (World Wide Web) The Internet's multimedia service containing countless areas of information, documentation, entertainment, as well as business and personal home pages.

WYSIWYG What You See Is What You Get—a description of computer software whose screen display is nearly identical to its printed output.

X

X Terminal A dedicated platform (terminal) that is designed to run X-server software; used with powerful machines that run 680 × 0 processors such as RISC computers. X is a standard base-level windowing architecture for UNIX machines and their terminals.

X Window A network-based windowing system that provides a programmatic interface for graphic window displays. X Window permits graphics produced on one networked station to be displayed.

X.121 A CCITT recommendation that defines general-purpose interface between data terminal equipment for synchronous operation on public data networks.

X.21 A CCITT standard governing the interface between DCE and DTE for synchronous operation on public data networks.

X.21 bis In PDNs, a CCITT recommendation that defines the most popular digital interface; it is equivalent to RS-232 and V.24.

X.25 The standard interface for packet-switched data communications networks, as designated by the CCITT.

X.25 Packet Assembler/Disassembler (PAD) A device that permits communication between non-X.25 devices and the devices in an X.25 network.

X.26 Same as V.10.

X.27 Same as V.11.

X.28 In packet-switched networks, a CCITT recommendation that defines the interchange of commands and responses between a PAD and its attached asynchronous terminals.

X.29 In packet-switched networks, a CCITT recommendation that defines the use of packets to exchange data for control of remote PADs.

X.3 In packet-switched networks, a CCITT recommendation that defines the parameters that determine the behavior of the interface between a PAD and its attached asynchronous terminals.

X.400 A CCITT standard for a messaging and document distribution protocol to connect different e-mail systems.

X.500 The CCITT designation for a directory standard to coordinate the dispersed file directories of different systems.

XENIX Microsoft trade name for a 16-bit microcomputer operating system derived from UNIX.

XGA (Extended Graphics Array) A video standard for Micro Channel 386SX or better PCs. Resolution is 1024 × 768 pixels, and 256 colors are supported. XGA's bus-mastering boosts performance by giving its graphics coprocessor access to system RAM, not just video RAM.

XModem An asynchronous communication protocol used to transmit files between devices such as personal computers and mainframes, and between personal computers and ITS.

XNS (Xerox Network Systems) A peer-to-peer protocol developed by Xerox that has been incorporated into several LAN schemes, including 3Com(R), 3+(R), and 3+Open network operating systems.

X-ON/X-OFF (Transmitter On/Transmitter Off) Control characters used for flow control, instructing a terminal to start transmission (X-ON) and stop transmission (X-OFF).

Y

Yellow alarm In T1, an alarm sent by the remote site to the local site when the second bit position is zero in every channel in a T1 frame. The yellow alarm is sent to the local site after a red alarm at the remote site.

Z

Zero insertion In SDLC, the process of including a binary 0 in a transmitted data stream to avoid confusing data and SYN characters; the inserted 0 is removed at the receiving end.

ZIF (Zero Insertion Force) socket A standard integrated circuit-socket design in which the user moves a lever to insert or remove the chip, instead of pressing and prying the chip manually. The lever reduces the chances of damaging the integrated circuit's pins.

Zero-slot LAN A LAN that does not require a network interface card.

Zone A logical grouping of devices in an internet that makes it easier for users to locate network services. Zones are defined during the router set up.

Bibliography

IEEE 802.3: Carrier Sense Multiple Access with Collision Detection

IEEE-802.5: Token Ring Access Method

Industrial Data Communications, Fundamentals and Applications, Lawrence M. Thompson, Instrument Society of America

Handbook of LAN Technology, 2nd edition, Paul J. Fortier, McGraw-Hill Publishing, ISBN 0-07-021625-8

Guide to Connectivity, 2nd edition, Frank J. Derfler, Ziff-Davis Press, ISBN 1-56276-047-5

TCP/IP and Related Protocols, Uyless Black, McGraw-Hill Publishing, ISBN 0-07-005553-X

DEC Networks and Architectures, Carl Malamud, McGraw-Hill Publishing, ISBN 0-07-039822-4

LAT Network Concepts, Digital Equipment Corporation, Order number: AA-LD84B-TK

Technical Aspects of Data Communication, John E. McNamara, Digital Press, ISBN 1-55558-007-6

Advanced Ethernet/802.3 *Management & Performance,* Bill Hancock, Network-1, Inc., L.I.C., NY. (800)NETWRK1

Recommended reading (via Internet) for TCP/IP is as follows:

Request For Comment (RFC) 793-Transmission Control Protocol

RFC 791-Internet Protocol

RFC 1011-Official Protocols

Index

About the Author

Daniel Capano is President of Diversified Technical Services, Inc., an independent consulting firm that specializes in providing Engineering and Management services to the construction industry. He has worked in the Instrumentation and Control System field for the last 15 years as both a consultant and a contractor, has written numerous articles about Instrumentation and Control Systems and Computers and has presented several papers dealing with Instrumentation and Control Systems. Mr. Capano is President of the NYC Section of the International Society for Instrumentation and Controls (ISA) and the National Director of the Water and Wastewater Division of the ISA.

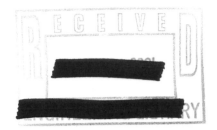
RECEIVED
RY